Blue-Footed Boobies

Drummond compares congenitally siblicidal brown boobies with only moderately less murderous blue-footed ones. Depending on circumstances, a blue-footed booby might occasionally allow a younger sib to grow up. Evolutionarily distant as these squabbling sea-birds are from our own order, primates, it's impossible not to be impressed by the parallels with other apes like bonobos among whom stress hormones go zooming up upon the arrival of a younger sibling, or the near universal manifestation of sibling rivalries across human cultures. In between lyrical descriptions of the habits and habitats of these consummate diving birds, and the adventures of the scientists studying them, Drummond provides an overview of sibling rivalry across the natural world and expertly guides readers through various evolutionary theories to make sense of behavior so seemingly counter-intuitive as lethal conflict between close kin. A fascinating, if sometimes harrowing, read!"

—**Sarah Blaffer Hrdy**, Emeritus Professor of Anthropology,
University of California, Davis

"*Blue-Footed Boobies* is a scientifically thorough, yet highly accessible, account of the evolutionary basis of family life. Although focused on the tropical marine birds Drummond has studied for decades, it does an expert job of relating his work to human behavior in a way that will appeal to professionals and nonprofessionals alike. Moreover, the meticulously dissected experiments provide a compelling counter to the 'just-so' stories sociobiologists have historically been accused of advocating."

—**Walt Koenig**, Research Zoologist Emeritus,
University of California, Berkeley

Blue-Footed Boobies

Sibling Conflict and Sexual Infidelity on a Tropical Island

Hugh Drummond

Departamento de Ecología Evolutiva
Instituto de Ecología
Universidad Nacional Autónoma de Mexico
Mexico City
Mexico

Illustrated by

Jaime Zaldivar-Rae

OXFORD
UNIVERSITY PRESS

Oxford University Press is a department of the University of Oxford. It furthers
the University's objective of excellence in research, scholarship, and education
by publishing worldwide. Oxford is a registered trade mark of Oxford University
Press in the UK and certain other countries.

Published in the United States of America by Oxford University Press
198 Madison Avenue, New York, NY 10016, United States of America.

Library of Congress Cataloging-in-Publication Data
Names: Drummond, Hugh, 1947– author.
Title: Blue-footed boobies : sibling conflict and sexual infidelity on a
tropical island / Hugh Drummond.
Other titles: Sibling conflict and sexual infidelity on a tropical island
Description: New York, NY : Oxford University Press, [2023] |
Includes bibliographical references and index.
Identifiers: LCCN 2023004674 (print) | LCCN 2023004675 (ebook) |
ISBN 9780197629840 (hardback) | ISBN 9780197629864 (epub)
Subjects: LCSH: Blue-footed booby—Behavior—Mexico.
Classification: LCC QL696 .P48 D78 2023 (print) | LCC QL696 .P48 (ebook) |
DDC 598.4/3—dc23/eng/20230308
LC record available at https://lccn.loc.gov/2023004674
LC ebook record available at https://lccn.loc.gov/2023004675

DOI: 10.1093/oso/9780197629840.001.0001

Printed by Integrated Books International, United States of America

To the Universidad Nacional Autónoma de Mexico,
a pillar of science and education in Mexico

.

Contents

Acknowledgments

I warmly thank the colleagues and friends who put an end to my years-long dithering over whether to write a book by responding to a synopsis of our research on boobies with these words:

> "Wow, Hugh. Thanks for sharing. What an outstanding encapsulation of your work!"
>
> "Thank you, Hugh! Who says it's not a page-turner? It's the entire booby soap-opera story, complete with the dramatic stuff that makes up their daily lives! In fact, the type of stuff that appealed to me when I went into the field of animal behavior...."
>
> "Remarkable stuff. It brings together all the strands that you have shared with us over the years and gives a complete picture of your life's work. Not only it is of great interest in scientific terms but fascinating in terms of all the blood, gore and cuckoldry! It would make a good series for Netflix following up on *Game of Thrones*!"
>
> "Hugh, thanks for sending. This is such a great story of avian love, deception, and murder. You need to write a book. Seriously."

Without this encouragement, I may never have taken the plunge. After all, I'd published the research in dozens of articles in scientific journals, so the blue-foot story was already out there. Why go to the trouble of repeating it all? I needed those friends, especially Hal Herzog, the author of the last response, to shake me up. And Hal went further—his advice helped me transition from writing technical, theory-oriented articles for specialists to entertaining naturalists of all stripes with the dramas of evolved family conflict. Similarly, I'm indebted to an anonymous reviewer of my book proposal who strenuously insisted on the book not just commenting on booby-human parallels but ending strongly with an explanatory account of the two species' similarities. They got me to write, finally, what I had always wanted to write.

Since I morphed into an evolutionary biologist 40 years ago, I have felt grateful to three American colleagues who showed me what it means to become involved in the lives and ecology of populations of wild animals. When, as a graduate student, I spent time with each of them in the field, I was already a nature lover and experienced in hiking, camping, and climbing, but I didn't know about garnering intimate knowledge of other species in their worlds. The examples of Steve Arnold (garter snakes), Stan Rand (green iguanas), and Brad Shaffer (axolotls) showed me what I must aspire to and how rewarding sustained fieldwork can be. And though I never actually spent much time with him in the field, the example of Harry Greene, my companion in graduate school, was similarly influential. Conversations with a thorough-going naturalist with a profound and unsentimental love of nature, an encyclopedic familiarity with reptiles, and the enthusiasm and patience to share his knowledge and passion with a naïve apprentice fostered my fascination with animals and my urge to delve into their lives.

I am grateful to a number of colleagues from different disciplines who took the trouble to critique one or more chapters in their areas of expertise. Sarah Hrdy generously provided salutary feedback on the chapter comparing family conflict in boobies and humans, taking time from her own writing to guide my outing into the literature of evolutionary anthropology. That chapter also benefitted from the comments of evolutionary/developmental psychologists Sybil Hart and Isaac Santoyo. Hal Herzog, a psychologist/anthrozoologist, kindly reviewed three chapters, and several ornithologically savvy behavioral ecologists reviewed chapters in their areas of expertise: Alejandro Gonzalez-Voyer, Lynna Kiere, Becky Kilner, Marcela Osorio, Judy Stamps, Roxana Torres-Avilés, and Alberto Velando.

My most staunch and supportive companions in this project have been the team of friends, not all academics, who read each chapter as it emerged from my keyboard and provided encouragement and advice for making chapters more readable and entertaining. Working in tandem, Roy Broughton and John Williams helped me to know when the prose and its organization were on track, rebuked my lapses into scientific journalese, nipped in the bud my obsession with data and details, and usefully encouraged my digressions into natural history and the incidental amusements of field research. Regina Macedo, Claire Drummond, and Cristina Rodríguez provided moral support as well as feedback on contents and style, and Patty Gowaty periodically startled me with encouragement so stirring it made me ashamed of my slacking and self-doubt. Finally, my wife, Sylvia Rojas, was a mainstay. In the early weeks, she helped me to grasp and clarify what sort of book I wanted to write; throughout the writing, she helped me keep up the momentum and gave valuable feedback, including judicious suggestions and criticism on every chapter.

We have managed to sustain decades of fieldwork on an uninhabited and waterless island far out to sea without any important accidents or setbacks because the manager of our camp, monitoring program, and database, Cristina Rodríguez, comprehensively coordinated the logistics, including the recruitment of volunteers and the support of fishermen, the Mexican navy, and the administrators of Isla Isabel national park. She also organized the book's bibliography. In addition, scores of volunteer students, mostly from the Universidad Nacional Autónoma de México (UNAM) and other Mexican universities, worked diligently for weeks or months at a stretch to sustain observations that were often repetitive and draining, and sometimes exhausting. Since I joined the UNAM in 1980, I have greatly appreciated and admired our students' willingness, commitment, and good humor.

No less important were the contributions to the booby research program of valued collaborators from universities in Mexico, the United States, and Europe who contributed their skills in research design, laboratory and field techniques, data analysis, thesis supervision, and manuscript preparation, either from afar or during visits to the island: Sergio (Cheko) Ancona, René Beamonte-Barrientos, Itzia Calixto-Albarrán, Jocelyn Champagnon, Diego Cortez, Cesar Domínguez, Brant Faircloth, Patty Gowaty, Robyn Hudson, Alex Kacelnik, Katharine Keogan, Becky Kilner, Sin-Yeon Kim, Sasha Kitaysky, V. V. Krishnan, Stacey Lance, Xinhai Li, Margarita Martinez-Gomez, Ale Núñez-de la Mora, Schyler Nunziata, Dani Oro, José Luis (Mogli) Osorno, Cristina Rodríguez, Miguel Rubio-Godoy, Salvador Sanchez-Colon, Hubert Schwabl, Judy Stamps, Mary Stoddard, Tamas Székely, John Wingfield, Rebecca Young, and Jaime Zúñiga-Vega.

Without finance, permissions, and logistics, there would have been no booby research program, so it's a pleasure to record my appreciation of the institutions that have steadfastly financed and otherwise supported us. These include, preeminently, the PAPIIT program of UNAM's Dirección General de Apoyo al Personal Académico, the Consejo Nacional de Ciencia y Tecnología, the Consejo Nacional de Areas Naturales Protegidas, the Secretaria de la Marina, the National Geographic Society, and the Conservation and Research Foundation.

Finally, two special mentions. I warmly appreciate the drawings of boobies that Jaime Zaldivar-Rae enthusiastically contributed to the book. While studying the social behavior of Isla Isabel's whiptail lizards, Jaime became so familiar with the blue-footed boobies that 10 years later, handed a set of mediocre photos, he could create the images that so faithfully capture the boobies' behavior and essence. And I am forever indebted to Gordon Burghardt. Who else would accept a lawyer-turned-language teacher he had never met to study in his lab for a doctorate in comparative psychology and ethology? And furnish him with litters of newborn snakes and laboratory facilities along with technical support for building experimental enclosures? Understandably, other universities brushed my applications aside, but Gordon took a chance and admitted me to the profession, the community, and the life in which I have thrived.

Introduction

When the second chick of the blue-footed booby hatches into the shallow pit under its parent's immense body, its sibling, 4 days older, bigger, and more coordinated, is just centimeters away. What these designated rivals get up to in that warm darkness is not known, but during the first several days each of them periodically arouses, eases its head from under the parent's breast or wing feathers, stretches its skinny gray neck precariously upward, and vibrates its head, chirping rhythmically. If the parent is ready with some predigested fish, and if the chick's begging goes on long enough to make it regurgitate, the parent will lower its daggerlike bill from on high and attempt an awkward mouth-to-mouth transfer to the naked, trembling creature one-thirtieth its own weight. Next, the chick will insert its head into the cavernous mouth and grope, usually unsuccessfully, for a mouthful of fish mush that it can swallow whole by withdrawing its head and tilting it back.

But the maneuver often fails well before the parental bill arrives, when the elder sibling's head crashes into the younger's, knocking the chick back into the pit. Or when repeated blows oblige the younger chick to crouch and cower. Nothing excites the rage of a booby chick as much as the lifted vibrating head of another chick; it must be toppled or bludgeoned into silence by a volley of pecks to the cranium, face, eyes, and nape, or indeed anywhere on its body. At the start of life, both chicks are equally motivated to attack in this way, but only the elder one has the height and muscle to pull it off. Then, over the course of 14 weeks of growth and development, new behavioral dynamics emerge as each family member adjusts its own strategy in the light of food availability, changing tactics of the others and whether its own interests will be best served by nurture, tolerance, or death of the second chick.

Sociobiologists, who study the evolution of the social behavior of animals, evolutionary psychologists, and anthropologists have argued extensively over whether humans are essentially good or bad. They get hot under the collar about whether evolution by natural selection has shaped them to compete selfishly and sometimes violently, or to empathize, cooperate, and work selflessly for the common good. In reality, natural selection has shaped species that live in families or social groups, including humans, to do both—compete selfishly and cooperate. Roughly 70 years of descriptive, experimental, and comparative studies of numerous species have revealed how animals treat their companions and their kin. We have come a long way in understanding how social behavior evolves in species over thousands of generations and develops in individuals over their lifetimes. And we understand better than ever

Blue-Footed Boobies. Hugh Drummond, Oxford University Press. © Oxford University Press 2023.
DOI: 10.1093/oso/9780197629840.003.0001

how the behavior of other animals can shed light on our own motivation, emotions, and behavior.

To be clear, the tropical marine bird that is the focus of this book, and which makes spectacular aerial plunges into the ocean to engulf small fishes and breeds in noisy colonies of hundreds, is worthy of our attention for itself, and it is in this spirit that my colleagues and I have studied it for more than 40 years. But the family conflicts of the blue-footed booby also merit our attention because they arise from the same fundamental evolutionary tensions and some similar behavioral mechanisms to those that drive conflict and cooperation in our own lives. To illuminate our evolved behaviors, we habitually turn to our closest relatives—chimpanzees and bonobos—but for family conflict, blue-footed boobies are more relevant models and provide starker insights into our emotions and behavior.[1–3]

Bonding of male and female to cooperate in a joint project of producing, nurturing, and defending offspring is simply not the chimpanzee or bonobo way. Each female raises a single infant by herself in a competitive milieu in which females mate with several males, and males compete with each other for opportunities to mate with females. Males may defend the whole group, including its infants, but they provide infants with little or no nurture. We certainly share emotional and behavioral characteristics with our closest primate relatives and even with many monkeys, including rhesus macaques. For example, there are some striking similarities in how adult males compete with each other and how mothers care for infants. But comparisons among primates can shed only limited light on social relations within human nuclear families. For that purpose, birds are often a better model because the majority of our planet's approximately 10,000 avian species reproduce in nuclear families in which highly dependent offspring that are full- or half-siblings compete for the care provided by two hard-working parents.

Among birds with nuclear families, blue-footed boobies are a particularly good species for making informative comparisons with key aspects of human social behavior because they live long lives and breed repeatedly. Every year they pair monogamously and both partners commit to sharing all tasks in defending and nurturing a highly dependent brood of one to three chicks until all of them fledge or die. In boobies and humans, the nuclear family sets the scene for intense and protracted cooperation and conflict among individuals who are important to each other. Consequently, the within-family emotions and social behavior of boobies and humans have been exposed, over geological time, to similar selection pressures. Given the huge differences between these two species, we expect those similar selection pressures to have channeled the evolution of different behaviors that ultimately serve similar functions. Blue-foots are considerably less intelligent than humans and, for all we know, may be strangers to social learning and culture, pivotal influences on our own endlessly malleable social behavior. But boobies face some of the same fundamental challenges faced by humans, including how to cooperate with siblings, partners, and offspring while prevailing in conflicts of interest with them.

Fig. I.1 Blue-foots loafing on a cliff's edge.
Photo by Hugh Drummond.

Fig. I.2 Female blue-foot.
Copyright Daniel J. Field, used with permission.

Competition between infant siblings, which occurs in most animal litters and broods that are cared for by parents, is a major component of booby social experience during infancy and supplies the first axis of this book. The raw, self-serving violence booby chicks use to gain the upper hand in competition and, *in extremis*, liquidate their siblings is the spectacle that first drew me to set up camp on a little tropical island and get acquainted with the nuclear families of a non-human animal.

I was not disappointed. Blue-foot chicks can go at their siblings so hard it hurts to watch, but it is by no means an unmoderated fight to the death. It is best understood as a partly caring relationship in which one chick imposes psychological dominance while carefully husbanding the dangerous but genetically valuable resource that is its sibling. That chick shoots for the jackpot of a double fledging in the service of its own reproductive achievement but is willing to kill if necessary. The dominant chick's restraint and the adaptiveness of the two rivals' competitive strategies are put in perspective by contrasting with a close relative that lives in leaner ecological circumstances. In the closely related brown booby, the younger chick's prospects for surviving alongside its sibling are so slim that it will give all it's got to attempt a violent coup, even against overwhelming odds. For this reason, the elder chick kills it preemptively in the first few days of life. In all this, boobies are following impulses shaped by natural selection, in response to their rival's behavior, their own experience, and current circumstances. As I will explain, I believe their competitive behavior, rather than being guided by comprehension of consequences, is driven by urges and emotion. They know not what they do; they just have to do it.

The second axis of the book is the reproductive cooperation and conflict between male and female blue-foots. Here there are no starry eyes, no ideals or moral codes, no guilt or shame. When a male and female commit themselves to a highly demanding and intricately coordinated project—producing fledglings while maintaining personal viability for subsequent breeding attempts—there can be nothing noble or disinterested about it. Natural selection does not readily tolerate unrewarded kindness or self-sacrifice. Rather, for this coordinated labor, both partners are economically driven and pragmatic about choosing the best available partner, calibrating their workload to circumstances, and allowing supernumerary offspring to be killed. Both members of each monogamous booby pair are also likely to engage in what in humans we call adultery. Males do it in order to produce additional offspring at another male's nest (and expense!), females supposedly to obtain more favorable genes for their offspring. Some females also dupe other blue-foot pairs into adopting their offspring by covertly laying in their nests. These tactics and the countermeasures adopted to thwart them involve selfishness and deception between bonded partners and exploitation of others, and they find parallels in human behavior. As do some booby tactics we would tend to admire. For instance, rather than divorcing, some booby pairs renew their monogamous bond over several seasons, resulting in better coordination, earlier nesting, greater hatching success, and the production of more fledglings.

Not that there are any straightforward moral lessons in the behavior of boobies or other non-human animals. Moral lessons are often sought though, and some of my colleagues strain to find, and are delighted when they uncover, social behavior in their study species that exemplifies what they consider praiseworthy in humans. This is fair enough when the worthy behavior is used casually as a motivational tool, but misguided when occurrence of the behavior in nature is taken to grant it moral authority in humans, a mistake known as the *naturalistic fallacy*.[4] In nature, there is breathtaking beauty and there are countless behavioral phenomena to admire and marvel over. For example, the pole-to-pole migration of Arctic terns, the complex community coordination of cooperatively foraging honeybees, and the paternal devotion of emperor penguins incubating in the long Antarctic night. But there is also much in nature that we are likely to find disturbing, such as rodent mothers eating their own offspring, packs of wolves disemboweling live moose, and female wolf spiders eating their mates after copulating (sometimes instead of copulating!).

So while blue-footed boobies can delight us with the familiarity and complexity of their family behavior, and study of it suggests persuasive explanations for our own similar behavior, we commit a fallacy if we look to boobies for moral justification of our sibling hostilities, parental favoritism, or marital deviations. Natural selection has dictated what boobies must unknowingly do, to promote the propagation of their genes. We humans choose what we will do, under the influence of genetically rooted urges and emotions shaped also by social experience and culture. We can rebel against inclinations that are morally distasteful. However, the logic of similar, naturally selected behavioral tendencies in boobies and humans sometimes explains *why* we are tempted to abuse our siblings or partners. Some of the unworthy feelings and inclinations that we struggle to suppress are quite natural, and we should not feel guilty for experiencing them, only for indulging them.

From Law to Boobies

Let me tell you how I got so involved with boobies and why I want to share my story with you. The personal voyage by which I transitioned from a legal career to research in sociobiology, a branch of behavioral ecology, and devoted 40 years to studying family conflicts in boobies is a tale of wanderlust, progressive awakening to nature and evolution, and intellectual stimulation by new theories of social evolution. When in the late 1960s I followed growing up in London by studying law at the University of Bristol and the College of Law in Guildford, I had already resolved to switch to language teaching. It was a heady time at the end of the 1960s, and I had no desire to inhabit offices and courtrooms or entangle myself in the paperwork of other people's conflicts. My game plan was to travel the world, experience exotic cultures, climates, and landscapes, and continue climbing cliffs and mountains (my undergraduate passion). After teaching American high school students in Montreux, Switzerland, for a year, and completing a 1-year graduate course in teaching English as a second language at the University of Leeds, I was offered teaching jobs in Norway, Costa Rica, and Mexico City.

The last of these had the most appeal. As I saw it, Mexico had it all, in the richness of pre-Colombian civilizations and their cultural and archaeological legacies; Atlantic, Pacific, and Caribbean coasts; climates ranging from sweltering heat in the tropical lowlands to permanent snow on the high volcanoes; dramatic diversity in landscapes; and a megadiverse flora and fauna. And it had not all been accessed by highways and condensed into maps and guidebooks, so travel felt like exploration. To my great surprise when I settled there, exploring natural habitats and landscapes inhabited by traditional communities largely supplanted my passion for climbing, and in my first 5 years I became familiar with Mexico's high plateau and coasts, and some of its volcanos and islands. Guided very often by Joaquin Albuerne, an extraordinary enthusiast who rejoiced and indulged in Mexico's natural and cultural treasures, I visited forests, deserts, grasslands, and coral reefs; dived in the Pacific Ocean, the Gulf of Mexico, and the Caribbean; and navigated rivers and lakes in home-made boats. I thanked destiny for landing me in a developing tropical country with warm-hearted inhabitants where I could satisfy my own mandate not to go to my grave without experiencing the glorious diversity of the planet. Delighted by brilliant humming-birds and coral reef fish, slothful tarantulas, and bustling army ant columns, as well as the imposing roars of howler monkeys and sweet gurgling of Montezuma oropendo-las, for several years my cup was brimming. Never much of an animal lover, I was ever more a wild animal admirer.

Through books, that fascination led to a new vocation. I devoured inspiring pioneer studies of wild animal populations, including Jane Goodall's chimpanzees and African hunting dogs,[5,6] and George Schaller's mountain gorillas.[7] I read three page-turning books by Robert Ardrey dramatizing academic accounts of conflict and co-operation in wild animals, and offering tentative theories to explain it.[8-10] Finally, with a nudge from my wife, Sylvia Rojas, a developmental psychologist, I alighted on the works of the three celebrated Europeans—Konrad Lorenz, Niko Tinbergen, and Karl von Frisch—who in 1973 shared a Nobel Prize for founding ethology, the science of natural animal behavior. I was captivated by the questions they addressed, the methods they used, and the concepts they invented to analyze the behavior of digger wasps and honeybees; cichlid fishes and sticklebacks; crows, herring gulls, and greylag geese. They were patently fascinated by animals in nature, and they explained the causation, development, and function of their behavior from a convincing evolutionary perspective.

So attracted was I by the elegance and astonishing explanatory power of evolutionary theory, and so eager to experience the behavior myself, that I set up twelve aquaria in my apartment in Mexico City. I wanted to see for myself what I had read about in Lorenz's popular book *King Solomon's Ring*.[11] And that did it! Watching the courtship, mating, and parental care of paradise fish, mouth-brooding cichlids, and Siamese fighting fish resolved me to transition from rootless voyeur of natural history to professional ethologist. After 5 years of teaching English in Mexico, this lawyer/

language teacher was generously accepted to work for a PhD thesis in ethology and comparative psychology at the University of Tennessee. In the laboratory of Gordon Burghardt, a specialist in the behavior of reptiles and black bears, and one of the first American psychologists to appreciate the magic and theoretical importance of ethology, I set about gleefully studying whichever reptiles came to hand.

Thus did I spend the last 5 years of the 1970s probing the causation, development, and function of predatory fish-luring by alligator snapping turtles, migration of hatchling green iguanas from their nest burrows, and predation of semi-aquatic snakes on fish. I watched litters of newborn snakes make their first attempts at feeding. I learned how to probe the lives of secretive reptiles in California's Sierra Nevada, the Little River in the Great Smokey Mountains, and Panama's Lago Gatún. For hours at a time, I watched garter snakes feeding on tadpoles and minnows in little mountain ponds, and hatchling green iguanas being picked off by toucans, greater anis, and basilisk lizards as they dispersed from a rainforest clearing. For my doctoral thesis, I compared the underwater predation of five species of snake.

Under Gordon's relaxed and benign supervision in Knoxville, Tennessee, immersed in a group of graduate students all working on their own reptile projects and discussing findings at enjoyable fortnightly seminars, I shed my legal and linguistic identities and became, as I had dreamed, a walking, talking apprentice ethologist. I also became a member of the community of ethologists and psychologists that converged annually from North and South America onto a university campus in the United States or Canada to present and discuss their research under the auspices of the Animal Behavior Society, my professional home, as I saw it. Those meetings were where, inevitably and almost unconsciously, I acquired the attitudes, values, and habits of thought of a community of people who delighted in observing and understanding what animals do in their natural habitats. Heaven! They were where I hooked up with experienced researchers I would eventually accompany and learn from at their field sites in northern California, Panama, and the Mexican altiplano. And they were where I made friends with admired colleagues who years later would host me in their laboratories or visit my field site to enjoy the island and bring their research skills to bear on blue-footed boobies.

But an important part of my mind and the minds of many graduate students in animal behavior at that time was elsewhere, focused on the unfolding sociobiology revolution. Understanding of the evolution of social behavior was undergoing extraordinary, effervescent rethinking after the publication of papers in the 1960s and 1970s by a new generation of theoreticians, most notably, by the father of the revolution William Hamilton,[12,13] followed by George Williams,[14] Robert Trivers,[15] Geoffrey Parker,[16] and John Maynard Smith.[17] Then in 1975 Edward O. Wilson published *Sociobiology: The New Synthesis*,[18] a landmark review of the field, and in 1976 Richard Dawkins followed it up with *The Selfish Gene*,[19] a masterful theoretical synthesis that rejoiced in explaining the treasure trove of social behavior of our planet's animal species. Sociobiology was the talk of animal behavior conferences and was getting headlines as legions of students and researchers jumped on board.

Meanwhile, some colleagues, leery of what they took to be distasteful political implications of genetic influences on behavior, denounced the theory and some of its most eminent proponents. Despite this unfair tarring,[20,21] the field prospered and has continued to prosper as part of behavioral ecology, effectively the current name of ethology, and the social theory reviewed by Wilson and Dawkins has stood the test of time and burgeoned into the present.[22–24]

Before the 1970s, scientists and nonscientists had mostly assumed that animal families are fundamentally harmonious, with all members of every family committed to a common project: producing healthy, well-nourished, independent offspring. Observed deviations from that, such as bickering by littermates or tantrums by offspring denied food by parents could be put down to imperfections in animal design. After all, you can't expect the division of food among the hungry or the trauma of weaning to go entirely smoothly; nothing does. But the emergent social theory, grounded not only in ethology and ecology but also in population genetics, was conjuring up quite different expectations.[25]

For example, it posited a fundamental conflict of interests between each caretaking parent and its offspring, with the parent better served by providing less care than would be ideal for the offspring. This penny-pinching would allow the parent to apply its remaining resources to other offspring (current or future), and thereby maximize the cumulative production of offspring over the parent's lifetime. Equally shocking at the time, the theory predicted conflict between siblings, with animals treating their siblings better than non-kin but not as well as they treat themselves, implying selfish and manipulative competition among brothers and sisters that could spill over into siblicide. Once you accept the insight of Hamilton and Trivers, that the fundamental competition in any population is not between the individuals themselves but between the competitive strategies they adopt, and even more fundamentally between the genes that influence the strategies, your understanding of family interactions is transformed. And the theory applied to all species that reproduce sexually, including humans. There was no reason to exempt humans, beyond our pride and species chauvinism.

Most radically, the new theory acknowledged the power of dependent offspring to advance their own interests, conceiving of them as active players rather than passive recipients of care. This opened the door to realizing that natural selection can favor the evolution of offspring wheedling extra care from parents, for example, by exaggerated begging, illness feigning, and aggressive suppression of siblings' begging. Naturally, the existence of such traits sets up conditions for the evolution of parental countermeasures such as diminished responsiveness to begging, or punishment of sibling suppression. The upshot would depend on the characteristics of the species and the age of the offspring. It could be overt behavioral struggles between demanding offspring and non-indulgent parents. It could involve infants attempting to kill their siblings while parents defend them, or embryos attempting to extract extra nutrients from mothers by physiological means while mothers resist with their own physiological tactics. Evolution by natural selection could no longer be conceived of as a broadly benign promoter of cooperation and harmony among family members. Incredibly, natural selection rooted for both sides of every family conflict and, paradoxically, every individual should seek parental indulgence when it's an infant and rein in infant greed when it becomes a parent.

This was seismic. If you've always understood that in Darwin's theory of evolution by natural selection is an almost omnipotent master craftsperson who, given enough time, can transform forelegs into wings, elaborate single photoreceptor cells into immaculate vertebrate eyes, and perfect all manner of physiological processes, corporeal structures, and behaviors, it's hard to grasp that it has a dark side—that it can also pit the closest relatives against each other, to the detriment of individuals, the population, and ultimately the species. I was elated to be joining the evolutionary biologists at just the moment when a new paradigm was sweeping their field.

After a pause, the unsettling ideas counted on the approval of some of the most distinguished specialists in animal behavior, and they were moored to a wealth of scattered observations on wild animals. In his seminal paper on the evolution of parent–offspring conflict, the Harvard wunderkind Robert Trivers cited numerous provocative examples of mammalian and avian families interacting in conflictive and inefficient ways that demand an explanation, and which seemed to fit the predictions of his theory.[15] For example, yellow baboon mothers and their infants in Kenya had been seen in protracted weaning conflict, involving daily tussles and loud cries from the infant, despite these monkeys being under strong selection to avoid predation by maintaining silence. Red warbler parents in the mountains around Mexico City had been seen attacking their own newly fledged offspring to silence their pestering food demands, and George Schaller,[26] the naturalist of mountain gorilla and snow leopard fame, had described extreme begging by white pelican chicks at Yellowstone Lake. To demand more food, one chick threw a fit that seemed to threaten self-harm. It ran to the parent, threw itself to the ground, and beat its wings wildly while waving its head from side to side, at one point seizing its own wing in its huge beak and growling continuously as it spun its body round and round. Even more dramatic was the black eagle chick in Tanzania that starved its younger sibling to death by harassing it fiercely during its first 6 days of life.[27] The mother could rarely transfer food to the younger chick because the elder one persistently pecked and tore at it, shaking it as a terrier does a rat, stifling its begging by squatting on top of it. Was it frustrating the mother's attempt to raise two chicks? Or was she complicit? These observations predating the new theory plainly contradicted the notion of family harmony.

Surely, you may think, it's a truism that family members often, maybe even mostly, support and assist each other. Yes, of course, but we always knew that! The new challenge was to uncover and understand the exploitation, manipulation, and deception that accompany support and assistance, to discover how these strategies are resisted, and to work out who wins. The behavioral ecologists who dedicated themselves to that mission were not pessimistic about animal nature or human nature. They'd simply become aware that it was time to incorporate family conflict in their studies because the theoreticians had opened a fascinating can of predictions.

The body of game-changing social theory that was causing all the excitement at the end of the 1970s, and also eliciting some healthy skepticism and repudiation, needed testing, and its implications needed exploring. The theory had run ahead of the data, and there was a sense that it was now the turn of the empiricists. To bring the Hamiltonian revolution to fruition, researchers needed to get into the field to measure and describe the cooperation and conflict between relatives and assess its effects on them and their progeny. They needed to identify suitable species for making persuasive tests, find ways of getting the critical descriptive data, and carry out experiments in the field or laboratory. In addition to confirming or contradicting the expected patterns of interaction, they would need to understand the processes by which appropriate behavior develops in the individual, tease out the mechanisms by which the behavior is controlled, and identify the ecological contexts and species characteristics that favor the evolution of particular patterns of cooperation and conflict. That was the agenda I would eventually address by studying the blue-footed booby and other birds on Mexican islands.

Booby Minds

We are currently experiencing a sea change in our perceptions of animals as sentient beings and in our attitudes to their welfare, treatment, and exploitation, so I need to come clean about my own understanding of booby psychology. I am not religious, but I have a reverence for nature, for species of plants and animals, for habitats and landscapes, and for the cosmos; I would describe this feeling as spiritual. I describe boobies in this book as I see them, as emotional but largely uncomprehending animals that do not mentally anticipate or plan. Rather, they follow what behavioral ecologists call rules of thumb: in situation A, do X. The manual of rules they follow, written by natural selection acting on millions of generations of their avian and dinosaur ancestors, is long and complex. It shepherds them reliably through the multitudinous challenges of growing up, staying alive and healthy, obtaining food, and producing viable offspring.

Forty years working in a blue-foot colony and hundreds of hours simply watching social interactions in neighborhoods of booby pairs have given me a personal perception of booby minds. Although I could not possibly know for sure, I suspect boobies have no intentions, experienced consciously as ours are, and have little or no ability to understand or imagine the probable outcomes of their behavior. Like most insects, I suspect, they may just do what comes naturally in the current situation, without in any way understanding the function of their deeds. I imagine that when a male booby dutifully scrapes a nest pit in the soil of his territory with his beak, he has no comprehension that this is a necessary step for producing offspring, nor even an intention of producing offspring. Scraping is just what he feels strongly inclined to do at the start of spring after courting and pairing with a female on his territory. Just as, days or weeks earlier, courting passing females was the irresistible urge of that male when the end of winter stimulated him to stake out a territory in the blue-foot colony.

I expect he feels enthusiasm and satisfaction as he does what he is inclined to do, finding pleasure in courting, copulating, incubating, seizing sardines, fleeing from intruding humans, arriving back at the colony with a stomach full of anchovies, and cheerily greeting his partner on arrival. If he has bred in a previous year, then I expect the sequence of reproductive actions feels familiar as it unfolds and experience enables him to pull it off with ever more panache (blue-foots with breeding experience breed earlier and more successfully), but I doubt that he mentally grasps that the whole long sequence of his behavior will lead via eggs and chicks to fledglings.

Some birds are evidently smarter than boobies. The poster children of avian intelligence, for their remarkable feats of making tools to access food and outwitting pilfering neighbors, are the Caledonia crow and the scrub jay.[28,29] I'm confident that more birds will eventually be found to have remarkable cognitive abilities, but I'm skeptical of boobies being among them,[30,31] because they are extreme ecological specialists, feeding on only one type of prey (fish) which they capture by only one method (plunge-diving), and they don't live in social groups, only colonies, in which social relationships are simpler. Flexibility is not their vocation. There is plenty of room in their lives for learning, for instance, because they must hone their weather-reading, shoal-finding, and mate-choosing abilities, but probably little scope for inventiveness. I have no doubt they sense and perceive the world around them, for example, listening attentively to their partner's calls and visually tracking fleeing sardines underwater, and I like to think their mental experience of the world is rich and varied in ways that are beyond me. But although I would be delighted to be proved wrong, I expect the vast domains of understanding, imagination, and intentionality that humans ordinarily inhabit are largely closed to boobies; even if their brains are adept, as they undoubtedly are, at unconscious learning and information processing in contexts such as plunge-diving and competing with neighbors for territory.

Caledonia crows and scrub jays may also be following rules of thumb and learning through experience without any insight, but the apparent cognitive subtlety of what they do suggests understanding and intentionality. Take food caching, for example. Unearthing some food and reburying it elsewhere because another jay witnessed you burying it the first time, and being more likely to rebury if you have previously pilfered another jay's buried food, speaks to foresight and wily understanding. It is difficult to imagine such flexibility in boobies. As we'll see in Chapter 8, a blue-foot female cheats less frequently on her partner when he is present, and before sneakily laying an egg in another booby's nest, a female booby waits for her to be absent or occupied by fighting with her neighbors.[32] But here rules of thumb along with learning by association seems a more parsimonious explanation than intention to deceive. Boobies are not simpletons, by any standard. Like all other vertebrate animals, they have immensely complex genomes, nervous systems, and endocrine systems, which respond in conditional and nuanced ways to the complex ecological and social environments they inhabit. And they undoubtedly learn throughout their lives. For example, an elder chick's response to its younger sibling's sudden peck might be determined by many experiences, states, and stimuli, such as the recent and accumulated frequencies of pecks given and received; the elder chick's energy reserves, current hunger level, and body temperature; and the younger sibling's body size, posture, and proximity. One way or another, all of that information is integrated, but the integration is experienced as an inclination to behave, not a calculation.

I don't mean to suggest that boobies are organic robots; they are sentient, emotional beings. When a chick responds to a sibling's peck by instantly rising up, growling, and redoubling its pecking, or by instantly turning aside, crouching, and burying its bill in its breast, I infer from this emotional behavior that the chick is *feeling*, respectively, angry or submissive. I am going way beyond any data I could possibly have recorded, because although I can see, record, and measure (e.g., with a ruler, stopwatch, or sonograph) components of that *emotional behavior*, there is no way I can experience a booby's feelings. They are strictly unavailable to me, in a domain that I can never hope to enter.[33] Whether the sensory organs and nervous systems of another species are very different to ours (e.g., in snails and vampire bats) or similar to ours (e.g., in chimpanzees), we simply cannot access their experience.[34,35] For all we know, some species such as amoebas, jellyfish, and cockroaches may not be conscious in any sense at all. Although they may perceive chemical, visual, or auditory stimuli, and respond appropriately to them, they may not *experience* the stimuli any more than a robot does.

It's easy to appreciate how this could be so. In principle, we could build quadrupedal robots with sensors and sophisticated software capable of responding to chemical, visual, and auditory stimuli in complex and subtle ways, and programmed to attack approaching robots at variable intensity, while puffing themselves up, squealing loudly, and turning red. They could also be programmed to avoid similar robots thereafter. Add all the design and programming subtleties you can think of, and you may end up with a robot so realistic that people mistake it for a real animal. But would it actually *feel* anything? I suspect not. It would, as far as we know, experience nothing: no vision, pain, anger, or sense of direction; no understanding, no awareness or self-awareness. As I have described it, it would detect and analyze stimuli and have observable emotional behavior, like variable anger (puffing up, squealing, and turning red) but be totally devoid of feelings.

The question is: couldn't some animals be like that? Maybe amoebas and spiders? Even maybe iguanas, snakes, and … boobies? Although neuroimaging and brain lesion studies are telling us which brain regions of humans are involved in their experience, no one truly knows which elements of our bodies (e.g., brain regions, neural networks, viscera) actually give rise to experience or how they do it, and this profound mystery is a subject of much conjecture.[35-37] Until we resolve it, we can only make guesses about which species have experience; and after we've resolved it, we still won't feel what they feel. Of course, from personal experience we know healthy awake humans have experience (but not at what age, pre- or post-partum, it starts!) because most of us are philosophically relaxed enough to attribute to other humans what we experience ourselves. We also assume, reasonably, that other humans' experience, including their emotions, is qualitatively similar to our own, except in exceptional circumstances such as brain damage and under influence of psychoactive drugs. But what about dogs and boobies?

Charles Darwin described the emotional behavior of dogs and cats and included drawings of their aggressive and fearful postures in his ground-breaking book *The Expression of the Emotions in Man and the Animals*.[38] From such emotional behavior as postures, bristling (hair-raising), and vocalizations, he readily inferred feelings and intentions, for example, a "humble and affectionate frame of mind" and "hostile intentions." He described "angry" postures and feather-ruffling in numerous species of birds, including chicks, swans, and tropicbirds, suggesting also that reptiles, amphibians, and earthworms are emotional. Some scientists have been disapproving or censorious about attributing feelings to other animal species because, as they correctly insist, we cannot experience their feelings or measure any more than their emotional behavior. But just as we cannot be 100 percent confident that dogs or any other animal species have feelings, neither can we be at all confident that they don't. To live responsibly, equitably, and productively in this world, we often need to make assumptions about things we cannot know. What then should we assume?

Feelings in some animals are actually expected because evolutionary theory tells us that any biological trait observed in humans is likely to be present in at least a rudimentary form in related species—otherwise, what did it evolve *from*?—and because there's a solid foundation of neurological evidence for the capacity for feelings in other vertebrate animals.[39] I have detected what I take to be emotional behavior in many of the mammals, birds, and reptiles that I have carefully observed, and tentatively inferred the existence of feelings similar to the feelings I have experienced, including anger, fear, and hunger.[40-43] My hunch is that, while they almost certainly experience such feelings, boobies and most vertebrates lack moral emotions such as empathy, sympathy, shame, guilt, and loyalty.[43] Moreover, the thoughtful observations of Frans de Waal and Sarah Hrdy on chimpanzees have largely persuaded me that some apes, in addition to emotions and feelings, have imagination and an awareness of others at the level of "a theory of mind"—conceiving of others as beings that think, experience, and act.[3,44]

Boobies, on the other hand, may simply experience. They may have no awareness of self,[37] just ongoing, present-tense sensation and perception, sometimes accompanied by feelings such as alertness, fury, submission, a panicky need to flee, or a chick's impulse to peck at a wobbling head or reach into a gigantic open beak. Boobies surely experience physical pain, and I will suggest in Chapter 7 that they experience the booby version of love, but I doubt that they feel empathy or self-pity. For these reasons, and because I want readers to appreciate boobies' minds as well as their behavior, I season my text with references to boobies' emotional behavior and feelings, although I rarely do the same in my research reports. I concede that there's a risk of misleading. Even people very familiar with a species can find it difficult to agree about which feeling an individual is experiencing in a particular instance.[45] However, because animal feelings are a fascinating realm that arouses our curiosity, and because we have a duty of care to animals that can probably suffer, I believe that when behavioral ecologists address a wider public, we do well to acknowledge the feelings our study animals seem to experience.

Meeting the Blue-Foots

When I was hired as a researcher in 1980 at Mexico's national university after finishing my doctorate, I did not specifically intend to work on social behavior. In fact, I set about traversing the Mexican altiplano to discover what the semi-aquatic snakes were eating and experimenting on their food preferences in my office (I had no laboratory). Then two colleagues invited me to accompany them on a reconnaissance of Isla Isabel, a tropical island off Mexico's Pacific coast, to get acquainted with its populations of reptiles and marine birds. In the early 1970s, I had happened to see a Philippe Cousteau television program, *The Sea Birds of Isabella*.[46] A scene of brown booby parents repeatedly feeding a 2-month-old chick while allowing its younger sibling to starve had stayed with me. I had not noticed, and neither had Cousteau, that the face of the cringing, enfeebled, and desperate smaller chick was scarred by siblicidal pecking. By coincidence, at the time of the invitation, two undergraduates, Cecilia Chávez-Peón and Alicia Castillo, asked me to supervise a joint thesis on the island's blue-footed boobies, which had won them over during a diving and kayaking vacation. A quick dip into the writings of Bryan Nelson,[47] the international gannet and booby authority, revealed that in the typical brood of two blue-foot chicks the elder dominates the younger and, if food is short, kills it with "hostile behavior."

That settled it; I was more than willing to help Ceci and Alicia work up a research proposal. Their pals from the university climbing and diving club would sort out the camping and logistics, while they themselves worked full-time getting basic descriptive data on blue-foot reproduction. They would also quantify behavioral conflict among chicks and between them and their parents.

When my two colleagues and I finally jumped from a fisherman's boat onto the black, wave-beaten basaltic rocks of the island's northeast shore in the spring of 1981, Ceci and Alicia had already recorded enough descriptive data on reproduction and social behavior to confirm that the blue-footed booby was a highly amenable species for analyzing family conflict. At that moment, the two students were in the fishermen's camp on the other side of the island, but right in front of me, beneath a multitude of soaring magnificent frigatebirds, there were booby nest pits all over the beach and adjacent forest floor, and dozens of blue-foots were incubating eggs, brooding chicks, and courting with their partners. Others were doubtless occupied with threatening neighbors, intruders, and predatory Heermann's gulls, or off on hunting trips. Some were just meters away and none was overly troubled by my arrival in their colony. If I could resolve the finance and logistics of working on an uninhabited island with no fresh water 50 km from the port of San Blas, these entertaining, expressive, and highly observable birds were potentially ideal subjects for delving into cooperation and conflict between relatives. If their fledglings remained permanently attached to their natal colony, we would even be able to tell whether sibling conflict in infancy marks a booby for the rest of its life.

Next summer we returned briefly to the island to check on the boobies and confirmed that some of the few adults Cecilia and Alicia had banded were again nesting in the same neighborhood of the colony. With that revelation, my herpetological career was over. I would phase out the snake research and learn more about birds and how to study them. Since then, with the help of colleagues and technicians, scores of student volunteers, and dozens of thesis students, together with logistical support from the national park administration (not yet in existence), the Mexican navy (not yet in the picture), and the ever-generous fishermen who camp on the island every spring and summer, we've maintained a camp on Isla Isabel during at least 5 months of every one of the last 42 years.

The Isabel blue-foots are excellent study animals for getting to grips with theories of family conflict, not only because they are so numerous, accessible, and tolerant of humans but also because they don't live in social groups. The pared-down context of a nuclear family facilitates identifying behavioral adaptations with greater clarity and confidence than when additional relatives or unrelated group members are socially involved, as they normally would be in group-living species like meerkats and turkeys. So boobies offer a straightforward model to compare with nuclear families

Fig. I.3 Cecilia, Alicia, and Hugh on wave-cut platform with fisherman Tapacho and his team in 1981.
Copyright Pablo Cervantes.

of other species, including humans. True, blue-foot families are always embedded in reproductive colonies of hundreds, in which chicks can be killed by colony neighbors and partners have sexual entanglements with other adults. But consequential interactions with other colony members during a reproductive season are otherwise confined to squabbles over territory and occasionally combining forces with neighbors to repel predatory Heermann's gulls. In marked contrast with group-living primates, there is no evidence of sibling relationships or parent–offspring relationships persisting beyond the year of birth: blue-foots do not nest selectively close to or far from their parents or siblings, and nearness of close kin in the colony has little or no effect on a booby's reproductive success.[48] If boobies do networking, it must be very limited.

Although we didn't plan it that way, Ceci and Alicia's undergraduate thesis project burgeoned into the first long-term *individual-based* study of reproduction, life histories, and social behavior in a long-lived tropical seabird. Both ornithologists and research funding are scarce in the tropics, so behavioral studies there have been comparatively few and usually short-lived. The foundation of the blue-foot program is the annual labeling of fledglings with engraved steel bands (23,000 to date) followed by the recording of their survival and breeding efforts throughout every subsequent breeding season until they die. And this work is complemented by single-season studies involving intensive direct observation of families, physiological measures, and field experiments. We have never taken a booby into a laboratory.

At the outset, following Konrad Lorenz's advice and my own firm inclination to start by just watching,[49] we simply described what families were doing; we watched their behavior and recorded the growth and survival of chicks. Later, to test various behavioral hypotheses, we cautiously and respectfully manipulated chick feeding rates, sibling age differences, brood sizes and compositions, circulating hormone concentrations, and the attractiveness, nutritional condition, and apparent health of male partners. Experiments inevitably affect the lives of animals, and our experimental treatments sometimes influenced the boobies' feeding rates or aggressive interactions. But we've been careful not to induce any states or behavior outside the ranges normally experienced by blue-foots, and to confirm afterward that normal growth, behavior, and survival resumed after experimental treatments ended.

The overriding motive of our research program has been to explore conflict and cooperation among family members—between infant siblings, between chicks and parents, and between female and male partners. But every finding leads to more questions, and student interests have tugged the program this way and that, so over the years, we've analyzed many other aspects of the boobies' lives and their interactions with the marine environment: the relative contributions of mothers and fathers to the care of offspring; the benefits and costs of prolonged pair-bonds; loyalty to the natal colony and nest site; relative production of daughters versus sons; male-female negotiation over the choice of nest site; how youth and old age affect a booby's reproduction and the quality of the offspring it produces; impacts of El Niño Southern Oscillation, global warming and hurricanes on boobies' survival, diets, and reproduction; and how the removal of invasive rats and cats from Isla Isabel exposed boobies to a wave of snake predation. I have woven some of these themes into the book to compose what aspires to be a uniquely intimate, thorough, and multifaceted portrait of a tropical marine bird. I have also included some of our research on brown boobies, brown pelicans, and cattle egrets, and compared all of these birds with other siblicidal species when comparison can illuminate the adaptiveness and evolution of blue-foot behavior.

But first, let me set the scene by introducing the two species of boobies, Isla Isabel itself, and the other marine birds and reptiles of the island.

Two Boobies and Isla Isabel

The blue-footed booby (*Sula nebouxii*) belongs to a family of medium-sized, plunge-diving marine birds (the Sulidae) comprising seven species of boobies (genus *Sula*), which mainly inhabit tropical and subtropical waters, and three species of gannets (genus *Morus*), which are larger and mostly inhabit temperate waters. Sulids are morphologically similar to pelicans but not as close to them in ancestry as was once believed. Their nearest relatives (also in the Order Suliformes) are the frigatebirds, cormorants, and anhingas. Their brilliant, patchy coloring, often including a bright face and feet, may have earned boobies their name. It has also been suggested that sailors thought they were clownish because of their big webbed feet, waddling foot-lifting gait, and tolerance of approach by humans, which allowed sailors to club them to death. In reality, like many animals that live on islands lacking mammalian predators, boobies have a high threshold for fleeing from approaching animals, and when they have a brood, they'll stand their ground even when humans are just meters away. We capture them on their nests by hand.

In all booby species, females are larger and heavier than males; in blue-foots and brown boobies females are 33 percent and 31 percent heavier, respectively. Blue-foots have a white belly and streaked white breast, thick neck, chocolate brown back, cocked tail, and, of course, brilliant blue feet. They inhabit the narrow band of the eastern Pacific Ocean where the food-rich California Cold Current and South American Humboldt Current brush along the west coast of the American continent, from southern California to northern Peru. They nest, among other places, in the Revillagigedo Islands, islands of the Sea of Cortez, and the Galapagos Islands, favoring areas with cold-water upwelling and abundant pelagic fish. The brown booby (*Sula leucogaster*) is relatively small and slight, and more mild-mannered when approached by humans. Its white belly and lower breast are sharply demarcated from its solid brown upper breast, and its feet are bright yellow. Brown boobies are found all around the globe in tropical latitudes, and they often share islands with other species of booby, including the blue-foot.

Boobies and gannets are the world's consummate plunge-divers. Many marine birds such as frigatebirds, terns, and gulls can snatch food from the ocean surface, and some, such as brown pelicans, can plunge in to grab prey beneath the surface.

Fig. I.4 Blue-foot female flying.
Copyright Daniel J. Field, used with permission.

Fig. I.5 Blue-foot captures unusually large fish.
Copyright Daniel J. Field, used with permission.

Only boobies, gannets, shearwaters, and, to a lesser extent, tropicbirds dive in head-first from the air and go deep. Boobies pursue prey to depths of 10 m by vigorous paddling with crooked wings and webbed feet. Plunging typically happens when a blue-foot, flying and gliding just meters above the waves, suddenly dips and slices through the surface, or when it joins a frenzied group of boobies, brown pelicans, frigatebirds, and gulls circling 10–30 m above a vulnerable shoal of fish forced to the surface by predators such as tuna and dolphins. Detecting an opening, it suddenly folds its wings and plummets vertically, holding a streamlined torpedo shape as it pierces the surface, sometimes accelerating its entrance by flying vertically downward before folding its wings. In among the panicking herrings, anchovies, or sardines, it lunges and grabs with a long, narrow bill with razor-sharp serrated edges. Blue-foots mostly feed inshore; they can feed near their colony in water just a few meters deep or make foraging journeys of at least 30 km, but their foraging trips, flying solo or in small groups, seldom last more than a few hours. Unless you go to sea or look out to sea from the shore, you will never see blue-foots, because they nest only on islands and rarely fly over land. Brown boobies feed similarly, although they may be more inclined to forage in large groups and nocturnally, and they typically feed on the open ocean, taking squid and a great diversity of fishes, including flying fish, which they grasp in the air.

On the face of it, annual reproduction by blue-foots is the marathon undertaking of a monogamous pair. Early in the season, the two birds typically pair up on a patch of terrain in the colony secured by the male (his territory). After a few days or weeks of additional courting and copulating, the female lays a clutch of one to three pale blue eggs in one of the shallow unlined pits they have scraped out with their beaks. Then over a 5- to 6-month period the two partners share all caretaking duties—territory defense, incubation, feeding of the brood—until the chicks fledge and transition to independence. All of this goes on in the essentially hostile environment of the colony, where pairs must deal with neighbors encroaching on their territories, boobies on transit through the colony making brief territorial incursions, and other species attempting to prey on eggs and chicks. The two birds take turns crouching on the nest or standing over it. The off-duty partner may remain there, too, especially at night, but usually departs on a foraging trip after a change-over ritual.

The ocean can be stormy and blue-foot offspring face many hurdles. Most offspring either fail to hatch, die before fledging, or fail to make the challenging transition to independence. Most fledglings simply don't have what it takes to survive a few years of juvenile maturation and learning, and then compete successfully for a territory and a partner. Adult boobies must therefore do their utmost to nest early and fledge well-fed high-quality chicks, all the while staying in good enough physical condition to survive into the future. The most successful boobies—those that produce the most offspring—are those that reproduce many times. After surviving the perilous first 4 years of life, annual mortality of adults is only about 10 percent,[50] so the most outstanding and lucky individuals live as long as 25 years.

Brown boobies reproduce similarly, except that the female lays only one or two eggs, slightly smaller than the blue-foot's, in a volcano-shaped nest the partners build on the ground out of grass, twigs, and seaweed. The partners phase-out continuous brooding of chicks after the first 12 days to allow expansion of their hunting schedules. By then there is usually only a single chick in the nest because, in the event of two hatching, the elder kills the younger.

Isla Isabel is a 95 m high cluster of eroded Plio-Pleistocene volcanic craters set on a bulge in the continental platform; it lies 28 km off Mexico's Pacific coast and 430 km southeast of the tip of the Baja California peninsula. The circular crater of one of the volcanoes is intact enough to house a green hypersaline lagoon that is rich in algae and microbial mats and 17 m deep. Usually, the crater is eerily quiet except for the diabolical calls and bill-rattling of magnificent frigatebirds. The rough guano-stained basaltic rock and tuff of the island is mostly black but weathered to brown or rusty red in parts. The shores comprise vertical cliffs, wave-cut or elevated rock platforms, the lower ones with tidepools, and three yellow-black beaches of sand, gravel, or boulders of basalt and bleached coral. There are several unvegetated islets within a few hundred meters of the island's shores, including Cerro Pelon (Bare Hill) and two imposing tuff spires, known as Las Monas (The Apes), standing just 200 m off the northeast edge of the island, opposite the main subcolony of blue-foots, where we camp.

Seventy percent of the island is covered by a deciduous forest of garlic pear and papelillo trees, its floor littered with boulders, fallen trees, and broken branches, the legacy of periodic hurricanes. In the 7-month dry season, the largely bare forest looks gray, for the lichens that clothe trunks and branches; during the June to October rains, it lights up emerald green with the flush of new garlic pear leaves. There are swathes of grassland dominated by waist-high bunch grass along parts of the northern and eastern shores, and on Cerro del Faro (Lighthouse Hill).

Nearly 4,000 blue-foots nest between December and July, mostly in the forest, the beach in front of Las Monas, and patches of grassland along some shores and on Cerro del Faro. They favor approximately horizontal land, with a preference for sites blessed by enough nearby wind and gradient to facilitate take-off and landing. Nests are grouped in neighborhoods of pairs whose laying is very loosely synchronous. Those nesting in the forest interior wear visible 30 cm wide booby trails as they trek to and from the shore for their fishing excursions, gingerly avoiding hostile neighbors as they pick their way through rocks and fallen branches. Some avoid the walk-in after a foraging trip by descending through a gap in the tree canopy, but blue-foots are fast straight fliers with little hovering ability, so gap descents sometimes end badly with crashing branches and torn webs, and can even be fatal when a booby gets snagged. For this reason, 20 to 80 m walk-ins and walk-outs through the colony are the favored option for many boobies, despite the rocks and debris, and the gauntlet of hostile territory holders. Maybe they learn something useful about the value of other territories or the goings-on of other boobies as they pass through.

As many as 6,000 brown boobies nest on the island and the Cerro Pelon islet in the rainy season, mainly on rocky or grassy slopes, where there are few blue-foots.

Fig. I.6 Isla Isabel in the dry season. The main blue-foot subcolony lies in the forest between Las Monas beach on the right and the crater lake.
Photo by D. W. Anderson, UC Davis.

This terrain minimizes contact between neighbors, who are often on another level or hidden behind tufts of grass, and allows take-off and landing close to the nest. The heavy summer downpours that brown boobies regularly experience don't form puddles in nests built on sloping ground, whereas the rare rains of February and March flood a small minority of blue-foot nest pits.

Isla Isabel teems with wildlife, most of it highly tolerant of nearby humans. Since the removal of invasive cats and rats in 1995 and 2009, respectively, there are no mammals, but marine birds are always visibly abundant, and terrestrial reptiles are more discretely abundant in all seasons except winter. All year round and all day long, clouds of *magnificent frigatebirds* soar above the island, circling ever higher on its thermals while hundreds of others incubate eggs and feed chicks on rickety treetop twig platforms, or glide over the ocean snatching prey and carrion from the surface, often following fishing boats. Lacking a preen gland, their feathers are not waterproof, and they are unable to enter the water without getting waterlogged and drowning. In the air though, frigatebirds are superb. Small-bodied and short-legged, and with longer wings for their body weight than any other bird, their aerobatic skills in swooping for food and pursuing other marine birds are spectacular. Clusters of males lure females to their treetop leks—communal displays—by inflating enormous crimson pouches on their necks and rattling their bills irresistibly, although only one of them will be chosen. The female lays a single egg, and while it takes at least 16 months to raise a chick to independence, the male usually deserts when the chick is just 4 months old, to prepare for his next nesting attempt with a different female.[31] Female and juvenile frigatebirds are the bane of boobies' lives because they pursue and harass boobies arriving at the island with full stomachs after lengthy fishing trips, sometimes pulling their tails and upending them into the water. Seldom do these pirates provoke the booby to regurgitate, but when they do, a gull or another frigatebird often seizes the booty before the pirate can swoop down.

Heermann's gulls are the island's streetwise gangsters. Unlike the other marine birds, which faithfully stick to stereotyped feeding tactics, these little, lead-colored gulls with orange bills are flexible and daring. They can make an honest living grabbing fresh prey in tidepools or snatching them from the surface during the boobies' plunge-diving frenzies. They are also brilliant opportunists who loiter around boats and campsites to take advantage of discarded offal and food and appear out of nowhere to seize any kind of food inadvertently dropped by dish-washing biologists and chick-feeding boobies. They descend suddenly through the tree canopy to stalk the blue-foot colony in small ad hoc bands, some provoking and distracting the boobies by staring or darting at their nests, others dashing in and seizing visible eggs. They nest on the ground on rocky islets and shoreline rock platforms at the time when blue-foots are hatching their eggs, each pair raising one to three precocial fledglings from a clutch of two or three exposed but superbly camouflaged eggs. They swoop on biologists who intrude in their colony, uttering menacing calls and pecking heads.

Brown pelicans, fully twice the size of blue-foots and the only pelicans in the world to feed in the sea, nest both in the treetops and on the forest floor of the island's highest ridges at the same time as the blue-foots. Like the blue-foots, each pair raises and shares the care of one to three chicks from an initial clutch of one to three eggs, and in every brood, an aggressive dominance hierarchy determines who will thrive and who will die of starvation, wounding, or expulsion from the nest.[51] Groups of adults and juveniles loaf on the beaches and islets and float in the shallows, herding small fish and scooping them into their capacious pouches before straining out the water. They also join in the boobies' plunge-diving frenzies, but their dives are shallow. Brown pelicans are most impressive when, before sunset, skeins of them ascend in file or V-formation from a bay or beach to a ridge top, alternating gliding with leisurely flapping, sober and serene as judges.

The birds that seem almost mystically "other," however, are the *sooty terns*, handsome black and white swallow-like birds that usually nest in a single synchronous flock over several weeks any time between March and December. The world's most numerous marine bird, its local numbers have declined during our project from several thousand to several hundred, just as a much larger population on Ascension Island in the Atlantic declined by 84 percent over a similar but earlier period.[52] Traveling in fast-flapping flocks, sooty terns snatch food on the wing from the ocean surface. They range far from land following tuna shoals over blue equatorial waters all year round, flying continuously, never sitting on water or perching on boats or buoys, on the wing day and night, sleeping, if they do, in the air. I always imagine they are fearful of islands when they come on land to breed, with cats, rats, snakes, gulls, and iguanas potentially roaming both grassland and forest. They move in only gradually, and after much probing. In March, a dense, coordinated cloud of wheeling and plaintively screaming birds flies ever closer to Isla Isabel's shore, small groups eventually breaking off by night or day to reconnoiter the grassland at the forest edges on the wing until finally, the whole flock takes over a swath and hundreds of pairs make nest burrows in the bunchgrass. There, in a seething colony that erupts in whirling, screaming flight when any predator enters, each pair raises up to three fledglings.

Least visible to us because they have minimal contact with the land are three bird species in which every pair of partners works together in the spring to raise a chick from a one-egg clutch. Two to seven pairs of *red-footed boobies* nest out of sight atop tree branches and either plunge-dive into fish shoals or catch flying fish in the air during day-long feeding trips. They come in two color morphs—brown or white—and the subtle pink and blue hues of their faces are alluring. Three hundred or so politely nodding terns called *brown noddies* nest colonially in pockets low in cliff faces, fishing by snatching prey from the surface or making shallow plunges during 1- or 2-day foraging trips, never flying higher than 4 m above the sea. In February and March, scores of screeching *red-billed tropicbirds* flutter energetically about the island's cliffs, trailing elegant white tail streamers. These brilliant snow-white birds nest in clefts, pockets, and small caves, where they lay a single egg; they hunt alone, making shallow plunge-dives for fish; they don't join feeding flocks.

Reptiles range from the tiny *brahminy blindsnake*, a subterranean creature only as long and thick as the lead in just half a pencil (with specs and good light you might just see its forked tongue flicking), to the omnipresent *spinytail iguana*. Thousands of these great omnivores haul their heavy bodies through the grassland and over the forest floor and tree branches of the island. They refuge in rock crevices and hollow trees and feed on garlic pear fruits. Swaggering through the blue-foot colony, they steer clear of nests, which are vigorously defended, but sometimes manage to grab fragments of fish that fall to the ground when boobies feed their chicks and, rarely, seize unprotected chicks. Our campsite usually harbors a posturing orange- or yellow-tinted male with a massive head and an imposing dorsal spine crest, defending a territory inhabited by six or seven females. That dominant male periodically performs head-bobbing displays or dashes into the undergrowth to chase off smaller and duller rivals. There are more iguanas than is comfortable in a campsite, and it is with difficulty that we avoid treading on tails and manage to keep insistent individuals out of our vegetable and fruit store and frying pans. Though we seldom see them, there are also some emaciated *green iguanas* eking out a living on the sparse vegetation of the cliffs, for some reason disdaining the canopy of the island's forest.

Thousands of small insectivorous lizards populate the forest. *Clark's spiny lizards* perch motionless on tree trunks during the day, waiting to pounce on insects, and *yellowbelly geckos* scale the trunks and branches at night, peeping to each other as they search for insect prey. Even more abundant, or at least visible, are the *Western Mexico whiptails*, sleek long-tailed lizards that busily pick their way over the forest floor in sunny areas, digging hundreds of tiny pits to unearth insects and arthropods, repeatedly freezing and peering upward to detect predators, and dashing off at breakneck speed when alarmed. Male whiptails that have inseminated females follow them around as they forage, to ward off other males and prevent them from getting their sperm in, too, foregoing feeding opportunities to assure their paternity.[53]

The most beautiful and, for both boobies, lethal reptile on the island is the ponderously named *Atlantic Central American milk snake*. Nonvenomous and known locally as the false coral snake, it is greatly respected by fishermen because it resembles the highly venomous Nayarit coral snake, a burrowing species found in their villages but not on the island. At dusk, nighttime, and dawn during the 3-month hatching period of the blue-foots, over 200 of these sleek serpents with broad black, yellow, and red bands cruise the floor of forest, beach, and grassland. Constantly tongue-flicking, they are ever ready to seize and devour whole any whiptail lizard, juvenile iguana or sooty tern, or hatchling booby they can grab and work their jaws around. To our fascination and dismay, milk snakes often enter blue-foot and brown booby nests at night and either slowly ingest the chick while the brooding parent shifts about looking nervous but doing nothing to defend its offspring, or drag the chick out of the nest and consume it a meter or two away. This dereliction of parental duty is due to a failure to recognize as a predator a long, limbless creature with slow, sinuous locomotion. It certainly is not due to incapacity or excessive caution because boobies attack humans that come too close with volleys of violent, wounding pecks, and when they recognize a milk snake, they repel it with pecks or seize and fling it away.

El Niño and Hurricanes

From the boobies' point of view, Isla Isabel is blessed by its location close to the nutrient-rich waters at the southern limit of the cold California Current that washes down the western seaboard of North America, and cursed by periodic food crises due to El Niño Southern Oscillation. Every 2 to 7 years, the air pressure differential across the central Pacific Ocean is inverted and westerly winds drag warm waters across the ocean to the eastern coast of the Americas, subduing the life-giving upwelling of cold waters along the coast. El Niño events, named after el niño (the child) Jesus by Peruvian fishermen because they arrive around Christmas time, affect oceanic and continental patterns of rainfall and severely depress marine productivity, resulting in depletion of fish populations.

Mexican coastal waters are at a considerable remove and affected far less than Peruvian coastal waters, but the consequences of local warm water for the blue-foots of Isla Isabel are serious. During El Niño events, breeding activity and productivity are depressed, callow females and males are more likely to die, and boobies in the colony have elevated concentrations of corticosterone, the stress hormone, in their bloodstreams.[50,54] During the strongest El Niño events, few boobies even get as far as pairing on a territory and laying a clutch, and those that lay do so late in the season. These daring or unwise pairs lay small clutches and suffer frequent hatching failure, and *in extremis* sometimes resort to simultaneous foraging by both parents while chicks linger in undefended territories. Every additional degree in sea surface temperature near Isla Isabel means an average 0.45 fewer fledglings per nesting pair; and in 1992, when the mean sea surface temperature rose to 26.7°C, not a single chick fledged.[55]

Another potential hazard is the tropical storms and hurricanes that develop annually in the eastern North Pacific Ocean between July and October and sometimes pass close to or, in the case of Hurricane Rosa in October 1994, over the island. The wild winds, heaving seas, and foam-flecked waves of these events complicate plunge-diving and may well make it impossible over wide expanses of the ocean. A direct hit on Isla Isabel in the nesting season would surely be devastating for nesting boobies. Surprisingly though, after the passage of eight hurricanes of categories 1–4 (on a scale of 1–5) from 1990 to 2013, adult mortality *declined* and graduation of callow adults into the breeding population *increased*. This pattern could speak to more than resilience. It may imply that blue-foots can successfully evade hurricanes that strike after their breeding season, and that hurricanes increase local marine productivity and fish abundance, conceivably by mixing ocean waters from different depths.[56]

Logistics

Working on an uninhabited and waterless island 28 km offshore is logistically and financially challenging for biologists working from the national university in Mexico City, 2,250 meters above sea level and midway between Mexico's Pacific and Atlantic coasts. But we started modestly and have kept things that way, and our program has benefitted from the generous support of the fishermen of Nayarit for 42 years and the Mexican navy for 30 years. At the start of the 1980s, we departed from Mexico City every year in a university pickup loaded with camping gear and fieldwork materials. Late in the dry season, we drove west along the federal highway for 10 hours across the altiplano, often detouring in Michoacán and Jalisco to see what garter snakes were eating around the shores and reedbeds of lakes Pátzcuaro, Chapala, and Cuitzeo, and then continued on to Guadalajara. After a night in a flophouse or dossing down under the stars beside a farm track, we would follow the highway out of Guadalajara northwest to the coast. This lovely drive on winding roads packed with trucks descended through the mountains of the Sierra Madre Occidental into Nayarit, passing through the arid agave fields of Tequila and the bright green sugarcane fields of Tepic, the state capital. After a food stop in Tepic, we descended a further 920 m through grey deciduous forest to Boca de Camichín, a village at the mouth of Rio Santiago, where about 350 families currently live mostly from fishing and oyster farming.

A couple of days later, after returning inland for supplies of food, water, and gasoline in nearby Santiago Ixcuintla, and spending a night or two sleeping on the floor of a fisherman's house, a couple of fishermen (often Bouleras, Chepo, Pablo, or Tigre) would ferry us in their *buguis* (narrow 5 m open boats, usually without compass, tools, waterproof clothing, or life jackets) across the standing 1.5 m wave at the river mouth, and on to Isla Isabel in about 3 or 4 hours. Whereas the fishermen camp in huts in a sheltered bay near the southern end of the island, we biologists set up tents every winter in a scenic site on the island's eastern shore, on the edge of the booby colony. Here the forest abuts a wave-cut platform where at high tide we unload supplies from the fishermen's boats, and every dawn for 5 months we watch the sun emerging from the sea by the guano-stained spires of Las Monas.

At the start of our program, the fishermen of Camichín specialized in catching sharks, mostly sharp-nosed and scalloped hammerheads, in gill nets and longlines they tended by the island at dusk and lifted after dawn, as well as "scale fish" (bony fish) such as red snappers and sea bass. One cherished memory is of the night when, 30 km out to sea, in a *bugui* tethered to a floating net, Bouleras kept us both up with tale after tale of colubrid snakes and boa constrictors on the mainland doing amazing things like suckling on sleeping women and cows, asking for my confirmation after each tale, declaring flatly that he or a friend had witnessed them with their own eyes, and dissolving into laughter. Shortly after dawn, in dense sea fog and visibility of less than 40 m, we hauled in the net with its meager catch. Then Bouleras, with no navigational instrument but his reading of the waves, fired up the outboard motor and gunned the *bugui* straight as an arrow across the ocean, directly into the mouth of Río Santiago.

Shark fishing at Isla Isabel has declined as shark populations have dwindled worldwide, and in the last few decades we have been ferried out to the island in *pangas* (longer than *buguis*) by fishermen from San Blas (often Emilio, Yamaha, Chamarras, or Rigo), who fish mostly for bony fish using longlines, gill nets, and scuba equipment (with onboard compressor and hose). Some fishermen, like Emilio, have become tourist guides, ferrying tourists from Camichín and San Blas to see schools of frolicking dolphins, whale sharks hoovering up plankton under surface slicks, and humpback whales migrating northwest with their calves. The navy has supported us with its vessels, transporting our food, drinking water, equipment, and personnel between the island and San Blas or Puerto Vallarta at the start and end of fieldwork, and in the course of monthly or fortnightly visits to the island. The fishermen ferry all our cargo between the ship and the wave-cut platform because only they can enter shallow water and dice with the waves and rocks. We bathe and wash our clothes and dishes in the sea.

Founded in 1768, San Blas is the port from which the Spanish colonists launched trading ships to Asia and military expeditions to Sinaloa, Sonora, Baja California, and California. Sited among mangrove swamps and lagoons and 50 km from the island, this relaxed little town with a sultry climate and a reputation for nearby surfing has grown only slowly because heavy silting prevents ships from approaching its quays and insect pests discourage tourism. Nonetheless, it has a naval base; a population of about 14,000; shops, hotels, and a gas station; and a new motorway between Tepic and San Blas is bringing faster development. It is to San Blas that our university pickup, loaded with camping and scientific equipment, travels every February, so we can cross the ocean in *pangas* or a naval vessel to the northeast edge of the island and pitch our tents.

The first six chapters of the book address the nature and evolution of sibling conflict and siblicide in blue-footed boobies and other animals. The first contrasts the conditional violence of blue-footed boobies with the unconditional, murderous violence of brown boobies and briefly outlines the basics of the theory that explains the differences between them. Chapter 2 is about dominance-subordination, the behavioral mechanism that enables one sibling to control another. It describes a number of unique field experiments on the blue-footed and brown boobies of Isla Isabel and Isla San Pedro Martir. These demonstrated that chicks use aggression to partially disable their competitors, and they led us to the insight that the *propensity* to be disabled—to learn subordination—is an adaptation. Chapter 3 describes the colorful diversity of sibling violence in different species of birds and offers explanations for why quite different modes of aggressive control evolved; it closes by considering sibling violence

Fig. I.7 Las Monas shore and spires. A fisherman's panga is tethered beside the wave-cut platform at top center.

Photo by D. W. Anderson, UC Davis.

Fig. I.8 Las Monas spires from the shore.
Photo by Hugh Drummond.

Fig. I.9 Dishwashing on the wave-cut platform.
Photo by Hugh Drummond.

in the few mammals that have been studied. Next, the issue of how aggressors regulate their behavior so that siblings are controlled and eliminated only when necessary is discussed in Chapter 4. That chapter highlights the ingenious experiments conducted in the field and laboratory by several researchers on a variety of bird species to test the three leading hypotheses. The following two chapters deal with two questions of particular interest to humans. Do booby parents attempt to manage sibling conflict in any way, and does the conflict between siblings mark the younger one for life? Although theory predicts conflict between parents and offspring, the evidence from observations and experiments on boobies points to parents colluding with their elder chicks over the control and conditional sacrifice of their younger chicks. And long-term data indicate that younger chicks who survive the nestling period emerge from the calvary of sibling oppression in remarkably good shape, as indeed do human younger siblings.

The next two chapters focus on the monogamous and possibly loving relationships that boobies form for the purpose of reproduction. Chapter 7 describes how monogamous partnerships form through mutual evaluation and bonding, how both sexes sometimes switch to a different partner, the practical benefits of sticking with the same partner year after year, and how partners make important decisions together. Then it considers the intricacies of this collaboration between two unequal partners with conflicting interests, concluding that the relationship is asymmetrical but egalitarian. Chapter 8 then looks at the dark side, detailing conspicuous conflict, deception and countermeasures to cheating, including infanticide, and asks why cheating has evolved.

Many aspects of booby behavior are reminiscent of human behavior, so the book finishes with a chapter describing the sibling conflict and monogamous relationships of humans. It flags numerous similarities between two taxonomically distant and very different animal species—*Sula nebouxii* and *Homo sapiens*—whose sharing of key biological characteristics arguably favored convergent evolution. Boobies and humans are both warm-blooded, long-lived vertebrates that reproduce monogamously, breed repeatedly, and have highly dependent offspring that simultaneously require the dedicated care of more than one caretaker. The coincidences in both basic biology and family behavior imply that similar cooperation and conflict in the two species could derive ultimately from similar evolutionary processes.

1

Two Approaches to Controlling and Killing Siblings

Though he attributed the cause to original sin rather than biology, St. Augustine put his finger on a fundamental and essentially selfish motivation of the animal kingdom when, in about 400 CE, he described the onset of human sibling strife:

> The feebleness of infant limbs is innocent, not the infant's mind. I have personally watched and studied a jealous baby. He could not yet speak and, pale with jealousy and bitterness, glared at his brother sharing his mother's milk. Who is unaware of this fact of experience?

Certainly not WebMD, a leading health publisher, whose 2020 webpage description of the nature and causes of sibling strife in American infants does not mention biology but roundly declares the generality of strife:

> Although siblings can be the closest of friends, it's rare to find a child who gets along perfectly with all of his or her siblings.... Brothers and sisters fight—it's just the natural ebb and flow of family life. Different personalities and ages can play a role, but siblings also often see themselves as rivals, competing for an equal share of limited family resources (like the bathroom, telephone, or last piece of cake) and parental attention.... Sibling rivalry is a normal part of growing up.

Humans are not alone. Although the costs and benefits of selfishness and kindness among siblings can only rarely be satisfactorily measured, studies of vertebrate and invertebrate species have broadly confirmed the theoretical expectations for widespread sibling competition in the animal kingdom that arose with the Hamiltonian revolution described in the Introduction.[1] Those studies have also uncovered the diverse ways in which siblings of different species compete for access to food and other resources such as parental protection. As we shall see in Chapter 3, these run the gamut from begging, negotiation, and scrambling all the way to fighting, dominance, and siblicide—killing a sibling.[2-4] The present chapter describes how we unveiled and got to grips with the controlling and conditionally siblicidal aggression of blue-footed booby chicks, and contrasts those subtleties with the unconditional, unflinching siblicide of the brown booby. It also suggests how the different strategies of the two species evolved.

Blue-Footed Boobies. Hugh Drummond, Oxford University Press. © Oxford University Press 2023.
DOI: 10.1093/oso/9780197629840.003.0002

Sibling conflict is better known in the blue-footed booby than any other animal species and should be of interest to anyone intrigued by the high prevalence of conflict in human siblings or disturbed by the conflict in their own family. As we shall see in Chapter 9, human sibling aggression is not an aberration; although more flexible than the boobies' dominance and killing, and often combined with cooperation and generosity, it arises from the same tension between conflicting reproductive interests of close relatives.

When we started looking at sibling competition in booby families, siblicide among avian chicks had already been reported in some birds of prey. It was called obligate cainism if the younger chicks were nearly always killed, and facultative cainism if their killing was variable.[5] This was a reference to the biblical story of Adam and Eve's firstborn, Cain, killing his brother Abel in a fit of sibling jealousy because the Lord gave his younger brother more attention. The roles of family members in the deaths of birds of prey were not well understood though, because the dedicated pioneers who observed isolated and inaccessible nests were seldom able to get data on growth, mortality, and behavior in large samples of broods.

The Blue-Foot Way

Inspired by the new theory of family conflict, our aim was to characterize cooperation and conflict among family members of an accessible species. We were particularly interested in knowing how aggressive and submissive behavior develops between chicks, and how it is influenced by the parental provision of food. Ultimately, we wanted to infer how the relevant social behavior of each family member—senior chick, junior chick, parents—could have evolved by natural selection. In other words, how could genes for sidelining or killing close relatives outcompete genes for tolerating and caring? Foregoing experiments until we had a clear idea of how family members treat each other, during four successive reproductive seasons at the start of the 1980s, we systematically registered development and behavior in blue-foot broods until chicks were a few months old. Comparisons of closely related species that differ in ecology and behavior can often be especially revealing, so we also observed the behavior and survival of brown booby chicks and dipped into the literature on their reproduction elsewhere. In the two species of booby we witnessed, and later probed with additional observations and experiments, markedly different lethal struggles between infant siblings in full view of their parents.

To document the growth and mortality of blue-foots, we monitored the contents of scores of nests on Isla Isabel labeled with numbered wooden stakes driven into the ground, briefly visiting each nest every 2–3 days to record its contents, and banding the chicks to be sure which was which. For behavior, we sat in the garlic pear forest or on the beach about 20–150 m from the sea and a similar distance from our campsite. Typically, we watched up to five nests at a time for 10 or 12 hours a day, every third day, over a period of several weeks. Observers sat silent and almost immobile,

Fig. 1.1 Blue-foot neighborhood at the forest edge with courting and incubating blue-foot pairs.
Photo by Hugh Drummond.

Fig. 1.2 Volunteer monitors blue-foot nests.
Photo by Hugh Drummond.

faithfully recording frequencies of parental food transfers to chicks, and of pecks, bites, and threats between chicks.

To reach your observation broods, you first have to pick your way through the defensive and mildly alarmed colony, evading the savage lunges of incubating and brooding parents, "sword-dancing" to save your ankles. When you arrive at your chosen cluster of nests and settle into a folding chair, the nearby boobies with clutches or broods stare at you with one or two beady eyes and kick up a kerfuffle, protesting and threatening. Females lunge and grunt, and males lunge and whistle, while solitary adults and bonded pairs that haven't yet laid back off a few meters or dash for the forest edge and fly off. A wave of alarm ripples through the neighborhood and promptly subsides, as happens if you simply sneeze during an observation stint. The colony is always alert and alive, even if some of its members are asleep, standing, or squatting with their heads resting on their backs.

Island-dwelling birds are tolerant of humans, so within a few minutes all is forgotten and the neighborhood relaxes into its background sound of communal grunting and whistling. The fliers are back on their territories, having completed a loop or two over the ocean and verified from the air that their neighborhood is safe. Nearby adults are busy on their respective territories, fighting with neighbors, repelling intruders, courting and copulating; chicks under their parents are sleeping or begging and feeding, or looking out at the world. Some nests are on the beach, cooled by an onshore breeze but exposed to the tropical sun, their occupants' back feathers sponged up and throat membranes vibrating, as they ventilate and disperse body heat. Most nests are in the forest among the fallen trunks and scattered branches, where the air is still but welcome tree shadows creep across the sand and leaf litter. Alternating in 2-hour shifts, we would sit there in swimsuit, t-shirt, and sombrero, scoring behavior, enjoying the rhythmic sound of breaking waves and the intimate atmosphere of

Fig. 1.3 Parent feeds 2-week-old chick.
Drawing by Jaime Zaldivar-Rae; based on photo by Hugh Drummond.

a colony alive with spinytail iguanas, whiptail lizards, and Heermann's gulls, taking it all in, untroubled by the light whiff of guano.

What we saw and scored on our clipboards did not disappoint. In every brood of two chicks, hatched 4 days apart, each one had clearly entered this world ready to attack on sight any little head that bobbed nearby. There was daily violent aggression between the two chicks throughout the 3 months it took for them to fledge. Pecking was infrequent at the start because the two chicks slept most of the time and their periods of activity seldom overlapped, but with time they were increasingly in contact with each other. It started when the senior one, naked but with emerging white down, was about 8 days old and finally had enough muscular development and

Fig. 1.4 Destined for subordination. Barely a week old, the junior blue-foot chick is learning its place. This tentative peck and later bouts of more forceful pecks and bites will train it into habitual submission within 3 weeks.

Drawing by Jaime Zaldivar-Rae; based on photo by Hugh Drummond.

Fig. 1.5 A lull in sibling conflict. The senior blue-foot chick relaxes briefly before rising up and pestering its parent with more begging. Junior is about to beg uselessly from senior, mistaking its sibling's head for a parent's head. Behavioral rules of thumb are more likely to misfire in the inexperienced.
Drawing by Jaime Zaldivar-Rae; based on photo by Hugh Drummond.

motor coordination to land a blow with its little pointed beak. It was still going on, at a lesser rate, months later when both down-covered chicks were sprouting dark brown feathers, approaching adult size, and walking clumsily about the family territory, some of them impudently threatening adult neighbors.

The first swinging thrusts glanced off the junior chick's elevated gray head, knocked it sideways, or toppled it to the sandy or damp earth floor of the nest. Pecks came singly or in groups and tended to continue until the intolerable little head subsided or disappeared from view. This could happen because the pecks did their job or because the brooding parent, aloof but moving restlessly on the nest cup, got in the way. Outgunned and seldom able to land a peck, juniors were punchbags, but by the time the brood was 3 weeks old, both chicks were able to deliver volleys of well-aimed, hard-hitting pecks and bites. Usually though, only one of them did. The senior one, armed now with a longer beak and angrier than ever at the sight of an oscillating downy head, battered its sibling's cranium, nape, and face, not sparing the eyes. Sometimes senior growled angrily with each blow or, when its sibling was out of reach, threatened by standing tall and growling. When especially incensed, even capitulation and prostration were not enough to satisfy it; it continued sinking pecks into its sibling's back, rump, and abdomen, biting, squealing, and twisting at the downy skin.[6] When they weren't knocked over, juniors sometimes ignored this aggression, sank down into the safety of the nest cup, or adopted a submissive posture, with head rotated sideways and tilted so far down that the beak pressed vertically against their neck or breast. This "bill-down-and-face-away" posture inhibits the other chick's aggression, although not always promptly. Sometimes ire wins out.

None of the blue-foot broods under observation suffered a death, but the significance of seniors attacking their siblings several times more frequently than *vice versa* was revealed by the patterns of growth and begging in the families we monitored.[6–8] The average senior chick increased in size and weight faster than the average junior chick throughout at least the first 8 weeks, despite their weights being similar at hatching. No matter whether broods grew slowly because it was a warm-water El Niño year, or fast because it was a cool-water La Niña year, senior chicks received more frequent feeds and grew faster than their siblings. And seniors managed this even though juniors begged just as frequently, if not more so. Effort was rewarded,

Fig. 1.6 Three weeks of battering is enough. The dominant blue-foot chick's threat call instantly elicits submission by its out-of-range subordinate sibling.
Drawing by Jaime Zaldivar-Rae; based on photo by Hugh Drummond.

Fig. 1.7 Marginalized junior blue-foot sibling. While an indifferent parent grooms itself, senior's threat call is sufficient to prevent junior begging and contending for the next regurgitation.
Drawing by Jaime Zaldivar-Rae; based on photo by Hugh Drummond.

though: the juniors that most outbegged their siblings were the ones that got the most feeds. Frequent weighing of 30 broods during the first 5 weeks of life revealed that a greater mass of predigested fish was transferred to broods at night than during the day and that senior chicks got roughly 20 percent more of the 24-hour totals than juniors of the same age.

This difference in feeding success we put down to aggressive social control: by dominating its nest mate, the elder chick confines how often, when, and where it can beg, and parents fall in line. Very often the senior chick's attacks clearly prevented junior from begging during a feeding bout or banished it from the spot where the parental bill was about to descend and offer food. However, hungry seniors could show conspicuous magnanimity: while begging next to and synchronously with its rival, a senior chick sometimes allowed it to feed from the parental mouth when it could easily have interfered. What's more, seniors frequently pecked at juniors when neither chick was begging and parents were neither offering food nor about to do so. The aggression of blue-foot chicks does not function exclusively for getting a larger share of food. Our hypothesis was that it also functions to *establish a dominance relationship*.

When you visit hundreds of nests every few days and only very briefly, predation by milk snakes and Heermann's gulls on eggs and chicks, and abandonment of nests by parents, can obscure how family interactions result in the death of chicks. Occasionally we witnessed a gull flying off with an egg in broad daylight or a snake on the forest floor swallowing a chick in the twilight. We attributed the disappearance of single eggs to gull predation, and the disappearance of chicks to milk snake predation if it occurred during the first 6 days of life when hatchlings are small enough to be swallowed whole. Frequently though, we simply found the dead bodies of chicks in or near their nests, or chicks disappeared between one nest visit and the next. However, over the first 4 years, patterns steadily emerged, and these told a story of competition among siblings for food and frequent sacrifice of the junior ones. Although pairs with two chicks commonly raised both of them to fledging, they often lost one or even both along the way, especially during the first month of life. Discounting losses to snakes, in all 4 years death fell selectively on juniors, and death of a single chick was nearly three times as likely to fall on junior as on senior. In 7,540 broods of two chicks hatched over many years, 25 percent of seniors and 35 percent of juniors died before fledging, indicating a tidy 10 percent survival pay-off for bullying.

A clue to this special vulnerability of juniors was found by comparing chicks' body weights with the weights of their peers, chicks of the same hatch order and age. In each of the 4 years, *seniors* were underweight just before their younger siblings died, with average deficits of 22–25 percent. Later, a sample of broods in which we knew the sex of each chick and could take into account the faster growth of females revealed three more clues. First, senior chicks recovered normal weight within 9 days of their siblings' deaths. Second, juniors were on average 43 percent underweight on the eve of their deaths. Third, nearly one-third of these juniors died with visible wounds on the head. This circumstantial evidence implied that senior chicks respond to meager rations and poor growth by facilitating the death of their siblings, maybe by pecking them more. And it also suggested that seniors benefit from the death of their siblings by eating more and growing faster afterward. Tentatively, seniors have the motive and the means to commit siblicide, and they benefit from it.

But how did death come about? Did senior chicks themselves or parents put juniors on starvation rations and inflict wounds on them? Did starvation provoke senior chicks to increase their attacking? Although in the colony you occasionally come across a chick giving its sibling an unusually sustained beating, constantly pecking, biting, and chasing it around the territory despite its pathetic and exaggerated submission, visible wounds from sibling aggression are uncommon. Chicks are immature; they can't pound heads and rip skin as adults can. Besides, natural selection for withstanding sibling abuse in the chick period has doubtless endowed chicks with resilience. An experiment described in Chapter 4 would subsequently test the role of starvation in sibling aggression, and our incidental observations over a period of 40 years have left little doubt about who does most of the damaging pecking.

Those observations pointed to infanticide by adult neighbors. No parent in scores of broods we have watched intensively and hundreds of broods we have watched casually has ever pecked or bitten one of its own chicks. However, from time to time we've seen a chick that strayed across a territorial border hammered by one or both adult neighbors with volleys of wounding pecks. These are mostly directed at its face and cranium and, when it adopts a submissive posture, its nape, too. As many as four adult neighbors can rapidly converge on a chick that intrudes on one of their territories, and if the chick doesn't manage to scurry back into its home territory, those adults drive it down to the ground with a barrage of pecks and may keep attacking its body and head until it is still—incapacitated or dead.

We've also come across many cadavers of chicks that were old enough to walk about the family territory with lesions around the face, cranium, and nape. And one morning shortly after dawn I came across a 1-month-old chick standing immobile several meters from its family territory, its nape and whole crown pecked to a continuous raw wound by whichever adult neighbors had sustained an extraordinary group attack on it. A chick could not do such damage. Judging by its body language, the wounded chick was chastened and full of dread. It stood alone in the colony all morning; next day it was nowhere to be seen.

Natural selection has surely favored this uncharitable, infanticidal repulsion of neighbors' chicks because boobies who raise the chicks of non-kin are raising reproductive parasites. Any gene that inclined adults to such charity would necessarily find itself in the genomes of fewer offspring than the alternative "infanticide" gene which ensures its own propagation by channeling parental care exclusively to own offspring. The hazard of unwitting adoption of neighbors' chicks arises after they become mobile enough to wander around their home territory. Presumably, they wander to develop their muscles and locomotor abilities and get to know the neighborhood, while not straying so far that they miss a parental feeding bout or come under attack by neighbors. Sometimes, by chance it seems, a chick accidentally strays into a neighbor's territory, and if one of the neighbors is attentive, it instantly and viciously attacks.

Adoption may be actively sought, too. We suspect that underfed blue-foot chicks seek adoption by neighbors when their prospects of thriving at home are so poor that it's worth running the gauntlet of a neighbor's defenses. Our monitors occasionally find individual chicks comfortably settled into a neighbor's nest, huddled or fighting with the resident chicks and effectively adopted into their families. Once I think I saw this happening. I was standing nearby when a 3-week-old junior chick that was underweight and being battered by its sibling, stood tall and walked unsteadily away from its antagonist. That's normal but, incredibly, it just kept going until it was out of the territory. It ran away from home! (I put the chick back, lest my presence had played a part in provoking its departure, so I cannot say where it would have ended up.)

So the high rate of mortality we'd witnessed in junior chicks could be driven by a chain of causation in which parents bring insufficient food for two chicks, underfed senior chicks aggressively commandeer an unfair share of it, and underfed juniors starve to death in the nest or seek adoption and get killed by neighbors.

Why Share with a Sibling?

The key to understanding boobies' treatment of their own family members was a concept proposed by William Hamilton in the 1960s, in what many biologists consider one of the most important doctoral theses in biology of the twentieth century.[9] Bear with me as I sketch out the idea, the only theoretical concept in the book that is not intuitive. It applies to all animals that reproduce sexually, including humans. It explains why you are inclined to be generous to your sibling, but only up to a point.

It was formerly thought that natural selection favors those behavioral traits, such as capturing fish and brooding offspring, that best enable animals to maximize their *fitness* as individuals, meaning the number of offspring they produce over their lifetimes, offspring who will carry forward half of the genes of each of their parents. Since the publication of Hamilton's dense mathematical arguments based on population genetics,[10,11] however, it is now widely accepted that where animals interact significantly with relatives it is *inclusive fitness* that is maximized.[12] Animals should maximize the number of offspring they produce plus a fraction of the offspring they enable their relatives to produce. The core idea was so radical that even Hamilton's tutor at University College London was unsure what to make of it, and most colleagues took years to appreciate its enormous implications. It is now widely understood to have provided the missing theoretical explanation for the evolution of sharing and helping behavior among relatives. We used to accept that animals help relatives, but through the concept of inclusive fitness, we now understand the genetic processes behind the evolution of such family "altruism" and can grasp the far-reaching implications regarding what sort of generosity can evolve.

Understanding the evolution in boobies of a *non-family behavior* such as grasping fish prey via the differential success of new (mutant) genes is relatively straightforward: a new gene that happens to enable its bearer to grasp sardines faster than bearers of the old gene will be favored by natural selection because the new gene boosts the number of offspring produced by the bearer. Copies of the new gene get into more bodies of the next generation than copies of the old gene, simply because the new gene enables every booby that bears it to capture more sardines and consequently make more offspring. Generation after generation, the new gene progressively replaces the old gene in the gene pool of the population until all boobies have the new gene in their genomes and all are fast graspers. That is the essence of evolution by natural selection: differential propagation in the gene pool of new genes that have positive effects on the number of offspring produced by their bearers. The means can be various. For example, new genes might replace old ones by enabling their bearers to grow faster, evade predators better, or court females more successfully, provided the improvement results in more offspring, on average.

Now consider the evolution of a type of gene-based *family behavior* that increases the fitness not of the bearer who expresses the behavior, but of its relative, by indirectly increasing the relative's reproductive success. A convenient example for us is food sharing among blue-foot siblings: the tendency to concede food to needy younger siblings. We'll assume that the senior chick aggressively controls how much food the junior chick receives from its parents. At first glance, we might conclude that sharing is unlikely to evolve because individuals that share will consume less food than more selfish individuals and consequently produce fewer offspring. In this way, Sharing genes will enter fewer offspring bodies than Selfish genes, and we expect them to progressively disappear, generation by generation, from the population. Self-sacrifice for the benefit of others—known as *altruism*—is generally not rewarded by natural selection; it is selected out.

But what if the sharer's help to its sibling increases the sibling's own fitness by more than it reduces the sharer's fitness? For the purpose of illustration, let's suppose that due to its altruistic sharing the sharer's own fitness is reduced by four offspring over its lifetime and that this loss is compensated by its sibling producing an additional twelve offspring. How will this arithmetic affect the propagation of the Sharing gene in the population? Hamilton realized that the answer depends on the relatedness of the two siblings because this specifies the probability of them both having the same new gene—here, the Sharing gene—in their bodies. Typically, every individual gets half of its genes from its mother and the other half from its father, so a new gene inherited by any individual has only a 50 percent chance of being in the body of any of its siblings. In the next generation, while the new Sharing gene has a 50 percent chance of being in any of the sharer's own offspring, it has only a 25 percent chance of being in the bodies of any nieces (for simplicity, let's suppose they're all female). As a result, each of the sharer's own offspring is twice as likely to bear a copy of the gene as each niece.

Applying these probabilities to our example, we find the Sharing gene actually enhances its own propagation by inclining its bearer to share with its younger sibling. Sure, it ends up in the bodies of two fewer of its own offspring (50 percent of four fledglings), but this loss is more than compensated by the gene's presence in the bodies of three additional nieces (25 percent of 12 fledglings). The arithmetic in this example demonstrates why the Sharing gene will eventually replace the Selfish gene in the population: in every generation, bearers of the Sharing gene produce one more fledgling with their gene than bearers of the Selfish gene. After hundreds of generations, all boobies will be sharers with Sharing genes in their genomes, and it's because these genes increase the number of copies of themselves in the next generation by inducing their bearers to facilitate the production of nieces. And this they do by helping their siblings to survive to maturity.

In fact, for the Sharing gene to propagate, the benefit to the sibling does not need to be three times greater than the cost to the sharer. According to Hamilton's rule, for altruism between siblings, the benefit only needs to be more than twice as great, and for altruism between half-siblings (who have only a 25 percent chance of inheriting the same new gene), it needs to be more than four times as great. In its broadest expression this rule specifies the conditions for propagation of genes for altruistic treatment of all degrees of kin.[13] As the rule is usually stated, a booby should altruistically help its sibling whenever the benefit to the sibling exceeds twice the cost to the altruist. Sometimes we say, a little less precisely, that a booby values itself twice as highly as it values any sibling, or that sons and daughters are twice as valuable to a booby as its nieces and nephews.

Such a differential between benefit and cost might sound unlikely, but it could easily occur in nature, including in booby siblings. A bellyful of sardines could be twice as valuable to junior as to senior whenever junior is thin and half-starved and senior is plump and well-fed. If boobies follow Hamilton's rule, we expect senior to allow junior to receive food from the parents when senior is well-fed and junior is needy, but take priority when both are equally well-fed or needy, and enforce its own priority more strictly when senior is needy and junior is well-fed. The key idea is that every senior blue-foot chick should strive to maximize, not its fitness, but its *inclusive fitness*—its output of offspring plus half of the additional nephews and nieces it enables its sibling to produce. Depending on the condition of the two chicks and parental food supply, a chick maximizes its inclusive fitness sometimes by yielding food to its sib, and other times by killing it.

Not that boobies need to do any calculating. We assume, rather, that senior chicks maximize their inclusive fitness by slavishly following inherited rules of thumb. These rules should be experienced by senior chicks as behavioral inclinations, such as "When feeling thin, peck your sib twice as much" or "When your sib's begging vigorously and you aren't, allow your sib to get the next regurgitation." Humans are not programmed with such specific rules, nor could they be because kindness needs to be expressed differently in each human culture and each context within a culture. However, humans do experience Hamiltonian inclinations that influence their generosity or selfishness in relation to kin. For example, how do you react when a brother asks for $500? You are probably inclined to help your brother out when he's hard-up and you are flush, but less so when you're both equally hard-up or flush, and definitely not when you're harder-up than he is. If that's how you would *feel* if your brother were to ask for cash, then your inclinations are in line with Hamilton's rule. What you would actually *do* may be different, because other influences, your moral principles, and your own quirks may override the inclination and ways of helping vary with the culture. On average though, members of sexually reproducing species, including humans, are expected to treat their siblings and other family members in ways that tend to maximize their own inclusive fitness.

Hamilton's rule tells how generosity and selfishness among kin can evolve through what is widely known as *kin selection*, and the rule has been hugely influential in setting the agenda for research on the social behavior of animals and humans during the last half-century. That research has extensively confirmed the predictions that arise from Hamilton's rule, but we should not exaggerate its scope. Helping kin and non-kin can also be favored by natural selection through other mechanisms—because altruism is reliably reciprocated or because it generates benefits for both parties.[14,15] What looks at first like kin-selected altruistic helping of relatives in service of the helper's inclusive fitness can turn out to be simple naturally selected behavior that serves selfish benefit.

For example, in a group of meerkats foraging for food by digging in the desert floor, it is common for one individual at a time to stand sentinel on an exposed perch while its companions feed. When the sentinel detects a predator and calls out, everybody scurries for shelter. Almost any behavioral ecologist would suspect that kin selection has favored individuals renouncing feeding opportunities and exposing themselves to danger because they are amply compensated by the increased survival of several kin. The increased survival and reproductive success of those kin, even devalued by the probability of bearing the same Sentinel gene, might well exceed the sentinel's costs in survival and reproductive success. Apparently not so! It turns out that meerkats stand sentinel only when satiated, that even solitary meerkats do it, that no risk is involved, and that sentinels are often the first in their group to run for cover. Sentinel behavior is cost-free behavior that benefits the sentinel and family alike.[16]

Conditional Tolerance

Controlled, conditional sharing by senior chicks consistent with Hamilton's rule was what our observations on growth, survival, and aggression in the blue-foot colony seemed to reveal. After getting those field data, our working hypothesis was that in the face of uncertainty over how much fish parents will continue to regurgitate to the brood, the senior chick maximizes its inclusive fitness through a strategy of *control, conditional tolerance, and siblicide* (*conditional tolerance*, for short). When parents are currently providing too little food for two chicks to grow normally, the senior one both increases its suppression of the other's begging and consumes a greater proportion of the food, to the point where the junior chick starves to death in the nest or seeks its fortune with another family, sometimes getting killed by neighbors in the attempt. Conditional tolerance is the part of the hypothesis that most clearly begs for an inclusive fitness interpretation rather than an old-style fitness interpretation. In our understanding, then and now, rather than promptly killing its younger sibling without more ado, as occurs in the black eagle and the brown booby, the blue-foot senior chick provisionally shares food with its rival, allowing survival and fledging of both chicks in the event of parents bringing enough food for both to prosper and fledge. Instead of playing safe and killing its sibling while its own advantage in size and maturity is overwhelming, the senior chick husbands its sibling as a potentially valuable component of its own inclusive fitness.

In the event of parents bringing insufficient food for both chicks to fledge at a good weight, senior sacrifices junior, thereby assuring adequate growth for itself. In the event of parents bringing enough fish for two chicks, senior allows junior to survive and feed, and fledge alongside itself. Most importantly, to ensure that it is junior and not senior that gets sacrificed when parents only bring enough fish for one chick, senior routinely establishes dominance from the very start of sibling cohabitation in the nest, administering daily thrashings, not to kill but to subordinate. Most of that early aggression is not about food; it's about dominance per se.

There is an element of risk to the senior chick's conditional tolerance because, in a brood where senior is unusually feeble or junior unusually feisty, junior might invert the dominance relationship and supplant senior in the assured survivor slot. Indeed, in the first 12 broods I watched for several weeks, one senior chick was something of a pacifist—it pecked its sibling half-heartedly. Unusually, junior became dominant by the time senior was 12 days old, outgrew senior several days later, and remained dominant and larger through at least age 6 weeks. Presumably, the cost of this risk to senior is outweighed on average by the inclusive fitness benefits tolerant seniors obtain via frequent survival of juniors in addition to seniors, and occasional survival of juniors instead of seniors (devalued in both cases by 50 percent).

Incidentally, we were intrigued by how, in the face of savage nest defense by adults, starving chicks sometimes contrive to get adopted and parasitize the neighbors. Can't adults distinguish between their own chicks and someone else's? On that discrimination could hang an adult's decision to attack versus care for a chick that attempts

Fig. 1.8 A millisecond away from a drubbing. Junior has not yet learned the score, but if it starts to beg, its sibling's open or closed beak will crash down and administer a bite or shake.

Drawing by Jaime Zaldivar-Rae; based on photo by Hugh Drummond.

Fig. 1.9 Senior blue-foot chick fiercely pecks sibling in submissive posture.
Photo by Hugh Drummond.

to enter its nest. In pilot trials in a part of the colony where nests were sparse, we removed single chicks from their nests and compared parents' responses when their own chick versus an unfamiliar chick was released 75 cm from the nest. One of us crouched a few meters away ready to gather up any chick that was threatened. On release, most chicks soon headed for the target nest, shuffling or walking forward on their big webbed feet, sometimes pausing and even napping on the way. As it approached, each chick cheeped while the brooding adult stared, craning tensely forward and sideways, studying the new arrival. Some adults walked toward the chick and then walked back to their nest as if to lead the chick home. But adults did not clearly discriminate: whether a chick was met with encouraging calls or threats didn't seem to depend on its identity. However, adults seemed more likely to threaten approaching chicks that were larger than their own, and unfamiliar chicks seemed to have a strategy for getting past adult nest defenses.

Compared with resident chicks, unfamiliar chicks cheeped less frequently and approached more slowly, making more pauses. On arrival, instead of walking in upright, cheeping loudly, some of them suddenly leaned forward and dashed into the nest cup with neck outstretched, skimming the ground in furtive haste. Once in the nest, whether resident or unfamiliar, test chicks were treated as part of the family, just as we have observed in naturally adopted chicks. Booby pairs are perfectly well equipped to repel and destroy potential parasites that trespass on their territories, and they hurry to do this whenever an obvious intruder approaches or crosses a border. In contrast, they are reluctant to discriminate against a potential parasite on the doorstep of their nests. I expect they hold fire in that situation because their ability to recognize individual chicks is limited and natural selection has eliminated attack tendencies that put the family at risk, leaving them vulnerable to parasitism. It may be better to occasionally raise a parasite than to ever kill one of your own chicks.

It may seem odd, but booby defense is all about space. To judge by the behavioral mechanism rather than its normal result, adult blue-foots defend their nest and territory, not their clutch or brood. For example, during colony monitoring, reactions of caretaking adults to our temporary removal of eggs and chicks for measuring and marking suggest that boobies follow rules of thumb in an uncomprehending way. Apart from the current partner, no booby or any other animal except the resident chick is allowed to enter a blue-foot's territory or closely approach the adult sitting on its nest. As an intruding booby or potential predator approaches an incubating or brooding parent, the parent grunts (females) or whistles (males) ever more rapidly and intensely. At closer quarters, the parent tensely sweeps its head from side to side until the intruder is close enough to expel it with volleys of pecks and bites, sometimes holding on and twisting, tearing the skin. So far so good, but if after gathering up a booby's clutch or young chicks from its nest pit, you crouch nearby to measure them, setting them down on the ground inside or outside the territory in full view of the caretaking parent, it shows no interest in them. The parent remains on its nest, ready to defend it against you and all comers, and not in the least inclined to approach you, its eggs, or its chicks, though it will welcome them back with excited "yes-headshaking."

And you can't move a booby's nest. On Isla San Pedro Martir, in the Sea of Cortez, where we camped among the collapsed stone walls of an old guano-mining settlement, we once tried to relocate a blue-foot brood that lay between our cooking area and our tents. That way the pair would be in peace and our cooks would be spared their "Morris dancing"—panicky hopping past thrusting parents with pans of boiling water. No problem, we gathered up the two chicks, too young for locomotion, and set them down in an improvised nest pit 4 m away. We withdrew, leaving the parents with free access to their brood and our cooks with dignified access to the cooking area. It was not to be. For an hour after the switch, one or both parents remained on and beside their nest, defending it against boobies and behavioral ecologists alike, yes-headshaking and pecking with a will, but totally ignoring their own chicks in full view several meters away.

We gave up and returned the chicks to their nest (and resumed our dancing). Steven Gould,[17] the eminent evolutionist, was right when he claimed that Galapagos blue-foots follow rules of thumb like the apocryphal sailor who obeys the military command "If it moves, salute it; if doesn't, paint it!" But Gould's claim that blue-foot parents attack any chick that tries to enter the circle of guano around their nest was incorrect.[18] Booby rules of thumb are sometimes more complex than you might expect, and we still haven't figured out their gate-keeping rule for dealing with approaching chicks.

Why such fierce attachment to space? The short answer for boobies, as for humans, is location, location, location! On Isla Isabel at least, location affects so many aspects of a booby's well-being and reproduction that choosing a nesting territory and nest site is as momentous as choosing a partner. For a start, location affects exposure to wind and sun, access to take-off and landing sites—important for a heavy-bodied bird—and risk of a nest being flooded at high tide or during rainstorms. It also affects relations with other animals, such as the neighbors the family will have to deal with, and exposure of parents and progeny to snakes and gulls and to parasites such as soft-bodied ticks, which emerge from cracks at night and swarm over the boobies' blue feet and their chicks. Ticks are so harmful they can oblige marine bird colonies to abandon nesting. Integrating cues on all these factors is quite a challenge even if boobies have innate habitat preferences to guide them, so boobies rely heavily on the tried, tested, and familiar. Isla Isabel blue-foots nest for the first time an average 31 m (males) or 37 m (females) from the spot where they hatched and grew up—a location that self-evidently works well for reproduction. For the rest of their lives, although they typically move several meters from one year to the next, and movement depends on how well they and their partner fared and whether they divorce, they always nest close to the spot where they nested the very first time. And this despite spending the offseason elsewhere: between August and November, they wander over the ocean and loaf on the cliff faces of the island.

The Brown Booby Way

In stark contrast with the controlled and conditional aggression suffered by blue-foot juniors, brown booby juniors hatch into a world of unbridled violence. The pairs of brown booby hatchlings we observed on Isla Isabel, Isla San Pedro Martir in the Sea of Cortés, and the Islas Marietas off the coast of Jalisco live by other tactics and rules of engagement, reminiscent of the bloody cainism of black eagles. As soon as it is mature and agile enough to launch attacks, which takes several days, the elder brown booby chick takes advantage of its 5-day head-start to attack its sibling every time it sees it up and about. This is not often because, like blue-foot hatchlings, both nestmates spend most of their time concealed under the brooding parent and probably asleep. Only very occasionally do they peer out from under the parent's breast to look around and beg by head-bobbing and chirping. However, if the two happen to peer out at the same moment and senior has a line of sight on junior and enough mobility to approach, it will launch an attack.

The senior brown booby chick is single-minded. It pecks and bites in desperate haste, targeting whichever part of junior is most accessible; when it can, it follows through with an energetic upward and forward push. Depending on where they are both located in the cup of the straw nest, the "expulsion push" may or may not displace the other chick. Eventually though, when the senior one is strong enough and can shuffle or walk forward as it seizes and pushes, it manages to thrust its sibling over the rim of the nest cup. From there, junior topples down the flank of the nest mound, frequently lodging near the bottom. The gradient is too steep for an immature chick to scramble back up, so it rests and writhes ineffectually until it dies of exposure or starvation, or is taken by a predatory bird or reptile. Significantly, we have never seen a blue-foot chick expulsion-pushing its sibling.[19–21] A brooding brown booby parent may nudge its expelled chick half-heartedly with the tip of its beak but, to my knowledge, never actually retrieves it. Bouts of expulsion-pushing tend to be frenzied: once when we paired survivors of siblicide from two broods in an abandoned nest to test for effects of hunger on aggression,[22] the *smaller* one pulled off six evictions from the nest cup in 30 minutes!

Nest monitoring of several brown booby populations, including the one on Isla Isabel, suggests that the junior offspring rarely, if ever, fledges along with its sibling but may substitute for it at the egg or chick stages if the senior one fails to hatch, is predated, or is inviable. Juniors generally die before they are a week old. Critically though, on Johnston and Kure Atolls in the Hawaiian Islands, the single surviving brown booby fledgling at a nest was derived from the second egg in 12–16 percent of cases. Although usually doomed, the second egg is not laid in vain.[23–25]

Fig. 1.10 Out and gone forever. After the senior brown booby chick heaves its sibling over the rim of the nest, it will topple down the nest mound and succumb to predation or exposure.
Drawing by Jaime Zaldivar-Rae; based on photo by Alina Cepeda.

Fig. 1.11 No prospect of retrieval. Ensconced on her nest cup brooding her senior chick, this brown booby mother will not attempt to fetch the expelled junior chick from the dark little pocket at bottom right. Unless a predator arrives first, junior will die where it rests.
Drawing by Jaime Zaldivar-Rae; based on photo by Alina Cepeda.

Overly Large Broods

In most bird species, some parents produce more chicks than they are capable of feeding, and one or more chicks die in the nest. This wastefulness was explained by David Lack,[26–28] an Oxford University ecologist whose insights derived from extensive studies of species whose chicks compete with each other not aggressively but by begging and jostling. Lack's explanation also applies, with some elaboration, to the minority of birds whose chicks compete aggressively, including South Polar skuas, kittiwakes, and several species of boobies, herons, and kingfishers, and raptors. The essential idea is that in the face of unpredictable food availability during brood care and other uncertainties, birds lay extra eggs and then dispense with their extra chicks if they're not needed.[29] In a world where some eggs are infertile and some chicks are developmentally unsound, and predation on eggs and chicks by snakes and other animals is commonplace, it often pays to lay more eggs than you can expect to raise all the way to fledging, just in case. Then, if for any reason an egg or chick fails, you have an "insurance" egg or chick to take its place; and if ecological conditions are benign during brood care (e.g., sardines are plentiful) or your partner is especially worthy (e.g., an adept plunge-diver), you may fledge the whole brood.

"Brood reduction" is the term for the brood-culling infanticide that has to take place whenever all goes so well that the number of chicks exceeds the parents' capacity or willingness to feed them. If parents cannot cope with so many, their chicks could fledge underweight and consequently produce few grand-offspring, or parents could overextend themselves feeding them and damage their own prospects for future reproduction. Something has to give, and in many cases, the best short-term outcome from the parents' long-term perspective is to discard one or more chicks. We don't expect any sentimentality or anything of the sort. The cold logic of natural selection favors the success of genes that incline animals to do whatever needs to be done to maximize their lifetime fitness, understood by Lack to mean the total number of viable offspring produced. No intentionality or even comprehension is required. Some small mammals cannibalize their litters of newborn pups when the environment is unfavorable for raising them, sacrificing current offspring all the better to raise more viable offspring in the future. Birds terminate their surplus offspring, too, usually by underfeeding the least competitive members of the brood and allowing them to starve.

Because of their tenuous food supply, brown boobies are rarely and maybe never able to simultaneously raise two fully viable offspring. Attempting to do so could require feeding them for up to 20 weeks and likely would end in underfeeding of at least one chick or debilitation of the parents. Their extra egg is probably for insurance only. The feeding ecology of blue-foots, on Isla Isabel and elsewhere, is different. They feed closer to shore and make shorter hunting trips and, depending on El Niño and other factors, the number of fish a pair can bring back to the nest varies hugely among years. In some years no pair of blue-foots on Isla Isabel raises more than a single fledgling, in other years scores of pairs raise two or three well-fed fledglings, and exceptionally some individuals even attempt a second brood. Since hatching failure and chick deaths are common, and gulls and snakes often take a single egg or hatchling, a blue-foot's extra eggs fulfill two functions: insurance for when an egg or chick doesn't make it, and production of additional fledglings when all goes well and food is plentiful.

Infanticide, Siblicide, and Suicide

But there must be more to it than I have been suggesting. After Trivers's theoretical proposals made us realize that each family member has its own fitness interests and may have the means to implement them,[30] those tidy stories of families creating and discarding extra offspring *in the service of the parents' fitness* had to be questioned. Researchers needed to explore the implications for brood reduction theory of each chick valuing itself twice as highly as its sibling, each caretaking parent being in conflict with the other and with its offspring, and each family member striving to prevail in its conflicts of interest with the others. Theoreticians were telling them they should suspect conflicting family members of resorting to physiological ploys, deception, and behavioral imposition. Under the Hamiltonian paradigm, researchers needed to delve into family relationships to call out the conflict, to detect how family members attempt to get their own way, and to work out who prevails.

Not that animal behaviorists were now attributing consciousness, intentions, or Machiavellian planning to animals. Some among them were investigating animal cognition and consciousness, but that was still a sideshow. To this day, despite often using the language of intentionality, strategies, and deception, what behavioral ecologists aspire to objectively analyze is observable behavior. We analyze the behavior of animals as if we were dealing with the software routines and artificial intelligence of robots rather than thinking protagonists. We use mentalistic language because it is tedious and awkward to work around it when you are talking about behavioral adaptations such as conditional routines that can conveniently be called strategies. We do this routinely, in the same spirit that most of us talk about our apps and laptops by saying that they "like," "choose," "allow," and "prefer" options, as if their software routines were conscious. Some behavioral ecologists—myself included—are intrigued by the possibility and nature of feelings, consciousness, and understanding in their study animals, or firmly believe their study species has those capacities, but in the main, we've cautiously stuck to studying what we can measure.

An up-to-date theory of avian sibling conflict was published just 2 years before I returned from the United States to Mexico to establish a research program in the UNAM's Instituto de Biología. In a provocative departure from Lack's parent-centered perspective, Raymond O'Connor elaborated a mathematical model applying Hamilton's rule and Trivers's notions of evolved family conflict to avian brood reduction.[31] The model made specific predictions for both cooperation and conflict among family members over the elimination of extra chicks when parents bring insufficient food. Its most important assumptions were that the survival of each chick depends on the amount of food it gets, that brood members share all the food brought by the parents and that the death of the youngest brood member frees up food for the survivors. What the model showed was that in the face of increasing food shortage, every family member can potentially benefit from that death in terms of its own inclusive fitness, especially when the brood is small. Applying O'Connor's model to a brood of two blue-foot chicks, under *moderate* food shortage, the elder sibling is expected to benefit from the death of the younger while the parents and the victim itself are expected to suffer an inclusive fitness loss. Under *substantial* food shortage, the elder sibling and the parents both benefit but the victim still loses. Under *great* food shortage, all three parties benefit, even the victim!

It's implicit in these predictions that under the more modest levels of food shortage there is family conflict: under *moderate* shortage senior should attempt to kill junior while junior resists and the parents strive to keep junior alive and combat siblicide, whereas under *substantial* shortage senior and the parents should collaborate to kill the younger chick while it struggles to survive. On the other hand, under *great* food shortage, there should be no conflict, and suicide by junior is predicted because its death frees up food and boosts the survival of senior, in which, as we have seen, junior has a 50 percent interest. The idea, logical enough, is that in such straits this boost to senior's survival (devalued by 50 percent) can exceed junior's own survival probability plus 50 percent of senior's probability in the event that they continue to share a meager ration. O'Connor suggested not taking the suicide prediction too seriously though, because natural selection for suicide by junior is not expected to occur if all the other more powerful family members are out to kill it anyway. Junior will die whatever it does, so there is little scope for natural selection to favor Suicide genes.

For those with a head for it, reviews of family conflict theory can be consulted in Mock and Parker's *The Evolution of Sibling Rivalry*,[1] Mock's *More Than Kin and Less Than Kind*,[32] and Forbes's *A Natural History of Families*.[3]

Our Questions

Questions immediately arise about how boobies, with all their informational, cognitive, and behavioral limitations could implement the conditional strategies predicted by O'Connor's model. Is the model realistic? How could each member of a family adjust its behavior to promote its own inclusive fitness interests at these three levels of food shortage (which in reality form a continuum) when no booby in the real world has the godlike comprehensive information that the theoretician has? Think about the challenges each member faces. How should you, as mother, father, senior chick, or junior chick try to influence food division between parent and brood, or food allocation among chicks, or choose between nurturing versus directly killing the younger one when you know so little? Each parent may only know how much food it personally has brought to the nest, and how it personally has been allocating the food among the chicks. Each chick may only know how much food it personally has consumed, today and recently, and how needy it personally is.

To complicate matters further, although family members can potentially estimate each other's nutritional condition using visual and auditory cues like plumpness and voice quality, and they can and do communicate with each other about their condition, for instance by begging, self-interested communicators are expected to attempt to deceive each other, and attempt to see through each other's deceptions. And it gets more complicated. Food may be abundant right now, and you and your sibling may be in good condition and well-fed, but in the weeks ahead the shoals of sardines may move further offshore, plunge-diving could be frustrated by a week of boisterous

Fig. 1.12 Blue-foot chicks poised symmetrically around their parent's vertical bill. Both siblings beg simultaneously, the senior one allowing its rival to compete for the next regurgitation, for now at least.
Photo by Hugh Drummond.

waves, or one parent might tear and disable a foot by landing through branches. Is it really a good idea to share food fairly with your sibling if it might turn against you later when there's not enough food to go around?

The control, conditional tolerance, and siblicide hypothesis has framed our general understanding of family relationships in blue-foots, but from the time we derived it from descriptive observations it needed fleshing out with confirmatory analyses and further explorations. One of the questions that most intrigued us concerned the control of a rival. If the senior blue-foot chick only contrives or colludes in its sibling's death when the parents are falling short on their food deliveries, *how* does it maintain control of the sibling meanwhile? After all, each of the chicks values itself twice as highly as the other, so surely junior cannot be trusted to accept its status as a disposable inferior. Given any chance, it should turn the tables, take control, and, if necessary, kill its tormentor. What stops that? And how on earth does a male chick keep the lid on a younger sister when she, being a member of the larger and stronger sex, is bound to substantially outgrow him well before they fledge. How does he pull that off?

Another question concerns the role of the parents. In relation to their puny, immature progeny, you would think that they are all-powerful and able to impose their will. Besides, it's mothers that create the clutch and brood, and they can potentially tailor it to suit their own designs, manipulating the age and power differential between chicks by adjusting the relative size, nutrients, and hormones of eggs, and the hatching interval between them. With rare exceptions during strong El Niño events, there is always at least one parent on the family territory, towering over chicks when they're young and powerfully outranking them as they approach fledging. The parents' will and the parents' interests should prevail, but do they?

Just as intriguing is the issue of how senior chicks actually calibrate their aggressiveness. A junior sibling is too precious an asset to wantonly discard because its successful fledging can contribute to the senior chick's inclusive fitness, but if parental food provision falls off to the point where senior's own growth trajectory becomes precarious, then junior must go. So adjusting their current pecking rate to the current ecological context is a vital and delicate matter. How do seniors get it right?

Finally, a quite different issue concerning long-term development. We all think we know that privation, abuse, or stress in infancy has consequences during adulthood. For humans, it's an apparent truism and many studies have supported the idea, and psychologists have recently turned their attention to possible long-term effects of sibling conflict between children.[33,34] Maybe other animals that raise competing offspring together can tell us something about vulnerability versus resilience to domination by sisters and brothers. How, for example, does growing up on lean rations and subject to daily threats and violence affect a booby's ability to survive, compete for a territory, attract a partner, and produce offspring? We don't know the answers for any non-human animal except the blue-footed booby, whose resilience may well surprise you.

These then are the four questions addressed in this section of the book: sibling dominance, calibration of sibling aggression, parental control of sibling competition, and developmental consequences of sibling competition. We'll also look, in Chapter 3, at how sibling conflict varies among species of birds and mammals, and in Chapter 9, at how conflicts of interest between siblings play out in *Homo sapiens*. But first, in the next chapter, we'll examine the totally different ways that blue-foots and brown boobies keep a lid on their sibs.

2
Beating Siblings into Submission

In the blue-footed booby, dominance is the cornerstone of the elder chick's strategy of conditional tolerance, which allows its sibling to share the food brought by parents only when there's enough of it for both chicks to thrive. Right after hatching, what designates junior as a subordinate chick and potential victim of siblicide is the 4-day age difference. That inequality allows the elder chick to overwhelm its sibling with irresistible pecking and biting. However, as the two of them grow up and the difference between them in size, maturity, and fighting ability wanes, the designation of roles is maintained by dominance-subordination. If it weren't hobbled by psychological subordination, the junior chick could tie up and wear down the senior one in repeated struggles for access to the parental beak. Even worse from a senior's perspective, in the event of a food shortage making brood reduction inevitable, it might be senior rather than junior that takes the fall. Establishing and maintaining dominance surely requires diverting energy from growth and maintenance of the body to pecking and biting, but this investment very likely pays for itself by saving expenditure on extra begging and pitched battles with an ambitious rival. But what precisely is dominance-subordination between two altricial—helpless, slow-maturing—chicks?

We were keen to address this issue because the psychology of sibling dominance is intrinsically fascinating when you understand it as an outcome of the evolution of two distinct strategies to be expressed by the same genome, depending on the social environment. The first-hatched chick has an egg for a sibling and grows for 4 days before that egg transforms into a tiny, frail hatchling that the first chick will shortly have the muscle to attack every day until it learns to instantly signal submission. By the time the second-hatched chick can peck aggressively, its large, ever-present sibling is repeatedly beating on it, denying it free access to the parental mouth, and ramping up the pecking and threatening to ensure that junior never gets the upper hand. Thus, radically different social environments—having an elder versus a junior sibling—turn creatures with similar genomes into the dominant and subordinate siblings they need to be if they are to prosper in the crucible of daily, inescapable competition.

When we started watching boobies, aggressive competition between mammalian infants was a familiar enough phenomenon but regarded more as a troublesome feature than an outcome of biological adaptation. Not surprisingly, experimental analyses of dominance in vertebrate infants were lacking, although unbeknown to us a researcher at the University of Oklahoma, Doug Mock, was already 2 years into an ambitious program of field research on siblicidal egrets and herons. Anecdotal

Blue-Footed Boobies. Hugh Drummond, Oxford University Press. © Oxford University Press 2023.
DOI: 10.1093/oso/9780197629840.003.0003

observations and reports suggested that dominance relations emerge weeks after birth in litters of dogs, wolves, foxes, and coyotes, and in the precocial (semi-independent and fast-maturing) broods of chickens, geese, and quail.[1] But could altricial chicks that are born nearly naked, unable to walk, and wholly dependent on parents for food, body heat, and protection entrain each other as dominant or subordinate individuals in just 2 or 3 weeks?

And if they can, then we also wanted to find out whether the behavioral mechanisms of infant dominance are similar to the "trained winning and losing" and deference to superior might that are common in adult animals; whether and how subordinates can recover from their imposed social disability; whether learning subordination reflects an adaptive strategy or just an organic vulnerability; and whether and how broods of three blue-foot chicks manage to structure a *dominance hierarchy*. These last three issues have not been fully resolved for adult animals, so booby broods would serve as a model of the simplest dominance-subordination, with wide implications for understanding the mechanisms, development, and functions of dominance in vertebrates generally.

Few behavioral concepts have been used so universally and applied so broadly by ethologists and behavioral ecologists. *Dominance* refers to asymmetrical social relationships between individuals of species whose members compete with each other for access to resources such as food, space, or mates.[2] It sometimes refers to a strong animal simply imposing on a weaker one, but more often it means that one individual yields, defers, or concedes access to another. Submission by one of the two is often the key event that determines which one dominates in a particular circumstance. Dominance occurs in species that are mostly solitary, species that live in tightly knit social groups, and all intermediates. When two competitors confront each other, depending on what is at stake, the stand-off can be nearly instantaneous, as when one chickadee supplants another at a feeding table. Or it takes some time and involves displays of capacity and willingness to fight, and in some species progresses to probing scuffles or fierce sparring. After which, critically, the less able fighter backs off, leaving the other to claim the resource. So two red deer stags will roar and wrestle with locked antlers, two male chimpanzees will bristle up to appear bigger, charge and hurl stuff around, and the male spinytail iguana in our island campsite will turn side-on to its rival to show off its colorful, massive body while nodding emphatically with wide-open jaws. Typically, such displays accurately reveal the true fighting ability of each individual because natural selection favors the evolution of signals that cannot be falsified, honest signals as they are known. Loud deep roars, puffed-up bodies, and huge red mouths actually show how healthy and strong the individual is.

Both competitors benefit by displaying rather than fighting because both, especially the weaker one, can thereby avoid the rapidly escalating energetic cost and physical risks of all-out combat. Importantly, this prudent approach to conflict resolution is not confected by any guiding hand or mystical principle of survival of the species, but by straightforward natural selection for self-serving behavior. In theory, genes for prudent engagement that incline their bearers to start by threatening impressively and then either back off or wait for the other to back off depending on who is more formidable or determined, find themselves in more bodies of the next generation than those that incline their bearers to automatic meekness or reckless violence. Because they generally expend less energy, suffer fewer wounds, and gain more resources, the bearers of genes for prudent engagement produce the most offspring, enabling copies of the genes themselves to gradually substitute for the alternative genes in the population's gene pool.

Backing off is self-serving behavior. It may not look so, but individuals that concede when the expected cost of confrontation is greater than the expected benefit of fighting, both measured in terms of impact on personal fitness, are working for themselves and their genes. It is natural selection that fine-tunes their decision-making on the basis of factors such as the opponent's relative size and ferocity, and their own level of need. In confrontations of male spinytail iguanas, young challengers are nearly always content to slope off sullenly or run for their lives when the resident shakes his great red mouth at them, and they do best for themselves by waiting him out. The day will likely come when the aging resident male doesn't look quite so formidable and can be bitten and chased off, leaving a group of six fertile females for the usurper.

In groups of social mammals such as spotted hyenas and social birds such as wild turkeys and Arabian babblers, individuals interact frequently with other group members and establish pairwise dominance relationships. By all accounts, individuals of these species can recognize particular members of their group, and on the basis of a history of pairwise interactions, every pair in each group sorts into a dominant individual and a subordinate individual. These *personal dominance relationships* in social groups tend to be stable, but if one of a pair is in decline due to illness, old age, or defeat in conflicts with others, or is growing, thriving, and ascending in the group hierarchy, a personal dominance relationship can be inverted. So it's common for dominants to repeatedly reinforce their personal relationships by periodic threatening and attacking even when no resource is being contested, reassuring themselves by demanding submission.

Trained Winning and Losing

The dominance-subordination between blue-foot chicks probably is not a personal dominance relationship, because when these infants develop their social roles in the first 2 or 3 weeks after hatching, the first quarter of the chick growth period, they are unlikely to be able to tell one individual from another. Although the precocial chicks of several species of wildfowl and waterfowl are thought to learn the identities of individual brood mates,[1,3] and this ability has been demonstrated in greylag goslings at the age of 6 weeks,[4] recognition of individual siblings may be beyond the ability of altricial chicks. In a few species with altricial young, chicks learn to distinguish between familiar chicks—the ones they grew up with—and unfamiliar chicks.[5-7] They can recognize brood mates as a class because they grew up with them, but none has so far been shown to distinguish one familiar chick from another. Experiments we'll be seeing in this chapter—swapping chicks between nests—suggest that while a young blue-foot chick is sensitive to the relative size or aggressive posture of another, it probably doesn't distinguish brood mates from non-brood mates, let alone distinguish one brood mate from another. So we thought that the conspicuous dominance-subordination in blue-foot broods, rather than reflecting personal dominance relationships involving individual recognition, was more likely based on two more fundamental learning mechanisms: *trained winning* and *trained losing*.

Psychologists first tested trained winning and losing by exposing lab rats either to defeat, by pitting them against superior rats, or to victory by pitting them against inferior rats. Then they observed their performance against unfamiliar rats. In those tests, rats that had experienced defeat tended to be submissive and lose; those that had experienced victory tended to attack and win. No individual recognition was involved or even possible in the experiments, so what the tests revealed were acquired aggressive and submissive tendencies in relation to other rats in general, not personal dominance relationships like those ascribed to hyenas, turkeys, and babblers. To this day, trained winning and losing are thought to play a leading role in the development of vertebrate and invertebrate dominance relationships and social hierarchies,[8] including relationships that involve individual recognition and relationships that don't.[9] Their role in the litters and broods of vertebrate infants was investigated for the first time in blue-footed boobies.

Natural Sexed Broods

To unveil the dynamics of size asymmetries, dominance stability, and dominance inversions in blue-foot broods, we obtained additional, more focused observational data before moving to field experiments. We observed the growth and behavioral development of dozens of broods in which, for the first time, we would identify the sex of both chicks and follow their growth and survival through fledging. Because adult males are smaller and 25 percent lighter than females, and the blue-foot's 4-day hatch interval initially determines which sibling dominates, we anticipated that prolonged scrutiny of sexed broods of two would help us tease apart the effects of relative size, sex, and social experience on dominance. If we could only sex chicks, then by monitoring growth every 2 days and observing behavior every 4 days, the relative influence of those three factors would be revealed and could be experimentally tested.

Among other questions, we could ask whether younger sisters outgrow their elder brothers and whether this inversion of size permits sisters to invert the dominance relationship. Alternatively, the prior establishment of dominance by elder brothers in such broods might enable them to maintain dominance over their younger sisters despite being outgrown by them, and even provoke elder brothers to suppress their little sisters' growth, leading to their stunting or death. At the other extreme, in the broods of an elder sister plus a younger brother, great and persistent size asymmetry might provoke severe bullying of males. On the other hand, in pairs of brothers and pairs of sisters, with minimal size asymmetry, sibling dominance relations might be stable, low in conflict, and energetically efficient, favoring fast growth in both sibs. In short, comparisons of growth, behavior, and mortality among all four brood compositions—male plus younger female, female plus younger male, male plus younger male, and female plus younger female—could yield rich insight into the roles of sex, relative size and learning in dominance relations of wild vertebrate infants. As it turned out, those descriptive observations on Isla Isabel prepared the ground not only for experimental tests of our interpretations but also for an exciting experiment on a distant desert island that would test ideas about why blue-foots and brown boobies evolved such different behavior.

And we *could* tell the girls from the boys! Molecular sexing was not available at the start of the 1980s when we got going, but, as we discovered, the size difference between the sexes begins to emerge from the very start of growth and by the time of fledging, at age 13 weeks, females are substantially and consistently larger than males. We could also sex the chicks that died before reaching fledging age. Fortunately for us, blue-foot parents are relaxed about domestic hygiene and do nothing to dispose of chick cadavers, often simply treading them into the nest floor. So chicks that died before fledging could be sexed months later in the lab by microscopic examination of their preserved gonads. Conveniently, we were able to scoop up most cadavers before they rotted or got dragged away by scavengers. Combining the two methods of sexing yielded a sample of 42 fully sexed broods that were monitored and observed. Deaths of chicks, overwhelmingly juniors that starved, reduced that sample to 27 broods that survived intact through fledging, split roughly evenly between the four possible sex-hatch order combinations.

Let's start with the broods where parents brought enough food to fledge both chicks. In these, stability was the name of the game. Independent of sexual composition, in all of these broods the senior chick was dominant and the junior one subordinate from the time they began to interact and, with only one exception, dominance stayed that way all the way through fledging, no matter which brood mate was larger.[10] Even more surprising when you consider that females are bigger and likely to consume more food than males, across the four brood compositions, the average growth of junior and senior chicks of both sexes was unaffected by their sibling's sex. For example, a male senior grew just the same whether his younger sib was male or female, and a junior female grew just the same whether her elder sib was male or female. It's hard to square this pattern with out-and-out competition; it smacks of tolerance and moderation in individuals capable of being more selfish.

The single exception to stable dominance was a brood where the younger sister outgrew her elder brother and 20 days later went on the offensive, progressively increasing her share of the daily pecking until 16 days later her dominance was uncontested. After that, she never looked back. In all the other broods of a male with a younger sister, although she outgrew him at an average age of 37 days, he remained permanently dominant over the sister that increasingly dwarfed him. Maybe this

Fig. 2.1 Will she kowtow? The male's threat call warns his younger sister, now as big as him and tempted to rebel, that he will rise up and pound her head if she doesn't signal submission.

Drawing by Jaime Zaldivar-Rae; based on photo by Hugh Drummond.

Fig. 2.2 On the cusp of outgrowing her elder brother, the sister nonetheless submits when he shouts a threat. However, her bill-down-and-face-away posture looks half-hearted and may be insufficient to stay his wrath.

Drawing by Jaime Zaldivar-Rae; based on photo by Hugh Drummond.

occurred not because he was a male, but because he had been dominant from the start, an advantage that clearly worked for both sexes.

Most surprising of all was that juniors in fully fledged broods of all sexual compositions emerged from their months as repressed punchbags physically unscathed. Despite slow growth during the first few weeks, especially in females, the growth achievement of juniors was in no respect inferior to that of seniors of the same sex—not in maximum weight, final weight, age of maximum weight, skeletal size, or age at fledging. On the face of it, although exceptions can occur, dominance in well-fed broods is stabilized by daily doses of trained winning and losing, and dominant siblings (or maybe their parents?) regulate competition for food in such a way that neither the smaller sex nor the subordinate sibling ends up undersized or malnourished.

However, this almost happy story of stable dominance, equivalence of females and males, and seemingly complete recovery of subjugated juniors must be tempered by considering the somber events in less well-off families of the 42 sexed broods. First, two of the fourteen males with a younger sister may have fatally stunted that sister's growth. In both cases, the sister was undersized, bore head wounds, and starved to death at the age when she was due to fledge, evidently victimized either by her elder—and continuously dominant—brother or by adult neighbors defending their territories. Second, the death of a single member of the brood fell far from equally on *one* senior chick versus *eleven* junior chicks. Seniors don't train their siblings in vain.

When the family food budget is sufficient, sibling conflict in blue-foots, for all its daily violence, does no lasting harm to either chick; when food is not sufficient, dominant status delivers privileged growth and survival. These are the bottom-line benefits of being dominant. However, we still needed to find out whether enduring subordination throughout the chick period affects a booby's ability to thrive during adulthood. You might well expect prolonged abuse at the earliest, most sensitive phase of development to leave a booby with emotional or physiological scars that handicap its ability to survive, compete for a partner, defend a territory, and raise a family. Such long-term effects of sibling persecution early in life have not been detected or even seriously sought in any wild animal because it's so challenging to monitor individuals over their lifetimes. We'll return to this question in Chapter 6.

To test the aforementioned interpretations regarding the primacy of trained winning and losing in stabilizing dominance and regulating competition, we needed to experimentally disentangle the effects of age, sex, size, and social experience on the behavior of both sexes. Without experimental control of factors that could be influential, some of those interpretations are no better than persuasive hunches which leave us with an insecure grasp of the behavioral dynamics of chick competition. For example, younger sisters may indeed remain subordinate to their elder brothers despite outgrowing them because younger siblings are trained as losers and elder siblings as winners, or because male chicks are stronger or more aggressive than females. After all, males grow more slowly so their developmental programs might privilege maturation and motor coordination over size, making them better fighters. Or younger sisters may remain subordinate because the junior chicks' 4-day maturity handicap makes them inferior competitors throughout the chick period. Or younger sisters may remain subordinate because differences in size are not detected by booby chicks or have little effect on their fighting ability. Here we have four alternative explanations, and all or any of them could be true! Detailed descriptions of growth and mortality allowed us to propose attractive hypotheses, but we needed to manipulate the causal factors to reveal which ones are relevant and assess the magnitude of their effects.

Are Boobies Trained?

To this end, we observed the behavior of pairs of chicks either placed temporarily on artificial territories in the booby colony or swapped temporarily or permanently between nests, taking care to expose them to no more harm than chicks ordinarily experience in their boisterous family lives. Nine years of intimate acquaintance with blue-foots enabled us to design experimental procedures that could answer our questions while minimizing disturbance and perturbing behavior and growth only within the range of what is natural in the colony. To this end, all of our experiments have followed the ethical guidelines of the Animal Behavior Society: for example, the design minimized the number of broods affected and strove to minimize suffering and harm, keeping them to levels that occur naturally; and we discussed with colleagues whether the suffering and harm were justified by the expected advance in understanding, and monitored chicks after experiments to assess possible impacts.

We thought the boobies would not cotton on to us substituting imposters after kidnapping their offspring or siblings because during the first few weeks after hatching of a clutch, booby parents seem unable to recognize their own young chicks and siblings seem unable to recognize each other. To verify the latter, several days before the first experiment we set down each of 12 kidnapped chicks midway between its own sibling and a similar-sized unfamiliar chick of the same dominance status standing 3 m away, on unoccupied terrain in the colony. During 20 minutes of observation, we detected no evidence that they favored or disfavored siblings, or even recognized them. These pilot observations reassured us that our experiments would test the way chicks react to nest mates (siblings) rather than to chicks of other families seeking adoption.

In the experiments, we temporarily kidnapped chicks just a few weeks old, using none more than once. They settled into their foster nests in seconds or minutes, and none was discriminated against in any way by foster parents. Rarely did chicks appear disconcerted by fostering or get repelled by the resident chick, and in these cases, they were promptly returned to their home nests. We transported chicks in individual sand-lined plastic buckets, covering their bodies with soft cloths for insulation and reassurance, and observed them after reintroduction to their families to confirm successful integration, which always happened promptly, even after parents had been left for a while with an empty nest. Follow-up monitoring of participants in experiments and nonparticipating peers confirmed in every case that fostering affected growth only temporarily and survival not at all.

I carried out these kidnapping experiments with José Luis Osorno, who completed three theses on the island—undergraduate and masters theses on blue-foots, followed by a doctorate on magnificent frigatebirds—before joining our faculty. His success investigating frigatebirds was no minor achievement. After capture and banding, the adults have an annoying tendency to abandon not only their nest but the island, and their legs are so short you can only read their leg bands by recapturing them. They further complicate observations and monitoring by disappearing for days at a time on feeding trips, leaving their single chick alone on the nest. Worse even, their chicks take well over a year to fledge and often starve to death before making it. But Mowgli (named for his Jungle Book cartoon look-alike) had the passion and the patience to pull off study after study while also studying shell interchanges between hermit crabs and keeping an iron in the booby fire. Tragically, Mowgli died in 2002, just weeks after returning to Mexico City from his fieldwork on Isla Isabel, leaving a legacy of congenial collaboration, exemplary stamina and stealthiness in the colony, and great gentleness in handling birds.

Our hypothesis for the first kidnapping experiment was that senior chicks are trained winners and junior chicks are trained losers, individuals whose previous experience of violent conflict inclines the former to attack and the latter to submit to any other chick. Either or both of those learning processes would be enough to explain sibling dominance relationships being so stable and enduring even after the subordinate sibling becomes larger than the other. We temporarily paired unfamiliar similar-sized chicks in three combinations: dominant with subordinate, dominant with dominant, and subordinate with subordinate. This design would also tell us whether the previous social experience of the other individual affects a chick's aggressive tendencies. For dominant with subordinate, we set the two chicks down 30 cm apart on shady terrain in the colony that was empty of boobies and watched them for 4 hours. They stood around waiting for their parents, just as chicks do in El Niño years when both parents are simultaneously absent from the territory, but also crouched down, slept, and interacted with each other. For the other two pairings, we swapped single chicks reciprocally between two broods of two chicks and watched the new pairings—two subordinates or two dominants—for 30 minutes. To control for possible effects of fostering, six of the fostered subordinates were larger than the resident subordinate and the other six were smaller; likewise for the fostered dominants. This was essential to be sure that effects of relative size and residence were not confounded.

After two seasons of sweating in the island's muggy heat shouldering a heavy driftwood pole hung with buckets of chicks, our journeys back and forth along a cliff-edge path were rewarded with compelling data and proved that booby chicks are highly amenable to experimental tests. I could've hugged them. Results of the three tests confirmed trained winning and trained losing in blue-foot chicks: they showed that chicks accustomed to domination are strongly inclined to attack and chicks accustomed to subordination are strongly inclined to submit, largely independent of the relative size, age, or aggressive/submissive tendency of the other individual.[11] There were a few exceptions, and those were informative.

In nearly all of the 10 pairs of *dominant plus subordinate* that interacted with each other (two pairs just stood around or slept), the dominants performed several aggressive acts and no submissive acts at all, and subordinates consistently responded to pecks with submissive postures. The remaining pair, in which the dominant was 14 percent lighter than the subordinate, surprised us: the subordinate performed 12 aggressive acts, all of which elicited submission.

All 12 pairs of *unfamiliar subordinates* behaved unaggressively, meekly cheeping at each other and touching beaks together, although in a few pairs after about 15 minutes the larger chick gently pecked or threatened a few times, eliciting some submission.

Fig. 2.3 Experimental pairs of blue-foot chicks that were subordinate in their home broods just stood idly cheeping; rarely did either threaten or attack the other.
Drawing by Jaime Zaldivar-Rae; based on photo by Hugh Drummond.

Fig. 2.4 Experimental pairs of blue-foot chicks that were dominant in their home broods locked bills and wrestled while growling fiercely; they had to be separated to prevent harm.
Drawing by Jaime Zaldivar-Rae; based on photo by Hugh Drummond.

Fig. 2.5 A natural adoption between neighboring blue-foot families reveals the effects of trained winning and losing. The adopted dominant chick elicits submission from the resident subordinate despite being smaller.
Drawing by Jaime Zaldivar-Rae; based on photo by Hugh Drummond.

In all 10 pairs of *unfamiliar dominants* both chicks attacked repeatedly, some of them growling and wrestling with mutually gripped beaks; each trial was truncated as soon as the pattern was clear, to prevent further violence.

The package of results not only confirmed that each blue-foot sibling acquires a dominating or submissive personality and that dominants are inclined to truculence and subordinates to meekness, no matter who is the opponent. The results also hinted, through the incipient belligerence of a minority, that subordinates, although seriously cowed, are by no means resigned or passive. Subordinate blue-foots continuously monitor the aggressiveness and relative size of their hostile nest mates, and they go on the offensive when they detect behavioral or size-related weakness. This behavior is a far cry from the "learned helplessness" that psychologists reported for rats left in hopeless situations like immersion in water with no place to climb out. Blue-foot submissiveness was beginning to look less like disability and more like part of an evolved strategy with a conditional component.

Sizing Up Rivals

Later we would test that attractive idea, which holds out hope for the underlings of this world, but first we needed to test whether chicks can detect differences in size and whether they respect those differences. The outcome of a physical struggle between two members of the same species often depends more than anything else on which one is bigger—a 10 percent advantage is usually enough—but can two contending chicks tell which of them is larger? Obviously so, you might think, but how could a young altricial chick whose family and home nest are its entire world (along with the background sounds and sights of the colony) compare itself with the nearby chick it sees and hears, and pecks, and whose pecks it feels? How would you program a robot to do that?[12–14]

To answer our two questions—can chicks assess relative size and do they respect it—we needed to get the effects of sibling experience (learning) out of the way. So the second experiment simply asked whether singletons—solitary chicks in one-chick broods—defer to larger singletons and attack smaller singletons. This led to more journeys along the cliff edge with Mowgli and the pole. We set down pairs of singletons, as similar in size as we could find in our study area, on unoccupied shady ground 30 cm apart, and watched them for 4 hours. The larger individuals were 5 days older and 22 percent heavier on average, varying from 6 percent to 57 percent heavier, similar to the range of sibling weight differences in natural broods.

Again, the results were remarkably clear. In all but one of thirteen pairs it was the larger chick that performed the most aggressive acts. What's more, nearly all pecks by larger chicks elicited submission. All smaller chicks adopted submissive postures when pecked, and only two of them were ever aggressive, one of which had no stomach for a battle. None of its pecks drew submission, and after the big guy pecked and threatened just once, the little guy submitted and was assertive no more.[11]

Infant boobies' sensitivity to size difference was impressive: even in the three most evenly matched pairs, where the big guys were only 1.0 percent, 1.1 percent, and 4 percent bigger than the little guys—and the little guys weighed the same or slightly more—the big guys dominated. We marveled at how half-grown, altricial chicks that had never even been near another chick could so readily size one another up and do so on the basis of so little social interaction. In only three pairs was the smaller individual the first to be aggressive, and after receiving a few pecks it became consistently submissive.

Even when tempted to rebel, a socially inexperienced booby chick defers to a larger size. So, yes, blue-foot chicks can detect and do respect relative size, and they don't need any experience with siblings to do so. How they pull this off we can only guess.

Training Versus Size

The idea that chicks trained into habitual submissiveness remain alert to opportunity to turn the tables on their antagonists was confirmed as an incidental but important finding of the next experiment, which paired dominants with subordinates that were bigger. We had already proven that a booby submits to others if it has grown up taking beatings and submitting, but also that a booby will attack others if they are smaller. But which factor is more important, trained winning and losing or relative size? A clear answer to this question could potentially confirm our inference that males outgrown by their younger sisters continue to dominate them regardless *because of the two siblings' prior training.* This result would allow us to rule out the three alternative explanations listed earlier, including the idea that male chicks may be intrinsically more powerful than females.

By swapping single chicks between pairs of two-chick broods, we created ten ex-perimental pairs in which a subordinate chick—one accustomed to subordination—was on average 4 days older and 32 percent heavier than the dominant one—a chick accustomed to dominating—and nine control pairs in which the subordinate chick was 4 days younger and 24 percent lighter, controls being similar to natural broods. In all cases, this reciprocal fostering was permanent, to test whether subordinates that take advantage of suddenly being bigger to dominate their dominant nest mates can keep it up indefinitely, or eventually cave in, reverting to type.

During the first 6 days, the size-disadvantaged subordinates in control pairs proved they were worthy controls by acting like regular blue-foot juniors: they performed no aggressive acts at all and responded to the frequent aggressions of their domi-nant pairmates with submission. In marked contrast, seven of ten *size-advantaged* subordinates in experimental pairs *were* aggressive. Evidently detecting their new-found advantage, these trained losers opportunistically turned on the aggression to subordinate their apparently shrunken nest mates. (Remember, they cannot recog-nize individuals.) The others may have done so, too, but our observations during just 2 hours per day did not pick up on that.

As you might expect, the bolshie behavior of the seven subordinates did not go down well with their experimental nest mates. Rising to the challenge of sudden re-bellion by their newly outsized nest mates, these trained winners upped the ante, more than doubling their rate of pecking and—you guessed it—exacting abject submission. Only one of the seven subordinates that gave rein to their inner aggressor prevailed over this stiff resistance. The others were overwhelmed by their smaller rivals pre-cisely because they buckled under pressure; trained losers all, they just couldn't resist capitulating, couldn't give more than they took, even one that was nearly two-thirds heavier than its diminished but unyielding nest mate! These six pairs, which settled into a typical dominant-subordinate relationship, gave eloquent testimony to the power of trained losing and trained winning, while the other four pairs demonstrated that for some individuals, or in some circumstances, a difference in size can prevail over that training.

That settled the question. We concluded that in blue-foot infants, trained win-ning and trained losing together constitute the main mechanism for establishing and maintaining dominance. These two learning mechanisms explain the stable dom-inance we had seen earlier in dozens of natural broods of all sexual compositions that fledged two chicks, including several broods composed of an elder brother and a younger sister who outgrew him.[11]

Exceptionally though, subordinates do invert dominance, and they make the at-tempt when their nest mates' size (and probably postures), always under scrutiny, re-veal a change in the balance of power. One experimental junior managed to invert dominance when paired with an undersized dominant, just as one of the six females naturally paired with a brother that she outgrew managed to invert dominance per-manently. As for the precise mechanism that governs inversion, I suspect it involves progressively undoing each of the two chicks' earlier training and substituting contrary training in its place. A trained loser triumphs over a trained winner by steadily increasing its ratio of pecks given to pecks received, and by sustaining that inversion of the peck ratio over several days. The evidence for this last part of the interpretation will become clear at the end of this chapter when we see how, in dominance hierarchies of three blue-foot chicks, the submissiveness of every chick increases with the number of aggressions it receives.

Trained winning and losing is, indeed, a cornerstone of the blue-foot senior chick's strategy of control, conditional tolerance, and siblicide. Remember, the senior chick has a 50 percent chance of sharing genes with its sibling and hence a 50 percent interest in the survival and eventual reproduction of its sibling. When parents are currently bringing sufficient food to the brood, the senior chick's inclusive fitness can benefit by allowing its sibling to share that food, fledge, and produce nieces and nephews. Extra nieces and nephews who owe their existence to the senior chick's generosity are a welcome addition to the senior chick's inclusive fitness, even if they are only half as related to senior as senior's own daughters and sons. On the other hand, when the food arriving at the nest is so meager that the senior chick's inclusive fitness would be greater by consuming all of it personally than sharing with its sibling, then senior should share no longer.

In a situation of uncertainty over how much food parents can bring, senior is probably best off hedging its bets. While there is hope, it should allow sharing to continue, in the expectation of those extra nieces and nephews. When hope evaporates, it should jealously defend its own growth and survival, and thence its personal production of daughters and sons, by commandeering the whole parental food supply and wishing its sibling well as it departs in search of adoption. Trained winning and losing allows seniors to retain their siblings as insurance backup and prospective sources of extra nieces and nephews until it's clear that their parents' plunge-diving output is simply not adequate to nourish two viable fledglings. Senior can postpone its decision with little fear of junior turning the tables because it has a double guarantee: senior has been trained to fiercely suppress insubordination and junior no longer has the wherewithal to sustain a serious challenge. Junior's strategy makes sense, too; being controllable is what buys its sibling's conditional tolerance.

Brown Booby Desperados

After witnessing the conditional tolerance of blue-foot siblings, as we understood it, the unconditional single-minded violence of brown booby chicks, when we finally got to see it in plain daylight, was riveting. We hadn't planned to invest much time watching brown booby chicks because the nuanced conditionality of blue-foot conflict seemed bound to teach us more about the evolution of sibling strategies. Moreover, it's not easy to witness what brown boobies do to each other. You make several visits to a nest to catch the chicks while they're at it, and seldom even see a chick because they're deep in the straw nest cup, snuggled under the brooding parent, you imagine. They could be slugging it out, but you wouldn't know. Once or twice, after confirming that both chicks have hatched, you get to see a reedy neck and pointy bill sticking out from the nest cup beside the parent's massive brown and white breast, maybe even begging frantically. Then the next time you prise up the brooding parent with a forked stick to inspect the cup there's only one chick there anyway! One day or night between nest inspections junior disappeared. Gone forever, its cadaver, too, or dead on the flank of the nest mound, shrunken by dehydration, its tiny mandibles stiffly parted.

Although many ornithologists have wrung their hands over the fact of this obligate siblicide, as if it were close to inexplicable that chicks dispose of their own flesh and blood, I have always felt that, sentimentalism aside, it makes fine adaptive sense to eliminate an extra offspring if parents can manage to feed only one. When the insurance chick—created only to substitute for senior in the event of senior's demise—*is no longer needed*, off with its head! That was the explanation offered at the start of the 1960s by the marine bird specialist D. F. Dorward.[15]

But wherefore such haste? That is the question! A 50 g chick eats so little, it could surely be fed for a week or three and then dispatched, just in case a raven or a snake makes off with senior, leaving the parents with no chick at all and a wasted reproductive season. Killing it in the first few days of life seems precipitate and wasteful, from the point of view of both senior chick and parents, and probably prejudices the inclusive fitness of both parents and both chicks. But what if junior has to be urgently dispatched because it's too dangerous to be tolerated in the nest? That was the thought behind an experiment that exposed the nature of the violent conflict between brown booby chicks by temporarily swapping them into nests of blue-foots that were twice as big. These swaps showed up the tragic flaw in brown booby insurance chicks that could have driven the evolution of their hasty elimination, and they helped us understand the trained losing of blue-foot juniors as an adaptation rather than organic vulnerability.

For the experiment, we traveled to Isla San Pedro Martir, 972 km north of Isla Isabel in the Sea of Cortés. This hauntingly beautiful sea was created 4.5 million years ago by the separation of the Baja California peninsula from the Mexican mainland. More than 1,000 km long, 100 km wide, and over 1 km deep on average, the sea houses one of the most ecologically intact archipelagos in the world. Isla San Pedro Martir is a 272 hectare lump of Sonora Desert that protrudes from the sea roughly midway between the coasts of Sonora and the peninsula. It is one of the most remote and logistically challenging of the sea's 900 islands and islets. Largely devoid of leafy vegetation or soil, it is covered with a sparse forest of massive elephant cactuses that nourish themselves on its bare slabs with the help of symbiotic bacteria and fungi among their roots. The island is also populated by uncounted thousands of blue-footed and brown boobies, as well as ravens and lesser numbers of brown pelicans and magnificent frigatebirds. Steep-sided and waterless, it was the site of a guano-mining camp at the turn of the twentieth century and has not been inhabited since.

We went there because that's where the nesting seasons of the two boobies overlap, allowing us to test hypotheses by swapping chicks between their nests. And working there brought a bonus we hadn't fully anticipated: an opportunity to observe junior brown boobies growing up naturally with their siblings. It's highly unusual for junior chicks of this species to survive more than a few days because their siblings are such urgent and efficient killers, but on San Pedro Martir some of them do. The surrounding waters are rich in nutrients, and 10 years before our visit unusually plentiful food enabled 7 percent of brown booby pairs to raise two chicks for at least several weeks.[16] However, whether these chicks became independent and were competitive enough to survive and breed, and whether their parents paid a high a price for raising them, is anybody's guess.

We arrived with our cargo on an isolated, wave-lapped shelf on the island's southern shore in the spring of 2001, after driving north in the university pickup to the naval port of Guaymas. Our team included Cristina Rodríguez, the tireless camp manager who masterminded our logistics for both islands that year, and six student volunteers, all of us freshly trained to administer antivenom should anyone step on one of the island's abundant rattlesnakes. We lashed our ten 60-liter drums of fresh water to the rocks of the platform and pitched camp on white, guano-stained limestone in the stone-walled remnants of the mining settlement, on a windy col atop a 50 m vertical cliff. From there we surveyed a tranquil seascape dotted with white islets that glowed pink at sunset and whirling congregations of boobies and pelicans diving into shoals of fish. In the blinding sunlight of the day, we were attended by pestering eye-gnats known locally as bobitos; on brilliantly starlit nights, our tents thrummed in the wind and California sea lions wailed at the shoreline. The sea lions kept us company when we bathed beside our shelf (*their* shelf really), but we never saw a snake, not even a milk snake. Too early in the season I suppose, but fear of rattlesnakes confined us to the camp at night.

During the first 3 days, besides portering our gear up the slabs, we searched for the broods demanded by our experimental designs by counting and measuring chicks in roughly 1,000 blue-foot and brown booby nests scattered over the island's boulder-strewn slabs and bone-dry gullies. Disappointingly, the available selection of broods was inadequate for either of the research projects we had in mind; we had arrived too late in the blue-foot nesting season. So improvising one night as this fact sank in, my gloved hands fumbling with data sheets in the wind streaming across our little col, I resolved to accommodate the supply of broods by tackling a different question. We would test the desperado hypothesis!

Proposed 14 years earlier by Oxford University theoretician Alan Grafen,[17] the desperado hypothesis holds that the wise-seeming rules that animals generally follow when resolving disputes over resources should be suspended when one of two contestants is a no-hoper. The usual rules mentioned earlier in this chapter, and analyzed using game theory by University of Sussex theoretician John Maynard Smith,[18] specify the contexts in which an individual should concede to an unfamiliar contestant who has some advantage, for instance, being stronger or resident at the site of the conflict. His models demonstrate how, given certain assumptions, strategies of squaring up to each other, assessing the other's fighting capability, and then either conceding or escalating are bound to prevail in evolutionary competition among strategies, winning out over other strategies such as unconditional aggression or routine capitulation. Competition among strategies, playing out over many generations, is expected to select for genes linked to sensible asymmetry-respecting strategies. This is why all-out fights are rare in nature except when there is a huge amount at stake, for example, when two red deer males or California sea lions compete for the opportunity to copulate with a *whole group* of females.

The theoretical exception discovered by Alan is the case where a contestant disadvantaged by some asymmetry in the current contest can expect to be similarly disadvantaged in all future contests, too, for example, because the contestant is unusually small or has a congenital or acquired disability. In this exceptional situation where the foreseeable future is one of endless concessions to more able contestants, conceding is a dead end, and in its place, the individual's best strategy can be to attack with all it's got. Desperados should attack more formidable opponents with almost suicidal determination when that is pretty well the only way out of their grim hole. By analogy, brown booby junior siblings are no-hopers doomed to starvation or urgent ejection from the nest followed by prompt death outside it, unless they can eject their older and more powerful siblings before that happens. Mission impossible! But every now and then a junior sibling could get lucky, and hypothetical genes for desperado violence would win out over genes for patient submissiveness provided desperado juniors fledge more often than submissive juniors.

And here's the thought that explains the urgency of brown booby siblicide: once the strategy for desperado violence by juniors has replaced the strategy for patient submission in the population, every senior brown booby chick would face a deadly challenge: cohabitation with an uncontrollable younger chick bent on siblicide. Not only that, with every passing day this implacable rebel would become more difficult to restrain and terminate, as it developed the ability to clamber back into the nest cup after ejection, and its age and size disadvantages progressively shrank to insignificance. This was our plausible but untested explanation for the hasty killing of junior siblings in the brown booby, and in other obligately siblicidal birds that nearly always kill the junior sibling. It could potentially explain urgent siblicide in the black eagle, the swallow-tailed kite, the Nazca booby, the masked booby, and the American white pelican. Seniors kill uncontrollable juniors as soon as possible because they're too dangerous to live with!

To test the desperado sibling hypothesis,[19] we needed to know what junior chicks of obligately siblicidal species get up to when they have the opportunity to express themselves—to behave. None had been systematically observed because junior chicks of such species are seldom seen in nature, and when they have been observed either they were too immature to do much more than stretch their necks up and beg, or they were undergoing an irresistible onslaught by their sibling. The next day they were dead or gone. The solution that occurred to me that windy night on the col was to temporarily foster nine junior brown boobies that were 7 days old into blue-foot nests containing just a single resident chick that was 5 days older. In these experimental broods, the blue-foot singleton, being larger, would naturally assume the role of senior chick. Critically, the brown booby would be unusually free to express its natural junior chick inclinations because, although blue-foot seniors dominate their nest mates, they do it with restraint. They don't automatically kill them as a brown booby senior would. So the behavioral tendencies of brown booby juniors would be laid bare.

This experimental design would cause an upset because it involved exposing blue-foots to the aggression of a more violent species. However, both species would adjust to fostering because chicks of the two species beg and feed similarly and are almost indistinguishable, and blue-foot chicks would be able to handle the risk posed by brown boobies because the blue-foots would be older, more mature, and nearly twice as big as their assailants at the outset. Naturally, we needed a control with which to compare the experimental brown booby juniors, and the most informative controls would be blue-foot juniors fostered into the same situation. For this, we could foster seven blue-foot juniors that were 7 days old into blue-foot nests containing just a single 13-day-old resident chick that was 6 days older and an average 117 percent heavier. The desperado sibling hypothesis would gain credibility if brown booby juniors with a blue-foot elder nest mate (and foster parents) proved to be unsubmissive and uncontrollably aggressive, compared with relatively submissive and unaggressive blue-foot juniors observed in the same context.

Next morning shortly after dawn we all fanned out over the slabs, swapped the junior browns and blue-foots between nests after weighing and measuring, and settled down in our folding chairs, mostly out of sight and sound of each other, to score the behavior. The experiment ended after we had sat and watched all broods for 3 hours almost every day on 8 successive days, each observer with their own little cloud of eye-gnats. Afterward, the brown booby juniors were fostered permanently into brown booby nests containing singletons of similar age, and the blue-foot juniors were left in their adoptive blue-foot broods, to spare them from having to integrate into another family. To our delight, the growth of seniors and juniors was normal in experimental and control broods, all seniors remained heavier than their fostered junior nest mates throughout their trials, and brown booby juniors put on just as much weight in blue-foot nests as brown booby singletons normally do in brown booby nests. This normality was important: we could reasonably attribute differences in the behavior of brown and blue-foot juniors to differences in their intrinsic inclinations, rather than to deficient feeding and growth of either species.

Confirmation of the main prediction of the desperado sibling hypothesis was swift, persuasive, and dramatic. Whereas junior blue-foots were no more aggressive than they ordinarily would be in natural blue-foot broods, junior brown boobies were little tempests of violence. On average they attacked their huge nest mates more than seven times as frequently as blue-foot juniors did, and four of them formed a distinct subgroup characterized by relentless attacking. As long as their imposing nest mates were awake, visible, and in range, these four continued their offensives until their targets counterattacked. Their attack rates on each one's most aggressive day were 31–81 attacks per hour or, if you scored only during the times when both nest mates were simultaneously active, 110–711 attacks per hour. From what we saw, these four brown boobies started these unremitting assaults when they were 11–17 days old, then maintained a similar rhythm of attacking daily. Juniors of both species pecked, bit, and pushed, but brown boobies did proportionately more pushing than blue-foots, and only brown boobies performed *expulsion-pushing*: systematically grasping the rival's neck or wing in its mandibles while thrusting upward and forward and, when possible, scrambling hurriedly forward. This is how senior brown boobies ordinarily expel their siblings to their death. Incredibly, these junior brown boobies sometimes managed to transport their nest mate several body lengths despite blue-foots being 91 percent heavier than them on day 1, and still 53 percent heavier on day 8. Due to the relative smallness of brown booby aggressors, they did no visible damage to their big blue-foot nest mates but, as the following two accounts make clear, the brown boobies' sustained harassment could disarm even a much larger blue-foot:

In nest A55, the brown booby junior (Brown) and the blue-footed booby nest mate (Blue) were paired on day 1 when Brown was 7 days old and Blue was 17.5 days old. When first observed on day 3, although Brown bit Blue three times, it received 30 pecks and bites and was frequently forced into submissive and prostrate postures; Brown was seldom allowed to even raise its head or beg properly. When next observed on day 5, except when asleep or under attack, Brown attacked Blue continuously whenever Blue was visible nearby. During the 41 minutes that both chicks were active that day, Brown delivered 224 violent pecks and 8 bites, as well as scores of pecks and bites at the attending adult's tail feathers and wing when Blue was out of sight. Brown prevailed over a much older and larger opponent through aggressiveness and by not being susceptible to intimidation. When attacked on day 5 (22 pecks and 1 bite), Brown briefly became immobile, cowered or hid, then resumed its frenzied attacking. On four of the next 5 observation days (days 7–11), Brown attacked incessantly whenever both chicks were active and in proximity, frequently seizing Blue by the nape or wing and propelling Blue toward the rim of the nest with an expulsion push. Blue mostly ignored the attacks and continued begging, but its demeanor was sometimes agitated and hurried when being pushed around. Occasionally Blue responded to aggression with a volley of pecks, once even giving a yes-headshake display in brief triumph when Brown crouched down in response, but inevitably Brown would renew its assault and Blue would absorb more punishment and adopt submissive postures or flee across the nest. Throughout this period of observations, Blue was roughly twice as heavy as Brown.

Fig. 2.6 Furious brown booby launches bout of attacking by biting its rival's wing. By day 5, the relentless attacks of the brown booby fostered into nest A55 were beginning to overwhelm the resident blue-footed booby despite it being older and 79 percent heavier.
Drawing by Jaime Zaldivar-Rae; based on photo by Hugh Drummond.

In nest B71, frenzied pecking, biting, and pushing by Brown were first seen at age 17 days, when Blue was not only obliged on occasions to remain at the nest perimeter but also briefly walked out of the nest in response to being bitten and pushed (despite weighing 43 percent more than Brown). Although when we watched the next day Blue was doing all the pecking and Brown was generally submissive, within 2 days a very submissive Blue was spending most of its time 50–70 cm from the nest, in the shadow of a boulder, while Brown remained in the nest with the attending adult. The next day, after repeated expulsions Blue declined to reenter the nest even when Brown was asleep. Blue approached the nest, and evidently could have shuffled in and joined its parent, but after a pause, it turned around, walked back to the boulder, and remained there.[19]

Brown booby juniors did not have it all their own way by any means, but they were more likely to dominate their large nest mate than blue-foot juniors: whereas only two of seven blue-foot juniors ever scored more daily attacks than their nest mate, eight of nine brown booby juniors did so, and only brown boobies tended to continue outscoring their nest mates on several days. Intriguingly though, the submissiveness of junior chicks of the two species was similar in form and incidence. The same submissive bill-down-and-face-away posture was used by both of them, although the

Fig. 2.7 Desperado brown booby usurper in nest B71. Under a blazing sun, a junior brown booby relaxes in the shade and care of a blue-foot adult into whose territory it was experimentally fostered, after forcing the adult's (older and bigger) chick to abandon home and shelter under a boulder.
Drawing by Jaime Zaldivar-Rae; based on photo by Hugh Drummond.

postures of brown booby juniors looked less stereotyped. And juniors of the two species were similarly ready to adopt the posture: on average, browns responded to 44 percent of pecks by submitting and blue-foots to 43 percent. However, in testimony to the effectiveness of brown booby aggression, whereas only one junior blue-foot ever elicited submission from its larger nest mate, six of seven brown boobies did so.

We concluded that brown booby juniors are considerably more aggressive than blue-foot juniors. Admittedly, brown booby juniors are equally inclined to submit when attacked, but their submission does not imply assumption of a subordinate role, because shortly afterward they resume attacking regardless. The brown booby's implacable aggressiveness nicely confirmed the main prediction of the desperado sibling hypothesis, with the surprising wrinkle that desperado aggressiveness can cohabit in the brown booby's psyche with the willingness to submit. That willingness could be a tactic for discouraging the senior chick's ongoing attacks, all the better for the brown junior to renew its own nonstop attacking as soon as senior relents. But wouldn't knuckling under and conceding carry a psychological cost over the medium term? Isn't submission to a bigger nest mate the royal road to aggressive neutering and permanent subordination?

That question—can brown boobies be ground down and forced to accept subordination—was answered in the negative without any need to swap chicks around. For this, we needed to find intact brown booby broods of two chicks and watch to see what they do to each other after the first 4 weeks of life. At this age, all pairs of similar-aged blue-foot chicks comprise a habitually aggressive individual and its submissive sibling, but what would brown booby siblings be up to? After the experiment, several days remained before the arrival of the naval vessel that would take us off the island. The ever-reliable Mexican navy had already resupplied us with water and fresh groceries, and our ragged team cheerfully signed up for spending the last days watching a whole slew of unmanipulated mature broods.

During our crisscrossing of the island, we had noticed a few broods of two brown boobies that were several weeks old, and by expanding our wanderings we could surely find more. So we combed the white slabs and sparse cactus forest under the glaring, bedazzling sun for pairs of naturally cohabiting brown booby chicks and, for contrast, pairs of similar-aged blue-foot chicks. They were easy enough to spot because the gangly, month-old chicks of both species are mobile and fluffy white until brown feathers erupt onto their tails, upper backs, and wing edges. They still sleep a lot, and they feed a few times each day when parents disgorge loads of whole fish into their throats. When feeling lively, they stretch or flap their wings, or shamble about the family territory, ventilating and drooping their wings in the heat, occasionally bursting into spasms of sib fighting. Our scouts found 16 broods of two brown boobies with chicks 29–60 days old and 17 broods of blue-foot chicks 28–64 days old, perfect for a fair comparison of the two species. In a few weeks, the oldest ones would be cloaked in somber brown plumage and ready for their maiden flights.

At the ages of the chicks in our sample, the desperado sibling hypothesis predicts that every pair of brown boobies should still be fighting it out, each chick battling to subdue its sibling and unwilling to submit, whereas every pair of blue-foots should have sorted into a dominant who attacks infrequently and a subordinate who attacks far less and routinely submits in response to attacks. Of course, we already knew that blue-foots behave like that on Isla Isabel after the first month, by which time the dominance relationship is well enough established to be sustained by less frequent attacking, using threats nearly as much as pecking and biting. However, comparisons of the two species would be more telling if both were observed in the same environment: same island, same dates. Once again, the broods we needed to watch were widely dispersed across the southern end of the island, but there were just enough of us to watch each pair of chicks for 3 hours before rapidly dismantling the camp, erasing our nest tags, saying adios to the obstinate blue-foot family beside our kitchen, and portering our gear down the steep slabs to the rock platform, before the naval vessel arrived.

Three hours was enough. There was no expulsion-pushing because juvenile boobies are far too big to carry each other around and anyway mobility makes expulsion pointless. But there was considerable pecking and biting, and the arresting contrast between the two species closely matched our expectations. Subordinate brown boobies were more than thirteen times as aggressive as subordinate blue-foots, and dominant brown boobies were more than seven times as aggressive as dominant blue-foots. In reality, subordinate brown boobies were so only in name (we had to call them something); what we were calling dominance between brown booby siblings was in reality just temporary or average ascendance in the aggressive see-saw between two combative near-equals. The desperado sibling hypothesis was supported because at age 28–64 days, brown booby juniors still had not learned to be subordinates: every pair of brown boobies was still fighting it out, each chick battling to subdue or kill the other. No relationship of dominance-subordination had been established.[19] However, contrary to one prediction, on the receiving end of attacks, brown booby juniors were just as willing to submit as blue-foot juniors, as we had previously seen in the fostered juniors. There's a basic difference between the two species, though. When a weeks-old blue-foot junior submits, it is renouncing aggression; when a weeks-old brown booby submits, it is taking a breather and forestalling further punishment before unleashing its next salvo. Junior brown boobies are not susceptible to trained losing; they are desperados, and their ungovernable aggressiveness is explanation enough for their hurried dispatch by elder siblings.

The sheer size of chicks in the natural brown booby broods of our sample, with individual weights approaching 1.3 kilos, made for the most damaging pecking and biting I've ever seen between booby siblings; some had bald patches and welts on the nape where their siblings had insistently jabbed, seized, and twisted. There were long periods when they rested or slept, but when they got into gear again, they were fierce. Victims that staggered away sometimes gained a respite; other times they were pursued. Often, they just stood there in total submission, for a while taking their beating when fighting back could draw more fire.

They made for a sorry sight, pairs of such large brothers and sisters growling nastily as they exchanged an average 10 pecks and bites per hour, each in turn digging at skin, scars, and wounds. Fluffy white Sesame Street dolls gone rabid. The hostility of brown booby chicks is not the stuff of natural selection acting for the good of the species. It's an example of how uncaring natural selection can shape selfishness and brutal discord among even the closest of relatives; just as the trained winning and losing and conditional tolerance of blue-foot siblings show that outcomes of conflicts of interest are not always so cruel. And there are touching nuances: just as blue-foot nest mates often huddle together on Isabel, no doubt during San Pedro Martir's frigid nights chicks of both species huddled with their siblings, because they need each other's warmth.

Oddly, you may think, in neither species does a youngster ever signal pain or distress by calling out. It doesn't have to be so; lost lambs and farmyard chicks bleat and cheep to call in the help of their mothers, and from the get-go, booby chicks signal hunger to their parents with a clicking vocalization that gets louder as they age. On the other hand, why signal if no one's ever going to help? Parents are mere bystanders. As far as sibling violence is concerned, each chick is on its own; it must take its medicine. What chicks actually *experience* is another matter, and it troubled me more than ever as I watched the brown boobies slugging it out on those bare slabs. Conceivably, they experience nothing, they may have no more feelings than robots, but I've never suspected that. My guess is that chicks under attack suffer ongoing fear, soreness, and pain but don't reflect on their own experience. I doubt they envisage themselves as victims; they just suffer, and get on with it.

After watching those last 33 mature broods we were laughably hard-put to dismantle our camp and porter it all down to the wave-lapped shelf from which, next morning at seven o'clock, a naval boat would ferry us to a vessel standing offshore. Partly, I confess, because in desperation we had painted hundreds of nest numbers on boulders and now we had to clean up our mess. The numbered wooden stakes we hammer into the sandy soil of Isla Isabel to mark nests were useless on San Pedro Martir. We couldn't even drive 6-inch nails into its slabs to pitch our tents. We had used paint on the pristine slabs and boulders, and now it took Cris and me nearly a whole day to rub it off with wire brushes. Meanwhile, Claudia, Adriana, and Gabi were diligently packing and carrying loads down the slabs, wryly amused by the sight of the gas tanks, boxes of food, and water bottles on the slabs that during a whole month we had failed to carry all the way to the camp, and now had to carry back down. With creeping shame, we had been leaving heavy items for "la proxima vez que subo" (next time I come up).

Then, after 10 hours of these exertions, and some ineffectual head-scratching for a plan that could save us from working throughout the night to keep faith with the navy, I experienced the second miracle of my 55 years on this planet! The first miracle came on an ascent of the north ridge of the Aiguille Noire de Peuterey in the Italian Alps, when staring down the abyss into which I must rappel, I beheld in the cloud below a tall Christian cross in the center of a circular glowing halo! (It turned out to be my own shadow in the cloud as I stood with the sun behind and arms stiffly outstretched to coil a rope.) The second miracle came on San Pedro Martir about an hour before sunset as I sat scouring rocks, my head and shoulders drooping with fatigue, my vision narrowed to the circle of pale rock at my feet. This time I beheld, not 2 m in front of me, a vertical pair of white, crisply ironed pant legs, and as I scanned incredulously upward, the erect, uniformed figure of a naval officer saying "Doctor, si me permite, mis marineros les ayudarán a bajar sus bultos a la orilla" (Doctor, with your permission, my sailors will help carry your loads to the shore). And so it came to pass.... By midnight, all of our gear including our precious data sheets was on board a carpeted ship (confiscated from a drug lord) and we were in the galley supping a heavenly seafood broth, thanking the Virgen de Guadalupe that the officer and his men were impatient to get back to Guaymas.

Blue-Foots and Brown Boobies Compared

By focusing attention on the victims as much as on their oppressors, and comparing two closely related species, our research program developed a fresh perspective on the evolution of sibling conflict and, in particular, trained winning and losing. In the blue-footed booby, whose food supply frequently allows both chicks to grow to a viable size, siblings train each other into two distinct personalities. One is the tolerant oppressor who allows food sharing while reserving the power to terminate its sib if necessary, and every day administers a dose of punishment to top up its sib's submissiveness. The other is the subservient victim who by renouncing aggression and showing daily submission offers a guarantee of its own controllability, betting on parents bringing enough food for both chicks to grow and fledge. In the brown booby, whose parents rarely bring enough food for more than one chick to grow to a viable size, two unconditionally intolerant chicks strive flat out to kill their sibling as soon as they can; they are both as congenitally aggressive as they could be, and incapable of learning subordination. From an evolutionary perspective, the junior chick's posture is pivotal in determining the difference in strategy between senior chicks of the two species. Senior blue-foot chicks can get away with allowing their sibs to grow up with them and possibly add to the seniors' inclusive fitness because seniors are in control. Senior brown boobies can't take that risk because their sibs are uncontrollable and conditional tolerance would allow mutiny and almost certainly penalize the seniors' inclusive fitness.

Feeding ecology, which here boils down to how much widely dispersed prey parents can bring to their offspring, is the external context that has channeled the evolution of these behavioral differences between the two species. And we can assume that genetic differences between the species are the internal forces that steer the development of different behavior in their chicks. Hypothetical genes in blue-foots predispose all chicks to attack their sibs as soon as they can, and they predispose chicks that dish out many beatings to become more aggressive (trained winning) and those that receive many pecks to become more submissive (trained losing). Equivalent genes in brown boobies are on a quite different developmental mission: they incline all chicks to be unconditionally and indefinitely aggressive no matter what their sib does, hunker down under attack, and heave hatchling sibs over the nest rim when possible. In both species, for the reasons outlined in Chapter 1, some of the genes pulling the strings on the chicks' behavior maximize their own transmission to the next generation via the progeny of the chicks themselves and also via the progeny of their siblings (who are only half as likely to bear those same genes).

The development of all animals, especially their behavior, varies with the physical and social environments they are immersed in, but genes sometimes specify narrow limits to the range of behavior that natural environments can induce. Undoubtedly, the behavior of brown boobies in our experiment was influenced in minor ways by growing up for a while in blue-foot nests, with blue-foot parents, blue-foot nest mates, and blue-foot food; but this experience did not turn them into controllable blue-foot-like subordinates. They fought like fury, as did the brown boobies that cohabited with brown booby siblings for several weeks. Brown booby genes are probably capable of inducing desperado aggression in any environment a brown booby is likely to grow up in. And blue-foot genes probably always build chicks that can learn to be dominant or submissive and, after radical change in their social environment, unlearn that and learn the opposite.

Comparison of the two boobies gives further insight into trained losing and strategies of sibling competition. Strictly speaking, the taming of junior blue-foot chicks does not occur by the violent *imposition* of a disability, but by activation of an evolved tendency to *adopt* a subordinate role in response to punishment. Trained losing is accompanied by physiological changes in some animal species,[20] and, indeed, we have found that subordination of blue-foot chicks doubles the concentration of stress hormone (corticosterone) in their veins, probably because subordinates get less food.[21] Physiological involvement smacks of imposition, but the example of the brown booby seems to show that boobies can resist, or evolve to resist, imposed disability.

If subjection to pecking imposes an unavoidable physical disability on blue-foot juniors, then how come the San Pedro Martir brown booby juniors that suffered more intense pecking over 4–9 weeks withstood it and remained fully belligerent? The two species of boobies are closely related, so blue-foot juniors presumably could have evolved comparable resilience if it served their inclusive fitness interests. Most likely, blue-foots have evolved a propensity to be tamed when convincingly outmatched[22,23] because this flexibility serves their own interests. By becoming progressively more submissive, blue-foot juniors that take daily beatings earn the tolerance of their powerful siblings. Their submissiveness is a strategic, reversible handicap, as we saw when a submissive female chick on Isla Isabel outgrew her elder brother and permanently inverted dominance. Those sexed broods also showed us that temporary subordination can be a gateway to success. When blue-foot parents brought enough food for the whole brood, junior fledglings began their transition to independence just as big and well-nourished as their oppressors. And subordination may conceivably reap a two-pronged reward: propagation of hypothetical Subordination genes into the next generation via the subordinate's own future offspring, and (with 50 percent less chance) also via the future offspring of its dominant sibling. Presumably, the sum of those twin prospects pays juniors better on average than the fraught alternative of trying to unseat a powerful rival and putting both sibs at risk. I believe that is why blue-foots have evolved a *propensity to be tamed*.

And being tamable, as opposed to faking tameness, is a winner in the evolutionary competition among strategies because it enables the evolution of conditional tolerance in the rival.[24,25,26] When our experimental swaps on Isla Isabel between blue-foot families made subordinate juniors larger and older than their nest mates, we saw that the juniors' previously acquired tameness was not fake. Nearly all of the juniors that responded to the sudden opportunity by launching campaigns of aggression crumpled and submitted under the hail of pecks they provoked. They were unable to promptly switch on full-blooded aggressiveness. They buckled because they couldn't help themselves. Fake tamability, if it evolved, would open the door to successful coups by subordinates that outgrew their dominant nest mates and violated their trust. And this would be answered by the evolution of unconditional lethal aggression in dominants. According to this interpretation, subordination involving authentic tamability and dominance involving conditional tolerance are two strategies that evolved each in answer to the other—a "coevolved" solution to a conflict of interests between siblings.

Sibling Hierarchies

If broods of two blue-foot chicks represent one of the simplest cases of dominance-subordination in a wild vertebrate, then broods of three blue-foot chicks are surely one of the simplest cases of *dominance hierarchy* formation in a wild vertebrate. Dominance hierarchies in social groups of animals can be linear, with A dominating B and C, and B dominating C,[27] or they can include complications like triangles, with B dominating C and D, who are of equal rank, or even circularities, with B dominating C, C dominating D, and D dominating B. Imperfect linearity is most common, with individual ranks reflecting the relative size or fighting ability of group members. It gets more complicated in the largely independent female and male hierarchies of some primates when offspring gain elevated rank through the nepotistic support of their mothers, or when two or three individuals form an alliance that outranks powerful individuals.[28] And the structure and relevance of a primate hierarchy can vary with the value of the resource that is being contested: even high rankers can be deferential around a subordinate who has some fresh meat to share out. In humans, the most socially variable primate of them all, psychological, sociological, and anthropological data converge to show that status-seeking is a universal motive.[29] Humans can attain hierarchical rank by power-mediated dominance or, more uniquely, via consensus-mediated prestige.[30] We will get to human sibling conflict in Chapter 9.

Here, we'll get down to the basics of the simplest vertebrate dominance hierarchy by analyzing aggression in broods of three blue-foot siblings. The genesis and maintenance of blue-foot sibling hierarchies are conveniently open to scrutiny because broods can be observed in nature from the time chicks start interacting and before they encounter any boobies other than their aloof and towering parents. Three chicks is the least common brood size on Isla Isabel, so three fluffy white chicks shuffling about a territory is always a heart-warming sight, and tells you their parents are an outstanding pair pushing the envelope. The clumsy, endearing youngsters approach and greet each other, pass sticks from beak to beak, and sprawl on the leaf litter for naps, but they also compete seriously for parental regurgitations. In 24 seasons on Isla Isabel, although 20 percent of 217 pairs with three chicks fledged them all, showing that not all ambition was misplaced, mortality rates were 25 percent for A-chicks, 31 percent for B-chicks, and 69 percent for C-chicks, figures revealing hierarchical control of the food brought by parents.

To identify the structure of blue-foot hierarchies and work out how hatchlings assemble them, in a single season an undergraduate thesis student, some volunteers, and I watched 18 broods of three during each brood's first 6 weeks or until its first fatality occurred. On average, A-chicks were 3 days older than B-chicks and 7 days older

Fig. 2.8 Blue-foot feeding hierarchy: A feeds, B begs, and C sleeps. C's turn will come after A and B fall asleep.
Drawing by Jaime Zaldivar-Rae; based on photo by Hugh Drummond.

than C-chicks and, so you'll know where this story is going, the proportions of A-, B-, and C-chicks that died before fledging were 6 percent, 16 percent, and 100 percent, respectively.[31] Without exception, As beat up heavily on Bs and Cs, and Bs regularly on Cs, and in each of these three dyads, there was minimal aggression in the opposite direction. A steady increase in pecking by the two eldest members of every threesome followed by a steady decline after 3 or 4 weeks told a story of maturing elders piling on the aggression until younger sibs were trained into maximal submissiveness, followed by relaxation of training effort to the minimum required to maintain siblings' subordination. In all three dyads of every brood, the staggered onset of pecking resulted in elder chicks training themselves as winners and their sibs as losers, and the outcome was 18 linear hierarchies composed of three dominant-subordinate relationships.

But there was more to it than that. In a year when food supply was inadequate for parents to raise three chicks, Bs received a worse thrashing than juniors in broods of two: attacks were at double the intensity and did not begin to ease up until after 33 days of continuous increase, compared with only 23 days for juniors in broods of two. What's more, the badly beaten Bs managed to beat up on Cs only infrequently. This regime turned Bs into fully trained losers (submitting to 80 percent of attacks), while largely denying them the opportunity to train properly as winners. Consequently, when an undergraduate thesis student, Tania Benavides, experimentally paired Bs from another sample of threesomes with similar-sized juniors from twosomes, to test their mettle, Bs failed to outperform juniors on any measure.[32]

Why was there so little sign of Bs being trained as winners? I now believe that As' intensive attacking of Bs functions to neutralize the long-term risk of insurrection. By giving Bs an extraordinary thrashing, As dilute the amount and effectiveness of Bs' aggressive enhancement (training as winners) by pecking C-chicks. Thus, three-chick hierarchies have a stabilizing strut that makes them more than just a collection of three dyadic dominance-subordination relationships. As curtail Bs' training as winners, effectively defanging potential challengers. In contrast, senior chicks in broods of two face a lesser risk of insubordination because juniors cannot build up their own mental muscle: there's no lesser sib to safely attack.

More eye-opening were the rates of submission we observed in broods of three; these disabused me of a wrong assumption about trained winning and losing that I had been making for about 20 years. Taken together with our observations on the development of pecking, the development of submissiveness in threesomes spoke to blue-foot submissiveness and aggressiveness being not, as you would expect, two sides of one coin, but two coins. The data suggested that submissiveness and aggressiveness are two largely independent axes of learning and a chick can learn both at the same time! The key observation was that all age ranks started their lives submitting to few of the attacks received and learned to submit to most of them, *including As*! Similar developmental increase in the submissiveness of all three ranks suggested that the submissiveness of every blue-foot chick increases with the number of attacks it receives. And that increase occurs whether the chick is, over the same period, pecking others as rarely as Cs do, as frequently as Bs do, or as abundantly as As do. Well after they established track records for greatly outpecking their younger sibs through at least age 30 days, As and Bs were submitting to those very underlings' (infrequent and ineffectual) pecks. Moreover, the submissiveness of all age ranks to pecks increased progressively from roughly 15 percent in the first 5-day block of being pecked to roughly 55 percent in the third 5-day block, and in Bs and Cs (the only ones observed longer) 85 percent in the fifth 5-day block. Might these increases in submissiveness be due not to learning through experience but to simple maturation over time? I doubt it, because the increase started later in Bs than Cs, and later still in As, on account of older chicks not getting pecked until their sibs were old enough to do it!

The neat strut added to the blue-foot's three-chick linear hierarchy—capping B's aggressiveness—could reflect the deft hand of natural selection operating on the aggressive tendencies of As to increase their inclusive fitness by defanging Bs, or simply be an unselected byproduct of the most senior chick having to juggle with two unequal inferiors at the same time. Either way, in a year when food was insufficient for threesomes to fledge intact, the simplest linear hierarchy in the book of wild vertebrates was founded on linear differences in age, size, and maturity among three infant siblings who probably cannot recognize each other as individuals. The hierarchy consisted of three dyadic dominance-subordination relationships based on trained winning and losing, one of which was fortified by extra pecking, possibly to keep the intermediate sibling from becoming insubordinate. And I counsel against assuming that training in winning is incompatible with training in losing. Blue-foot chicks manage to do both at the same time!

A comparison of blue-foots and brown boobies has shown how chicks of closely related species that differ in a single important aspect of their feeding ecology—food availability—can differ greatly in their tactics and strategies of sibling competition. To enrich and widen our perspective on sibling conflict in vertebrates, the next chapter introduces another six species of birds and three types of mammals. Comparison of their offspring will illustrate how, in response to particular species' characteristics and particular ecological circumstances, highly varied and bizarre modes of aggressive sibling competition can evolve.

3

From Wild Violence to Courtly Rituals

In this chapter, we widen our perspective by scanning across the diversity of sibling dominance relationships documented in field studies of birds, by far the best-known vertebrates in this respect. After Chapters 1 and 2, you may be thinking of blue-footed and brown boobies as curiosities. I would prefer you to see them as two well-studied points on an intriguing spectrum of variation among avian species in how chicks compete aggressively with their siblings. They compete for the reasons we considered in those earlier chapters, basically because evolution by natural selection impels each offspring to maximize its own inclusive fitness. In every species, each chick values itself twice as highly as any sibling, so if food is in short supply—the typical situation—brood mates will compete for it. How they do this depends on the characteristics and ecology of the species, and behavioral ecologists busy themselves working out which factors have favored the evolution of particular forms of competition. Comparing several avian species—particularly cattle egrets, brown pelicans, oystercatchers and crested ibises, game birds, and waterfowl—will give us some sense of the ingenuity of evolution by natural selection and prepare us for thinking about sibling conflict in other animals.

Mammals have been neglected because they are difficult to observe, but the three mammals whose offspring have been studied, mostly oddballs, will give us an idea of the variation in sibling conflict that awaits discovery. Domestic pigs have been studied because of their commercial importance, spotted hyenas are distinctive enough in their social behavior to have merited extensive study in African savannahs, and a few species of canids have been watched long enough (alas, mostly in captivity) to take a tantalizing peek at how their pups form hierarchies and get serious with each other when parents start delivering solid food. *Homo sapiens*, more studied than any other species, but by social scientists rather than evolutionary biologists, will be saved for the final chapter, which discusses the relevance of booby family conflict to understanding human family conflict.

Cattle Egret Resistance

Between the brown booby's lethal, unconditional ferocity and the blue-foot's conditional tolerance lies a continuum of aggressive intensity; and beyond the blue-foot end of that continuum lie such delights as ibises that "attack" their siblings with the

Blue-Footed Boobies. Hugh Drummond, Oxford University Press. © Oxford University Press 2023.
DOI: 10.1093/oso/9780197629840.003.0004

ritualized manners of Elizabethan courtiers and owlets that negotiate vocally with their siblings. Sitting somewhere near the middle of that continuum are cattle egrets, those compact white herons you see on road trips across the Mexican altiplano and tropical lowlands and the southern United States. Dotted across lush meadows and pastures in loose flocks, statuesque or walking deliberately around the cattle that scare up their prey, cattle egrets seize and devour small creatures such as frogs, grasshoppers, and crustaceans with their long slender beaks. By chance, I got my first look at their broods in the early 1970s while boating from island to island in Oaxaca's Lake Temascal, in the lands of the Mazatec Indians. Interspersed among glossy black cormorants, the egrets were nesting in the low bushes of an islet, dozens of pairs caring for broods of two or three gangly white chicks perched on rickety twig platforms. Marveling, as I was, over obsidian arrowheads exposed on wave-lapped shores, swards of sleepy plants clenching their fronds as we stroked them, and herds of black toadlets hopping down grassy banks to the water, I had no idea at the time—I was still a language teacher—that these cattle egret nests are hotbeds of family violence.

It was not until 10 years later, at Animal Behavior Society meetings, that I learned the basics of their sibling antagonism from studies by colleagues—Doug Mock, Bonnie Ploger, and Curtis Creighton—on a dredge island in Texas and clusters of small oak trees in Oklahoma. Years later, to answer unresolved questions about the development and causation of their aggressiveness, my doctoral student, Alejandro Gonzalez-Voyer, made behavioral observations on a Mexican colony at a brackish lagoon on the Gulf coast. From floating observation hides tethered near an islet of red mangrove bushes, Alejandro and a team of student volunteers watched a throng of cattle egret broods for 6 hours every day for 6 weeks.

When I joined his team in the summer of 2003 to finally see the contending sibs for myself, it was with high expectations. After a night's rest in a nearby field station, we puttered out to the islet after dawn in an inflatable raft with a four horsepower motor. Smaller than many bath toys, it was just the right vessel for a languid tropical lagoon alive with birds. The floating hides with folding chairs before a circle of bushes crowded with nests was our theater of sibling hostility, and on the stage chicks were exhibiting neither the murderous resolve of brown boobies nor the measured tolerance of blue-foots.

A cattle egret's three chicks hatch at 1- or 2-day intervals, and this staggered hatching gives the elder member of each dyad enough of a head start to win most fights, get more of the food provided by parents, and survive better. Yes, unlike boobies whose aggression is normally one-sided, these chicks actually fight—they face each other and trade blows like boxers. During bout after bout, day after day until they fledge, the three siblings slog it out for status and food. More often than not it's the younger member of each dyad that gets out pecked or dislodged and the elder that manages to scissor higher on the parent's beak and catch more boluses of regurgitated food. Younger siblings are subordinates in the limited sense that they yield more often, but they don't signal submission, not at least with the theatrical clarity of a junior blue-foot's bill-down-and-face-away display.

Indeed, to all appearances, trained winning and losing are minimal in cattle egret broods, and each chick periodically assesses the other's relative brawn or willingness to fight when the two of them face off, rearing up, stretching skyward, and tilting their beaks upward to outreach each other. If this test of height satisfies them, they subside and rest; if it doesn't, they swap blows at full stretch until one of them (usually the younger) yields. It does this by temporarily averting its head, crouching on the nest platform, or hanging its head over the edge. Sometimes an overwhelmed loser flees, even at the risk of falling or getting tangled in branches. But however it yields, it'll be game to fight again when parents next offer food. Egret chicks lose battles but not the war, which continues until one of them dies or their parents no longer feed them.[1-6]

As we finished our observations of the cattle egrets, an incident occurred that illustrates a key ethical challenge in this type of fieldwork. Just as our stint in the hide ended, we spotted a 2 m boa constrictor a couple of meters above the water, stealthily weaving its heavy body through the mangrove branches barely a body length from the nearest focal nests. The limbless, sinuous locomotion of a snake slowly approaching its prey can be hard to detect because it includes few moving edges, and the egret colony seemed unaware of the danger. The boa was a plump, magnificent specimen and fully capable of engulfing a few chicks, even if terrified parents screamed and attacked it, though it might, for all we knew, delay its moves until nightfall. What to do? Whether chicks were asphyxiated by constriction and swallowed afterward or simply engulfed alive by flexible jaws slow-walking over their bodies, lives would surely be truncated and chicks and parents would *experience* the horror of their deaths. On the other hand, boas are native members of the ecosystem who need to eat, and they evolved to consume only one type of prey—live vertebrates. Should humans, the most destructive and effective predators on the planet, be arbiters between predators and their prey? Should humans meddle in natural ecosystems? Predation is an integral ecological process, and in North America snakes are the leading predators on birds' eggs and broods; and birds have evolved countermeasures such as mobbing predators and nesting on inaccessible islands. Ought we to protect our warm-blooded friends against dangerous uncaring reptiles or allow the natural contest to play itself out?

I waded to the islet, gently unraveled the boa from the branches, cradled it gingerly in my arms, and returned with it to the hide. Then we puttered to the nearest shore, from which the boa had probably swum to the islet in the first place, and gently lowered the glossy serpent into the mangrove. It glided serenely away.

You can seldom study wild animals without affecting their behavior, but you are usually confident that the alterations are brief or minor and will not significantly prejudice the animals' survival or reproduction, or substantially alter the behavior you are watching. Although behavioral ecologists sometimes manipulate study animals in important ways—for example, adding a chick to a brood—we do our best to minimize impacts on them and their environments, including other members of the plant and animal communities they inhabit. Indeed, scientific societies have ethical guidelines for the treatment of animals, and most field biologists feel affection, awe, and a duty of care for the species they study. Should they sometimes give needy individuals a helping hand then? On Isla Isabel, we often come across magnificent frigatebird chicks that have fallen from their unstable nest platforms to the forest floor, several meters below. Unharmed, they crouch there on the leaf litter, looking lonely but composed; waiting. Although sometimes tempted, our team doesn't return the chicks to their platforms, even knowing that their parents will never risk entanglement in branches to descend and attend to them and that, abandoned on the forest floor, the youngsters will take many days to starve to death. But I did save the cattle egrets when the boa was about to feed on them. And once on an islet in Panama, I protected hatchling green iguanas that were warily waiting to emerge from nest cavities before migrating to an adjacent rainforest. I hurled sticks at the overhanging branches to keep greater anis from perching there and swooping on my research subjects. If I'd allowed the boa to eat the egrets, and the anis to eat any more iguanas, my samples of egret and iguana behavior might not have sufficed to answer our questions about egret dominance and iguana orientation mechanisms.

Putting it baldly, I save animals to study them, but not for themselves. And you may think it gets worse because ordinarily I am fascinated to watch how predators capture, subdue, and consume their prey, as well as how prey animals evade, resist, and escape. The dramas of wild lives and the ways animals rise to challenges are enthralling! In the channel between the iguana islet and the forested island itself, I once watched in awe as a juvenile American crocodile sped behind a hatchling iguana streaking across the water surface, seizing it with the last snap of its jaws. At a cattle pond in California's Sierra Nevada, I was shocked and fascinated to watch the larva of a diving beetle seize, envenomate, and kill a tadpole-hunting garter snake, then suck the nutrients from its body. Whatever the animals may be feeling, I feel privileged to witness the dramas of their lives and happy to better appreciate the webs of ecological relations they inhabit, whether particular encounters bring me joy or sadness. I was sad when that little snake died; for several days, I had been watching its first excursions over the pond with some tenderness. Mowgli was not heartless in these matters and sometimes cared for injured birds on Isla Isabel, once endeavoring to splint a booby's broken wing. I admired and respected his empathy and compassion, but my own inclination is to minimize intrusion and interference with nature, altering lives only in the service of understanding them better. And I believe that the more we silently contemplate species, habitats, and ecosystems the more we respect, cherish, and want to conserve them.

Comparisons among the cattle egrets and other aggressive birds helped us understand, in functional (adaptive) terms, why youngsters of different species evolve to struggle differently with their siblings.[4] A low-ranking cattle egret violently resists being dominated because low rank incurs such disadvantage that it should never give up struggling for a better deal—a greater share of food, or promotion to the next rank, after usurping or killing a sibling. In contrast, a blue-footed junior's low rank ultimately incurs so little disadvantage that it is better off submitting to low status than engaging in costly struggle. And a brown booby junior is at such extreme disadvantage that it should desperately go all out to kill its rival. Estimates based on colony monitoring have shown that occupying the lowest rank depresses the junior blue-foot's probability of surviving to fledge by a mere 10 percent, the youngest great egret's survival by 22 percent,[7] the youngest cattle egret's survival by a massive 75 percent, and the junior brown booby's survival by fully 83 percent. Those percentages—the cost of low rank—express, very roughly, *what a low-status youngster stands to gain by struggling rather than acquiescing.*

Moreover, in all three species, it's the low-ranking chicks' aggressive strategy that drives the evolution of the high-ranking chicks' strategy of containment. Elder blue-foots first beat their highly tamable siblings into submission, then coast through the rest of the chick period by pecking just enough each day to maintain their dominance. Elder cattle egrets govern the resistance of their almost untamable siblings by repeatedly beating them down. Elder brown boobies have no choice but to kill their dangerous, desperado siblings as soon as they possibly can. In each species, the amount of food typically available and its typical allocation among chicks determine how vigorously the youngest one needs to stick up for itself, whether by submitting, resisting, or killing; and it's the youngest one's typical behavior that determines whether the eldest one should tame it, struggle perpetually against it, or do away with it. The chicks themselves aren't born knowing any of this, and nor do they learn it. Rather, natural selection equips every species with the genes that will steer each chick's development toward the most convenient levels of baseline aggressiveness and appropriate flexibility in the face of experience. I believe the factory setting on any avian species' sib aggression dial increases as the survival probability of youngest brood members declines.

Low-ranking chicks of the brown pelican, the great egret, and the grey heron are also intermediate between the blue-footed and brown boobies in their survival prospects, and they resist dominance much as cattle egrets do. For her undergraduate thesis project, my student, Dalila Pinson, studied the brown pelicans of Isla Isabel for 3 months in 1987, when the mild El Niño event of the previous year was drawing to a close. Every morning at 0630 h, after consuming her signature breakfast of cornflakes with chopped cabbage (the secret of her success?), Dalila set off through the tall grassland beside our camp, struggled up the forested, boulder-strewn flank of the pelicans' ridge, and spent 3 hours on the crest in a cloth blind, watching 10 broods on the forest floor. In the afternoons her volunteer assistant took over for a further 3 hours. Treetop broods were left out because climbing trees to weigh and measure chicks would have upset the colony adults and possibly spooked them into abandoning.

A pelican specialist who hadn't yet taken on board the Hamiltonian revolution described in the Introduction had told me that we wouldn't find systematic aggression in pelican broods; there was no phenomenon worth investigating. "Yeah sure, chicks peck at things. That's what young birds do. They'll peck at your hands and might peck each other. There's nothing to it." But what Dalila observed on the ridge were clear aggressive hierarchies in all broods of the huge-headed chicks, with the senior chick of each dyad, which was 2 days older, outpecking, outbiting, and outpushing the other. A-chicks in broods of three delivered an average of 11 pecks per hour, and only 39 percent of those were during feeding bouts, implying that most pecks were for training the sibling into submission. And, indeed, junior dyad members sometimes signaled submission, using the same bill-down-and-face away posture used by blue-foot juniors. However, pelican B-chicks were more pugnacious than blue-foot juniors, delivering roughly half as many pecks as they received. Junior pelicans fed less frequently than seniors and died of starvation rather than expulsion, but some of them attempted to invert dominance. In two out of ten broods, the B-chick managed to out peck the A-chick, and of these two rebellious B-chicks, the one that made a complete switch from submitting to pecking eventually triumphed: A died with visible lesions at 7 weeks, and rebellious B went on to fledge. The other rebellious B-chick didn't have enough mojo to truly dominate: despite going on the offensive and outpecking A rather decisively, it persisted in submitting; eventually, it succumbed, leaving A to fledge alone.[8] The lowest rank in a brown pelican brood implies a 62 percent reduction in the probability of surviving, intermediate between 75 percent in a cattle egret brood and 10 percent in a blue-foot brood. So it's not surprising that the tamability of brown pelican chicks is intermediate between that of cattle egrets and blue-foots, and that they meet with an intermediate level of repression.

Dalila eventually moved on from pelicans, and from her ground-breaking diet, becoming an assistant professor at the Universidad Autónoma de Queretaro.

Oystercatcher Avoidance

Markedly different in their ecology and early development, and consequently in their sibling aggression, are the chicks of oystercatchers, those jet black shorebirds (some species with white breasts) with jutting orange beaks so brilliant they seem to glow. Socially monogamous like boobies, egrets, and pelicans, oystercatchers nest solitarily on rocky and sandy shores, where they snap up invertebrates such as shellfish, bristle worms, and sea urchins. Their chicks are unique among birds in being fully mobile on land within 24 hours of birth while remaining entirely dependent on their two parents for food. Shortly after the brood of two or three chicks is complete, they're all walking out of the nest scrape, scampering around like baby chickens and escaping from predators but receiving all food from the parental beak. To compete for these transfers, chicks hurry to meet their mother or father when they return to the family territory with food. That, of course, is when sib fighting would be expected, but, strangely enough, it doesn't happen.

Although oystercatcher chicks are in serious competition with their siblings—poor growth prejudices not only the chance of fledging but also the future reproductive success of survivors—and are capable of pecking forcefully, even if you watch a brood for hours, you probably won't see obvious aggression. Because chicks are not nest-bound—not forced into each other's company like slow-developing boobies, cattle egrets, and brown pelicans—the low-ranking members of a brood can elude their oppressors! At least in the two most studied species—the Eurasian oystercatcher and the American black oystercatcher—low-ranking chicks simply avoid contact with their siblings, managing this so effectively that it took a while before observers were aware of their aggressive hierarchies. Chicks avoid approaching parents for food when a high-ranking chick is already there, they depart from parents when a high-ranking chick approaches, and, when risky proximity is imminent, they make themselves scarce by fleeing or hiding, taking advantage of the mottled feathers that camouflage them. Here, too, the strategy of high-ranking chicks is an evolved response to the strategy of low-ranking chicks: seldom does a dominant threaten or chase its sibling, and only rarely does it punish it with a peck on the head. It doesn't need to.

It took an inadvertent field experiment on the Welsh island of Skokholm 40 years ago to reveal the subtle hierarchies of Eurasian oystercatcher broods and the *potential for violence* that creates and sustains those hierarchies. When Uriel Safriel enclosed four nest territories in 50 m^2 corrals (to facilitate observation),[9] although parents could still fly back to their territory with food, subordinate chicks no longer had enough space to keep out of the way of dominants. Forced into each other's company, pairs of siblings contested prey items by tugs-of-war and fought ferociously. When subordinates began to lose weight, Safriel dismantled the corrals to rescue them from sibling aggression and starvation. During the fighting, submissive postures were not reported and may not have occurred. Why should a mobile youngster learn to submit to its antagonist if in the natural context it can avoid clashes almost completely? I suspect that all two or three oystercatcher brood mates are born roughly equal and each dyad quickly sorts into dominant and subordinate, based on their relative size and the outcomes of brief fights over access to parents. Thereafter, a tense inequitable peace reigns as subordinates steer clear of dominants, allowing them privileged access to food. What precisely oystercatcher subordinates learn may well be less onerous than what blue-foot juniors have to learn: trained avoiding rather than trained losing. *Avoidance*, enabled by mobility, could be preferable for subordinates because they don't have to go through the physical and psychological attrition of learning to submit.

Another two variants of sibling avoidance have been uncovered, one delightful for its simplicity, the other grim and often tragic. Sandhill cranes, spindly grey birds with crimson foreheads that congregate in vast flocks for migration, forage in grasslands and wetlands along with their two self-feeding chicks, taking insects, amphibians, and parts of plants. Zookeepers long ago learned that within a few days of hatching brood mates attack each other so violently that it's fatal to leave them in the same pen. By all accounts, neither sibling yields, and they peck at each other until one of them dies. In at least some wild populations though, the chicks live in peace by foraging separately, one tagging along with Mom, the other with Dad. However, the priority of the elder one can emerge in lean times, when the younger one receives fewer food offers from parents and may be evicted from the family circle by its sibling.[10,11]

The sad variant of sibling avoidance is practiced by a species of skua. Skuas are large, gull-like birds notorious for seeing off humans and other mammals that trespass on their nesting areas by swooping on them and violently striking their skulls with their beaks or feet. They are versatile, generalist predators, and thieves, too; they often frighten other marine birds into dropping or regurgitating prey by threatening them, attacking them, tugging at their wings in flight, or stooping on them like a peregrine falcon.

In the raw cold of Ross Island in the Antarctic, where one of Ernest Shackleton's expedition huts still stands, brood mates of the South Polar skua fight it out on the ground. A day or two after the second egg hatches, a couple of days after the first, both chicks are at it, pecking, biting, and shaking each other furiously. They're at it tooth and nail for several days, with the senior one usually prevailing and the junior increasingly moving away, keeping its distance and hiding, often missing out on the food and brooding offered by parents. Juniors are seen hunkered down out of sight as far as 30 m from the nest, occasionally approaching a parent to beg for food, but often unwilling to try, for fear of more battering. At such a remove, the subordinate chick can still be attended by parents, and there's a slim chance of either being adopted by another pair or, if parents bring enough food, fledging in addition to its sibling. Either way, it's the skua's precocious ability to walk, run, hide, and generate body heat just a few days after hatching that enables younger sibs to avoid aggression and survive a little longer.[12–14]

Crested Ibis Rotating Dominance

Blue-foots, for all that I have lauded their measured aggression and vocation for sharing with siblings, are not at the peak of sibling civility, which may be occupied by the crested ibis, an aggressive species so gracious it makes blue-foots look like barbarians. You would think that slow-developing, nest-bound young and staggered hatching of several chicks at 2-day intervals would set its broods up for hierarchical dominance and status-based access to food, with elder individuals brutally enforcing their priority. And that would make perfect evolutionary sense; no healthy chick should forego the privilege of high rank, because natural selection inevitably favors infants being more kind to themselves than to any sibling. Genes for unselfishness are not expected to compete successfully with genes for selfishness. And yet, in the *rotating dominance* of this ibis, careful observation by a Chinese team striving to rescue the last extant population from the brink of extinction has shown that although there are trappings of violence, every chick's watchword seems to be equity.

The crested ibis is a stout, white wader with a slender downcurved beak that sifts through water and silt for loaches, eels, and invertebrates. Monogamous pairs nest in trees on forested slopes near water. When my colleague, Xinhai Li, and his team observed feeding bouts of broods in the Shaanxi province of Central China, they beheld chicks in nature *taking orderly turns* to aggressively dominate their siblings and receive regurgitations from parents.[15] I've watched their field videos of two- to four-chick broods over and over again, fascinated by the chicks' manners. When a parent returns from a foraging trip and prepares to offer food, *any of the chicks* can rise up and start pecking the heads and napes of all the others. One or two chicks may briefly peck back at the start of the bout, but within seconds all are submissively hanging their heads to nest floor level while the pecker, the *current dominant*, continues to peck at all of them and receives food from the parent's mouth. And then, a marvel of marvels, after feeding the current dominant hangs its head and a different chick assumes its mantle, pecking at all the others, who meekly hang their heads, and takes food from the parent. And so on, with each individual taking its almost uncontested turn to be the boss and get the food. What's more, all deliver similar numbers of pecks and receive similar amounts of food, despite there being a patent age-based size hierarchy.

And if that weren't bizarre enough, can you also believe that more than half of the pecks by each ibis chick are fake? Rather than striking the hanging heads, fake pecks sweep in a long arc to the nest floor, almost glancing the pecker's own flanks, alternating between left and right and striking nothing, no matter that the submissive victim is right in front and clearly accessible. Tellingly, fake pecks can also occur when the nearest victim is out of range and, during the brood's transition to independence, even when all potential victims are out of sight, seeking food elsewhere.

Xinhai's detailed measures disclosed three important clues to the function of crested ibis aggression: after real and false pecks parents often feed the current dominant, fake pecks usually occur after a parent has failed to feed the current dominant, and fake pecks never occur in the absence of a parent. Like the ritualized bows of Elizabethan courtiers, fake pecks are displays. Their form belies their function, which is to beg for food. Evolutionary ritualization of behavior that originally served a function such as body care, in the service of a new function—communication—is widespread in vertebrate animals and was much studied by the Austrian ethologist Konrad Lorenz. Thousands of generations ago, crested ibis pecking was assuredly genuine and favored by natural selection because it enabled peckers to subdue competitors and gain access to the parental beak. Then when mutant genes by chance inclined parents to feed chicks that pecked, those genes were favored by natural selection because parents who fed in response to pecking fledged more, or more viable, offspring. It's conceivable that this parental responsiveness increased the efficiency of food allocations or reduced damage to chicks' beaks and skulls. And once the motor coordination of pecking was functioning as a communicative stimulus, it could evolve to be more conspicuous to parents and even become partially liberated from its original function, resulting in pecks at nothing.

The crested ibis's fake pecks can be speculatively explained in this way, but its rotating dominance—orderly turn-taking itself—is mystifying when chicks of so many other birds clearly take advantage of their age and size to forcibly secure their priority in feeding and survival. More recent monitoring of the Shaanxi population has revealed that later-hatched brood members sometimes grow poorly and die more frequently than their sibs, raising the possibility that, in unspecified circumstances, older brood members can somehow achieve feeding priority while taking their turns.[16,17]

A single species with anomalously unselfish competition would not be too surprising, but there are others. In both aggressive and non-aggressive nestlings of several species, either rotating dominance or rotating access to food has been documented and attributed to nestlings communicating with each other to identify the most deserving recipient. According to this hypothesis, after identifying the individual that is hungriest or oldest by comparing all of the nestlings' begging calls or pecking frequencies, the others allow it access to the contested food and may even punish queue-jumpers. This is not an easy story to sell to a behavioral ecologist because, in theory, the evolution of such communal responsibility is highly vulnerable to invasion by mutant individuals who cheat and make off with their ill-gotten gains. Genes for cheating are expected to confer such large benefits on their bearers that, as soon as they crop up by chance, they outperform genes for communal responsibility and steadily replace them in the gene pool. That would be the death knell of communal responsibility.

However, communal, turn-taking responsibility has been reported for nestlings of the crested ibis, the bald ibis, the common kingfisher, and notoriously the barn owl, whose owlets negotiate by holding group beg-ins before a parent even arrives with prey for the brood.[4,18] Moreover, theoretical cheat-proof scenarios for the evolution of such virtuous competition among siblings, founded on selfish gene theory, have been devised.[19] Clearly, we need some additional probing of communal turn-taking, and to overcome skepticism it will need to confirm that turn-taking *in nature* consistently results in equitable food consumption, growth, and survival in natural broods. Meanwhile, I suspect that subtle manipulation is going on behind a façade of equal opportunity.

In any event, whatever the explanation, what cannot be written off is the empirical phenomenon of "polite" turn-taking competitors, exemplified in one ibis by sisters and brothers who graciously stoop to the floor while a sibling strikes them and feeds, and aggressors who graciously simulate violence. Brown boobies, eat your hearts out!

Hierarchy in Game Birds and Waterfowl

Our last variety of avian sibling dominance manifests in two groups that distinguished themselves 66 million years ago by surviving, almost alone among birds, the impact of a colossal asteroid that thudded deeply into Mexico's Yucatan peninsula, transforming the planet's weather and ecosystems and extinguishing all the dinosaurs. It may have been their ground-hugging and aquatic habits that saw the game birds and waterfowl through that ecological cataclysm. We have been much more given to hunting and eating these birds than studying the social behavior of their large broods. However, a generic portrait emerges when we pool scattered observations on infants of junglefowl and domestic fowl; red, black, and willow grouse; Japanese quail and western capercaillies; redhead and canvasback ducks; and Canada, barnacle, and graylag geese. This portrait of infant social conflict and organization is different from anything we have so far seen but remarkably similar to the adult "peck orders" of the domestic chicken, the touchstone species for animal dominance hierarchies.[4] Maybe their adult-like social organization is not surprising because these infants are *fully precocial*: capable of walking, running, and feeding by themselves within hours of hatching a brood, which happens roughly synchronously.

The behavioral maturity and self-feeding ecology of chicks of game birds and waterfowl resemble those of adults, and they compete with each other much as flocks of adults do. All brood members can benefit from protection and guidance, so they hang out with their parents in a mobile flock in which every individual is mobile enough to avoid sibling persecution. Their food is not concentrated at a nest or a parent's mouth but distributed across the habitat, so it doesn't usually pay to fight for it; it's more efficient and probably safer to ignore the sibs and go straight for the food. How different when siblings don't grow up in a nexus of cut-throat competition for food issuing from the parents' mouths! Compare fully precocious chicks with blue-foot chicks. Any parentally regurgitated food swallowed by a blue-foot chick is one item less for its sibling, but the dispersed food eaten by a baby quail or duckling may not have been discovered by its sibling anyway, and there's often plenty more in the vicinity. Competition is therefore milder in precocial chicks than in blue-foots. Finally, it's probable that fully precocial chicks recognize other individuals of their natal brood, although this ability has only been demonstrated for 6-week-old grey-lag goslings.[20] Therefore, after a certain age, dominance between precocial chicks is likely to be a personal relationship between individuals.

Each flock of immature game bird siblings or waterfowl siblings sorts out its linear dominance hierarchy over a few days or weeks, after a series of one-on-one dust-ups. In every dyad, one individual threatens or attacks and the other signals submission, and once that personal relationship is established, violence is uncommon and ranks are maintained by mere displays, to the benefit of both players. Naturally, siblings can benefit by hanging out with each other because companions can inadvertently decoy predators and help discover patches of food. But why, then, is there a hierarchy at all? Well, in almost all social groups, be they chickens, macaque monkeys, or human adolescent gangs, there is potential for competition, whether it be for food, space, parental attention, or mating opportunities. And, for all we know, high rank in the dominance hierarchy of its brood could make an animal more competitive in adulthood, either because it gets more food and grows better, or because it develops a dominant personality.

This critical role of the spatial distribution of food in determining whether young birds fight their siblings for it or ignore their sibs and simply go for the food, was put to the test by a student I taught at the Universidad Autónoma del Estado de Mexico

(UAEM), as part of a government program for getting faculty of the national university to give classes in the provinces. Once a week for 3 years, I drove northwest from Mexico City to a campus beside the village of El Cerillo and taught courses in Ethology to students of the UAEM from the surrounding villages. For her undergraduate thesis, Reyna Hernandez presented 2 g of ground corn to 24 isolated dyads of common quail brood mates in four spatial concentrations: homogeneously spread over 75 percent, 50 percent, 25 percent, or 1 percent of the cage floor. With little finance and a lot of improvisation and wheeling and dealing, Reyna borrowed chicks from a commercial hatchery and lodged them in her home. To assure that their scratching and fluttering in the test cage would not disturb her food distributions, she pasted the corn to the cardboard substrate, using corn flour paste. She ran 288 trials in her home and analyzed the data with the statistical package "R," using generalized linear mixed models.

The research conducted by Reyna, now a prize-winning high school teacher in the Estado de Mexico, supported the hypothesis that aggression increases with the spatial concentration of food. At the highest concentration—a single blob of corn—the two rivals pecked, kicked, hopped, and pushed at each other, sometimes both of them simultaneously, and the more dominant rival sometimes managed to drive the other away. At lower concentrations, both chicks walked around the cage, pecking at the corn and only occasionally pecking the other's head, if it came close. But they weren't equal: often one chick fed freely while its companion foraged less fruitfully away from the main patch of corn. Most important for the formal test of the hypothesis, aggressive pecking was virtually absent at the lowest concentration and progressively increased, up to a 16-fold maximum, at the highest. More tests with other fully precocial species and larger groups of chicks need to be made, but Reyna's experiment provided the first demonstration, for any animal species, that infant siblings compete for scattered food by going for the food and for concentrated food by attacking each other.[21]

Dominance In Mammalian Litters

Aggressive sibling competition has been studied in few wild mammals because in the great majority of mammalian species intimate family goings-on occur out of sight. Most mothers are secretive, nocturnally active, and conceal their litters in burrows. Even dogs and cats, which live in our homes, are poorly known in this respect and have only recently begun to be observed by colleagues at the UNAM.[22] After birth, besides a warm secure place in the litter huddle, litter mates typically seek maternal milk during nursing and solid food provided by parents—usually mothers—after weaning. Competition for food and space must be commonplace but may often amount to little more than each cub single-mindedly striving for the resource it seeks, just as most nestling birds compete for food by begging loudly and jostling to get close to the feeding parent rather than by attacking their nest mates.

In the absence of aggression, it's a tall order to disentangle the influence of individual pups and their parents in determining how much food is available for the litter, how it is shared out, and which pups—if any—should be sacrificed to benefit the growth and survival of the survivors. However, in some mammals, aggression among cubs can be very conspicuous, and the three mammals whose (very different) sibling conflict is best known—hyenas, domestic pigs, and canids—can help us appreciate not only the variety of ways that mammalian siblings compete but also some similarities in the sibling conflict of mammals and birds.

Spotted Hyenas

Highly intelligent and behaviorally flexible, spotted hyenas resemble some primates in their social organization, living in matriarchal clans in which dozens of individuals respect a rigid dominance hierarchy complicated by temporary coalitions of allies and nepotistic transmission of rank to offspring through social influence. Formidable predators, spotted hyenas hunt in groups of all sizes, commonly running down and slaying wildebeest and antelopes, and a great variety of smaller animals. They also consume carrion and famously drive outnumbered lions from their kills but, when raised in captivity, are sufficiently emotional to develop affectionate relationships with humans. Some chauvinists have conjectured that, similar to humans and chimpanzees, hyenas are aware of the thinking minds of their companions, which would put them on a high shelf. Nonetheless, when it comes to infant sibling competition, they're remarkably similar to blue-footed boobies.

Typically, a hyena dam deposits her litter of two cubs (rarely, three) in an abandoned mammal den with an entrance too small for her to enter. That reduces the probability of cubs being predated during the dam's long absences on hunting trips; it also means newborn cubs are left to their own devices at an age when they are already fierce and precociously armed with fully erupted incisors and canine teeth. Observations of captive litters showed that within days or hours of birth cubs attack each other, biting and bite-shaking in a fierce struggle for dominance, even when no food is on offer. It can take only a few days for a dyad to sort into dominant and subordinate, and that relationship, reinforced by regular threatening and attacking, tends to persist for weeks or months. Dominance confers privileged access to the milk and prey provided by mothers, as well as superior growth and survival, and if the maternal supply of food dwindles, dominants attack more intensely and subordinates sometimes die.[23-26]

So far, so nearly identical to the aggression between pairs of siblings in blue-foot nests, except that hyenas control access to the mother's nipples and dropped prey morsels, while blue-foots control access to the food that issues from their mother's and father's mouths. In both species, the weaker sibling is beaten into submission very early on by sustained attacking followed by regular, milder aggression to maintain its subordination, with potential for escalation if food is scarce. But whereas each booby chick learns through trained winning or losing either to attack or submit to other chicks in general, hyenas are thought to develop *personal relationships* between pairs of individuals that recognize each other.[4] Either type of learning will do to ensure that the individual who is more powerful at the outset will be the one most likely to survive a period of starvation, and boobies have no need to recognize their siblings as individuals anyway because they won't be cohabiting with them in a clan.

But are hyena subordinates as manageable as blue-foot subordinates? It depends on their feeding ecology. Observations of contrasting populations in the savannas of Tanzania and Kenya indicate that the pugnacity of spotted hyena subordinates increases with the risk of food shortage and starvation, rather as the pugnacity of avian subordinates is greater in species whose low-ranking individuals have poor prospects of fledging. In the Masai Mara reserve, where prey are abundant and subordinate hyena cubs tend to survive as well as dominants, subordinates graciously accede to low status in their litter: they almost never attack dominants. In the Serengeti reserve, where long commutes to hunting grounds oblige mothers to feed their cubs less frequently, 10 percent of subordinates die and subordination is resisted stiffly. Rather than responding to attacks with submission, Serengeti subordinates often back away or retaliate, and they attack one-third as often as their dominant sibs. Only amid the scarcity of the Serengeti do dominant cubs bear the scars of sib attacks on their bodies.

Domestic Pigs

For maximum contrast with hyenas, we turn to the sibling dominance of domestic pigs. Here, too, infants are born fully armed with dentition customized by natural selection for attacking siblings, but their dominance in no way resembles that of any infant birds we have considered. Oddly, it closely resembles the classic territoriality of mated pairs of adult songbirds, which divide the habitat into contiguous territories, each occupied by a pair, and reserve the territory's resources (mainly food) for themselves by threatening and attacking any neighbor that intrudes. As it is with birds and their patches of land, so it is with piglets and their teats.

As each of about eight to twelve shut-eyed piglets plops out of the sow just minutes apart, it stumbles about, then totters and sniffs uncertainly around her recumbent bulk until it locates a teat and can fasten on to it. There are usually enough teats for each piglet to get one, and there they all hang, nicely spaced, meal after meal, across multiple milk outlets that are synchronously productive in bouts. After each hourly meal, all detach and go about their business; when the dam grunts to let them know another meal is imminent, every piglet makes for its own teat. It has been suggested that a sow's teats may differ in value: anterior teats may produce more or better-quality milk, and teats on the upper row may distance suckling piglets from the risk of getting crushed between sow and floor. In this view, teat competition is all about getting the best teats.

Whatever the teats' relative merits, piglets own them: from day one, individuals develop attachments to the teats on which they first suckled, and vigorously resist the intrusions of neighbors, pushing sideways with their snouts and slashing neighbors' faces with their protruding teeth. Similar to so many songbirds on their territories, aggression is largely confined to the defense of teats, almost all aggressive encounters are with neighbors, and owners nearly always prevail over intruders.[27–30]

Piglets' *site-specific dominance*, also called teat ownership or teat order, could indeed be a mechanism for retaining exclusive access to high-value food outlets, but there may be a more important benefit. I suspect that independent of variation in teat quality, teat ownership reduces the costs—in time, energy, and injuries—of competing anew for a teat at every mealtime. Within a few days of birth, even the piglet that got the worst teat on day one might be better off latching onto its own teat and suckling than searching, fighting siblings to upgrade, and then suckling. But what sort of relationships and learning mechanisms are involved in teat ownership? Probably not trained winning and losing, nor learned individual dominance relationships. Site-specific dominance probably doesn't enhance or diminish an individual's competitive ability in other contexts. As in birds, the site-specific dominance of piglets is all about home advantage; it would be surprising if it affects the relationship of teat neighbors when they are away from the udder.

Site-specific dominance may be widespread in mammals whose teats offer multiple food outlets with coordinated milk availability. Disappointingly though, although teat ownership or fighting has been reported anecdotally in peccaries and boars—the domestic pig's wild relatives—and some species of cats and hyraxes, no experiments have yet worked out how ownership affects food allocation, growth, and survival. But what seems clear is that when the coveted resource is simultaneously available at multiple separate outlets, sibs don't need to fight for dominance; they're better off just going for the milk. Natural selection on dams seems to have saved newborn pigs from bickering by furnishing spaced-out teats that outnumber piglets. For hyenas, it's different because a single cub can aggressively monopolize access to the dam's pair of adjacent nipples.

Canids

Canid pups are essentially similar to spotted hyena cubs in that all dyads at some stage go through severe ranking fights and then settle into a supposedly personal relationship of dominance-subordination that persists while they are juveniles. The big differences between these two types of carnivores concern the age, severity, and outcomes of sibling conflict. Whereas hyenas delay only hours or days before attacking each other ferociously, canid dyads live in peace for a few weeks and then fight it out shortly before mothers start feeding them meat, as if their dominance mediates only access to solid food. In coyote dyads, for example, violence suddenly erupts after the third week and promptly subsides after status is established. Thereafter, fighting is consistently moderate, with the subordinate cub submitting when attacked or threatened. In captive litters of several canid pups, these ranking fights generate an approximately linear dominance hierarchy in which only the lowest-ranking and most persecuted individual is at serious risk of dying. Pups often suckle at different times and don't appear to own or contest access to teats,[31] which could imply that mothers ration milk provision in such a way that pups are unable to monopolize access.[4],[32–34]

How dominance enables weaned canid cubs to prosper in nature was revealed in 1986 by an ethologist who had the good fortune and commitment to spend thousands of hours watching red foxes in Canada's boreal forest. And what J. David Henry discovered was rather different from what you may be expecting. Fox parents returning to a den deliver prey, not preferentially to kits of higher rank, but to the first kit that begs. Even so, dominant kits get more food and grow faster, because dominants can get away with stealing prey from subordinates, and often do so. In sum, canid pups generally compete pacifically during lactation by going straight for the nipples, and aggressively during solid feeding by establishing a hierarchy of dominance-subordination and levering their status to gain privilege.

Competition among animal brood mates and litter mates is nearly universal and often ends in death or at least some degree of runting of the least competitive individuals. It goes on so secretly or subtly in most species that we are unlikely to appreciate how strenuous and often lethal the competition is. Even in domestic rabbits, where no aggression has been detected when the whole litter suckles frantically during the mother's single 5-minute visit to the burrow each day, scrutiny of suckling and growth reveals consequential competition among pups. Robyn Hudson and I found that pups which consume more milk grow faster, pups in larger litters grow more slowly and are more likely to die, and after a death, survivors grow faster. We concluded that food shortage limits the growth of rabbit pups, and the milk consumption of each one depresses the growth of the others. But *how* pups outcompete each other is not obvious.[35]

The great merit of violent species such as the blue-footed booby and the spotted hyena is that their conspicuous, unmistakable aggression gives us a window into sibling competition's seriousness and strategic nature. Why the offspring of so many avian and mammalian species compete non-aggressively is still something of an enigma, but likely has to do with many immature animals simply being unable to use violence effectively or economically. Certainly, there is no moral inhibition about using aggression. For most species, aggression probably is not a practical option, so infants are obliged to make do with peacefully outbegging or outscrambling their sibs for food.

Statistical comparisons across scores of species in seven taxonomic families of altricial birds showed that aggression is most common in species with small broods, long periods of nestling cohabitation, and infrequent parental feeding, and more likely in species where parents deposit food on the nest floor than species where parents insert it into nestlings' beaks.[36],[37] Those factors may favor the efficacy or profitability of using aggression. But those comparative studies left much variation among avian species unexplained, and I suspect the aggressive potential of nestlings—their armaments and ability to physically impose their will—will turn out to be important. Someone needs to come up with a methodology for measuring that potential.

Widening our perspective by focusing on several avian and mammalian species whose offspring compete aggressively, along with quick consideration of two comparative studies, revealed the engaging variety of sibling aggression and some of the factors that conspire in shaping it. Sibling aggression in humans, which mostly produce their infants in succession rather than in litters, will be addressed in Chapter 9. More predisposing factors will eventually be added as behavioral ecologists work out how to study additional wild species.

The factors include, first and foremost, the distribution of parental food outlets in space and time. Avian offspring are more likely to compete aggressively for food when parents drop it on the floor than when they pass it from mouth to mouth, and least likely when searching for scattered food. Mobility can allow younger chicks to largely avoid aggressive altercations over food provided by parents but doesn't save them from aggressive subjugation. Dominant cubs can aggressively control access to teats that are sufficiently clustered for one cub to deny access to others, but when there are enough synchronously active teats to go round, every individual may aggressively own one of them. In birds, we've seen that aggression is greater when immobility or confinement in a nest makes it impossible to stay away from brood mates, when broods are small (as Raymond O'Connor's model suggested), and when subordinates' prospects of surviving are poor; and aggression can vary from damaging violence to courtly rituals. The capacity for individual recognition affects the nature of infant dominance hierarchies; game birds, waterfowl, and some mammals seem to form personal dominance relationships, whereas altricial nestlings recognize the status but maybe not the identity of siblings.

Holding out hope of an eventual synthesis for vertebrate broods and litters, sibling aggression and siblicide are remarkably similar in blue-footed boobies and spotted hyenas, two species that differ hugely in their morphology, feeding ecology, social organization, and locomotion. In both species, infants feed exclusively from a single or clustered outlet, newborns establish dominance-subordination as soon as they can, dominance assures better access to food, aggressive resistance of subordinates is greater when their risk of starving is high, and the ultimate sanction—siblicide—is common.

Once we had a good grasp of the incidence and adaptiveness of sibling aggression and siblicide, and the nature of aggressive struggles between siblings, a natural next step was to identify the mechanisms by which such damaging behavior is regulated. We wanted to know how dominant chicks calibrate their aggression in such a way as to allow siblings to survive when parental provision is sufficient and terminate them when it isn't. The criterion, of course, is for the dominant chick to maximize its own inclusive fitness. It's a delicate matter and complicated, too, because supernumerary siblings are not going to go quietly when they would actually be better off turning the tables on their oppressors. How, then, do dominants, who can benefit from their siblings' survival but be prejudiced by it on other occasions, prudently regulate their own aggression? Comprehension and planning are not options; what are needed are rules of thumb. Which information would you use if it was your decision? Researchers came up with several hypothetical rules of thumb that might enable dominant chicks to get it right, but, as we'll see in the next chapter, we didn't always agree on the interpretation of our observations and experiments.

4
To Kill or Not to Kill

The senior blue-footed booby chick has life-or-death control over its sibling. After a few weeks of beating on the sibling on a daily basis, senior's control is complete. At any moment it can deliver a jolting peck to the head or rise up and scream a threat; junior will assume a submissive posture or crumple down and desist in its begging. According to our hypothesis of conditional tolerance, this control is what allows a senior chick to receive more of the food regurgitated by parents, enjoy a better chance of surviving to fledge, and dictate whether its sibling will survive or die. But how can the senior chick of any avian species whose siblicide is conditional "know" at any particular moment whether its inclusive fitness interests are better advanced by continuing to share with junior or by facilitating junior's death? That is the question, and addressing it will take us into divergent interpretations of a medley of inventive field experiments on several bird species. These will illustrate some of the challenges of designing practical and telling field experiments and interpreting the data they yield. We'll begin by considering two proposed mechanisms by which a chick calibrates its aggressive intensity in response either to food shortage or large brood size; we'll finish by considering the proposal that a chick's decision to compete aggressively, rather than simply begging, could depend on how parents transfer food.

Remember that, depending on how sharing affects the prospects of each chick, in some contexts, it pays to keep on sharing, while in others it pays to be ruthlessly selfish. To kill or not to kill is the momentous decision that every uncomprehending senior chick faces every day the two chicks cohabit in the family territory and share in the food their parents bring.

The Food Amount Hypothesis

In practice, the senior chick just needs to limit its siblicidal behavior—intensified pecking—to contexts where killing will increase its own inclusive fitness. We expect evolution by natural selection to have ensured that senior's decision is based on the most relevant and reliable information available to it, although there are sensory and computational limitations on the data an immature chick can process. The way that most animals deal with complex inputs of information is not by reasoning but by following rules of thumb, and the *food amount hypothesis*, favored by researchers working on birds whose chicks compete aggressively, provides such a rule. It holds

Blue-Footed Boobies. Hugh Drummond, Oxford University Press. © Oxford University Press 2023.
DOI: 10.1093/oso/9780197629840.003.0005

that *as a chick's food ingestion falls off, it should intensify its aggression*. The rule is plausible because it combines obvious relevance with simplicity.

The hypothesis was born of both logic and tell-tale correlations reported by biologists studying a variety of wild birds. The circumstantial evidence was that chicks peck each other more often during their feeding bouts and less often after feeding, that aggression is more frequent in poorly fed colonies than in well-fed colonies, and that protracted poor weather increases chick mortality. None of these observations is conclusive, but together they raise the suspicion that food shortage incites chicks to increase their attacking, to assure their priority in feeding competition or to go to the extreme of killing their competitors.

Broadly understood, the hypothesis doesn't specify precisely which cues will trigger enhanced attacking, but one can imagine plausible cues. For example, senior brood mates could increase their attacking to lethal levels when their own rate of swallowing falls off, when parents spend less time at the nest (because they are having to hunt more), or when juniors beg insistently (because they're hungry). All of those situations are likely to be correlated with reduced food delivery to the brood, the environmental variable we expect to govern the growth and viability of chicks. And it is not much of a stretch to imagine a booby chick being able to detect one or more of those situations and adjust its aggressive behavior accordingly.

The food amount hypothesis may sound like a pedantic name for expressing a relationship between ecology and behavior that seems almost bound to be true, but natural selection can work in unexpected ways, and you don't know about causation until you make an experimental test. What's more, there are reasons for doubting the central role of reduced food ingestion. First, it connotes less available energy to fuel activities such as pecking at siblings, and second, food shortage does not in general provoke animals to be aggressive. If they encounter frustration when seeking food (or doing anything else), frustration itself can make animals aggressive, but food deprivation and hunger don't by themselves elicit aggression. If you think about it, food shortage may make blue-foot juniors more likely to die not because it provokes seniors to attack them more often but because the reduced rations received by a subordinate chick make it more vulnerable to dying of starvation, expulsion, or illness. So there's a real question whose answer may have implications for many birds and mammals, including humans: Does starvation make an infant more aggressive to its sibling?

Many hours observing feeding and dominance interactions in the blue-foot colony gave me a strong hunch that this was true. But a hunch is not an answer. You need an experimental test in which the amount of ingestion is manipulated upward or downward, and the predicted change in aggression is measured and confirmed. That can nail it. And for high confidence in the hypothesis, it would be hard to do better than a package of observations including (1) experimental increase in chicks' ingestion followed by a reduction in their pecking, (2) experimental reduction in chick's ingestion followed by an increase in their pecking, and (3) field observations on unmanipulated broods showing that chicks' pecking increases when they get less food. The trick is to devise ways of manipulating ingestion that work, and which don't unduly upset or frighten the chicks or their parents.

However, when such a package of observations was reported, the results, although described with appropriate caution, appeared to flatly contradict the hypothesis. Here's what the first serious study, embracing three species of herons, concluded:

> In total, then, we have found that ardeid chicks did not fight more when food supplies were artificially reduced or naturally insufficient (i.e., when fighting would seem most useful to them), nor did they fight less when their food was artificially supplemented or naturally abundant (when fighting might be a waste of effort). Indeed, the recurring general relationship between aggression and food amount in these birds is weakly positive, as if fighting rate may be governed more simply by the birds' current energy levels. We conclude that the Food Amount Hypothesis is unlikely to explain much about the rates of sibling aggression in these species.[1]

Such a statement, published in the prestigious journal *Ecology* by a team from the University of Oklahoma, automatically put this booby researcher on the back foot. Could it really be that chicks don't take food amount into account when deciding whether to kill their sibs? Doesn't this conclusion cast doubt on the whole idea of siblicide functioning to adjust brood size to a key ecological variable—available food? Maybe blue-foots are different from herons? What other variable could serve chicks better or be easier for them to measure? I accepted the conclusions of the heron study, but only for a while. My doubts grew as contradictory data accumulated on other species, and as I reflected on how the flightiness of herons must inhibit researchers from approaching colonies for observation, manipulating ingestion, and getting critical measurements on chicks' growth and behavior. At that moment I took a harder look at the research methods, data, and statistical analyses, and I found them less convincing than when I first read them. Let's take a look at the evidence for great white egrets, the leading players in the *Ecology* article.

Great white egrets are those tall white herons you see standing in shallow water along shorelines, making lightning strikes to grasp and spear aquatic prey like fish and frogs. The very symbols of avian siblicide, they are elegant, improbably spindly birds, and thousands of them overwinter every year in Mexico. On our annual descent through the Sierra Madre Occidental to San Blas, we know we've reached the coast when we start to glimpse crowds of brilliant white egrets in the mangrove bushes, hanging out and feeding in tidal lagoons before they migrate north to breed. On their breeding grounds, they mostly nest colonially in branches overhanging water or reedbeds. Each monogamous pair attempts to raise a brood of three to five chicks, but many young die of starvation or sib fighting over the course of a 6- to 9-week chick period.

To find out whether *adding food* reduces aggression, every day experimenters tipped 70 g of fish into nests on a dredge island in Texas, until the broods were 25 days old. All broods were still fed by parents, of course, so supplemented broods were considered very well fed and control broods normally fed. Contrary to expectation though, supplemented broods did *not* fight less intensely or frequently than control broods. So extra food does not snuff out aggressiveness, right? That was the conclusion.

But look again at the results, this time considering A-chicks, B-chicks, and C-chicks separately. Whereas the weight increase of provisioned Cs was 82 percent greater than that of control Cs, the increase of the provisioned As and Bs was a paltry 7 percent greater than that of controls. Oddly enough, provisioning did not substantially increase the amount of food ingested by the very chicks that do most of the attacking—As and Bs. It seems almost bizarre, but the supplements, which were usually gobbled down in less than 15 minutes, seem to have been consumed mostly by Cs! If that's right, then because the two aggressors in each brood were scarcely supplemented, the food amount hypothesis does *not* predict noticeably less fighting in supplemented broods than in control broods, and the results fall short of clearly contradicting the hypothesis.

To test the effect of *reducing food* on aggression, at another Texas dredge island colony, researchers took broods of three great white egrets into custody and installed them in stick nests on laboratory racks. Here, they fed them for 12 days using hand puppets that mimicked the heads and necks of regurgitating parents—a complicated and ambitious enterprise! In the total absence of real parents, every brood received daily either a "low amount" of fish (46 g) or a "high amount" (92 g, estimated to be one-third more than in nature). Surely the broods that received half as much food every day would fight more than the others? However, despite suffering more deaths, the low-amount broods did not fight more at each meal than the high-amount broods, so the food amount hypothesis was not supported.

Or so it was thought. However, there was ample evidence in the study that both high- and low-amount chicks starved, grew poorly, and fought with unusual intensity, and that this happened because the amount of food consumed in nature was underestimated. How else can we explain nearly all Cs and half of Bs in the experiment starving to death (including individuals expelled from the nest, rescued, and scored as dead)? It's easy to underestimate the food consumption of wild animals; for years we didn't realize that blue-foots also feed their chicks at night, let alone guess that *most* of their food is given to them at night—not until we measured food consumption.[2] So a credible alternative explanation for these egret results is that starvation on meager experimental rations pushed both treatment groups to the same aggressive ceiling. The low-amount broods surely suffered more hunger than the high-amount broods, but both groups may have been pushed by hunger to the maximum aggression of orphaned egrets in captivity.

Finally, there were also descriptive field observations of unmanipulated broods. After watching great white egret broods during all daylight hours (!) every day during their first 2 weeks of life, the researchers found, in a whole battery of statistical tests, no evidence whatsoever of a relationship between fighting and recent food ingestion, whether these were scored per individual, per dyad, or per brood. However, this negative result was inconclusive, too, because the mix of brood sizes, the incidence of brood reduction, and the variation in ages were not adequately taken into account, and failure to control variables known to greatly affect aggressiveness could easily result in data variance obscuring correlations. Furthermore, scoring brood mate ingestion and aggression in up to 15 broods at a time, in nests up to 30 m away, and keeping this up for several hours at a stretch, may have overloaded observers and yielded less-than-optimal data.[3,4]

At first glance, the two experiments and descriptive field observations on herons, along with other descriptive data, made for a compelling package, but under sober scrutiny, the huge research effort that had observers heroically spending 24 hours at a stretch in blinds (to minimize spooking the colony during changes of shift) left the issue unresolved. The study neither confirmed nor provided solid grounds for rejecting the food amount hypothesis. Proving that a variable such as food ingestion does *not* affect behavior is a demanding enterprise, because negative results of any test may only mean that the test failed, for whatever reason, not that ingestion never has any effect. Additional tests on herons are needed.

The blue-foots of Isla Isabel were more amenable candidates for testing the food amount hypothesis because they nest on the ground and tolerate human approaches and manipulations. We already knew that juniors tend to die when seniors are 20–25 percent underweight and, while walking through the colony, we had noted the occasional dominant giving its nest mate an unusually energetic pasting, as if submissiveness was no longer enough to calm its anger. These descriptive observations suggested that juniors die when food is short and that the aggression of seniors is highly variable.

We decided to reduce ingestion of parental feeds by manipulating chicks in their own nests, where we could monitor their condition by weighing them frequently and observe whether aggressive pecking increased as feeding diminished. To prevent chicks from swallowing the fish that parents dropped into their mouths, we fastened adhesive cloth tapes around the tops of their necks. We did this to both chicks in each brood, for greater realism and to avoid upsetting dominance relationships by making subordinates outgrow dominants. Tapes didn't damage or constrict necks in the least, but they prevented throats from expanding to allow the passage of food; liquid food and some mush got by. After a few attempts at swallowing, a chick would open wide and shake the food out of its mouth onto the sand and soil, where the whiptail lizards, Heermann's gulls, and spinytail iguanas scavenged it.

To our relief, the booby chicks and their families took the taping protocol in their stride, paying no attention to the tapes themselves, so the experiment could go ahead. Parents seemed a little disconcerted by the insistent begging of chicks they had recently fed, but families maintained their routines of brooding, feeding chicks, and overflying the ocean to plunge for prey. In their home territories, taped chicks shuffled around and begged, and parents continued feeding them, and their neighborhoods continued to buzz with grunts and whistles as the boobies courted and copulated under the tropical sun. All felt close to normal, except that selected broods in a couple of neighborhoods were going hungry and researchers were sitting nearby, quietly scoring the begging, feeding, and pecking of up to four broods at a time.

In experimental and control broods, both chicks wore tapes for 2 days (young broods) or 3 days (old broods), to simulate the food privation that comes with a spell of poor weather, but we briefly removed the control broods' tapes whenever parents were regurgitating and during the whole of every night. So experimental chicks consumed little food and ended up about 30 percent underweight, whereas controls fed and grew almost normally. After the experiment, chicks recovered normal body weight and suffered no more mortality than chicks that weren't included in the experiment. Long-term effects on them should have been minimal because temporary food deprivation is normal in the ocean environment, and during the experiment, deprivation lasted only 3 percent of a chick's total growth period. These considerations did not entirely save us—Mowgli and me, and Cecilia García, a dedicated thesis student—from guilt over disturbing the boobies' family lives and peace of mind, but we felt that their experiences in the experiment were within the range of hardships boobies experience in nature.

It took two spring seasons to accumulate observations on a large enough sample of broods because in the 1980s when research budgets were lean and we still had no support from the navy, we had few volunteers to help with the labor-intensive work. However, the experimental design worked like a charm, and the data told a convincing story.

During the days when they couldn't swallow, both nest mates greatly increased their begging, and aggressive pecking by dominants climbed to a fourfold increase in young broods, and a sixfold increase in older broods. Taped subordinates also tried to up their pecking but, as trained losers under fire, they had no chance; they were taking too much punishment to assert themselves. Dominant chicks in control broods did not increase their pecking, so the experiment demonstrated that restricted ingestion itself—rather than taping and handling of chicks—was responsible for the steady increase in aggression. What's more, aggression paid off: the taped, fiercely pecking dominants increased their share from 40 percent to 60 percent more regurgitations than were received by subordinates. Had we not removed the tapes, I suspect the death of subordinates through starvation or failed attempts at adoption into neighbors' nests would probably have followed.[5]

The precise cue for increased pecking under taping is still a matter of speculation. It could have been infrequent ingestion itself, hunger, emaciation, reduced intestinal distension, or even increased begging by the brood. However, whichever cue actually stimulates a chick to peck more, we know it's tied to diminished ingestion.

The experiment also revealed that this behavioral mechanism is conveniently reversible. After the removal of the tapes at the end of the experiment, dominant chicks resumed feeding and begged and pecked less often. This implies that dominants who intensify attacking to potentially lethal levels when food goes short can draw back from the brink of siblicide if feeding conditions improve. When you come across a dominant chick in the colony attacking its sibling mercilessly, sometimes words like "unhinged" come to mind, but that description would be wrong. In just a day, domi-nants can switch from lethal aggression to tolerant sharing of parental regurgitations with their sibs, and they probably make this switch whenever threatening feeding conditions improve to the point where they can increase their inclusive fitness by sharing. At least, that is what O'Connor's model (Chapter 1) leads us to expect, and the idea fits well with our descriptive and experimental observations. How well the aggression of senior chicks has been calibrated to their rate of recent ingestion and hence to the survival prospects of the two chicks is open to question, but seniors may well get it right on average because natural selection has probably been fine-tuning this calibration for thousands of generations.

Other researchers have field-tested three unrelated species since then, using a dif-ferent method of experimentally reducing ingestion in each case.[3] Black-legged kit-tiwakes, those cliff-nesting gulls that raise broods of one to three chicks on narrow cliff ledges above the sea, were starved in their nests by David Irons for 10 or 20 hours by fastening loops of wire around their closed beaks. Controls were treated likewise, but every 4 hours their wire loops were briefly removed to allow hand-feeding with chopped herring. The food amount hypothesis was supported when the proportion of broods showing sibling aggression increased with the duration of bill-wiring, and their aggression intensified as they lost weight. And there were other similarities to blue-foots: pecking reached its peak when chicks were 20–25 percent underweight, and reversibility was confirmed when pecking tailed off after wires were removed and chicks resumed feeding.

Lingering doubts over whether boobies and gulls may have increased their ag-gression in response to frustration over being unable to swallow food rather than di-minished ingestion were dispelled by the next two studies. Ospreys, among the first facultative brood reducers to be studied seriously, are handsome, bright-eyed brown and white raptors that lay their eggs on exposed treetop twig platforms and stoop on large fish at the water's surface. They are one of the glories of the lagoons and river at San Blas, where it's always reassuring to see an osprey flapping slowly overhead with a heavy fish hung from its talons. Their three chicks are highly aggressive, but the actual killing of sibs varies among populations and from one year to another, quite possibly as a function of natural variation in the scarcity of prey.

Osprey nests are generally too far apart to observe more than one brood at a time, but Marlene Machmer and Ron Ydenberg found a site at Kootenay Lake in south-west British Columbia where, with the aid of binoculars, they could watch a couple of observation nests atop two pilings less than 100 m from the researchers' perch on a railway bridge. Over two seasons they successively brought 10 broods from other nests by boat and inserted them one by one into the temporarily vacated observation nests, after manipulating their food ingestion without resorting to taping or wiring. For this, they detained each brood without food for 3½ hours, and then either hand-fed all two or three chicks to satiation (the controls) or merely pretended to hand-feed them (the experimentals), before settling the chicks in an observation nest along with a dead fish. Then, while the unsuspecting resident adult tore off strips of fish and fed them to its adoptees, the researchers counted their aggressive pecks, bites, pushes, and threats until the meal ended. As expected, unsatiated broods were more aggressive than satiated broods, consistent with short-term food deprivation increasing aggression.[6]

An even more cunning research operation probed for effects of short-term food deprivation on a circumpolar marine bird. The black guillemot is a jet-black blob with an equally black beak, a large splash of white on each wing, and coral-red feet. Less than one-third the size of blue-footed boobies, these marine bids are also consummate underwater hunters but, rather than plunging into the warm waters of the tropical Pacific, they dive from the surface of cold northern waters for fish and, including in the high Arctic. Mark Cook and colleagues studied black guillemots on the marvelously named Holm of Papa Westray, the smallest of the Orkney Islands north of Scotland, where the guillemots nest out of sight in boulder caves, rabbit burrows, and clefts in cliffs. In this marine bird's seldom-observed broods, the older and larger of two chicks pecks, bites, and tousles the other, which responds submissively by crouching and rarely pecks. The younger sib often dies before fledging, suggesting that it's a designated victim who will be raised through fledging only if parents bring plenty of food.

The researchers devised a way of manipulating the food ingestion of both guillemot chicks without taping necks, wiring beaks, kidnapping broods, or feeding chicks. They simply tethered a scaring device beside the burrow entrance, a balloon with painted eyes or a dummy great black-backed gull (a predator), to dissuade parents from entering the nest with food. Pecks, bites, and parental feeding were scored off video recordings made by cameras installed in the burrows before laying. Once again, the food amount hypothesis was supported. Senior chicks doubled their rate of attacking during the 3-hour block when the device was in place and parents were too scared to feed them, and then tripled that rate during the next 3 hours after parents resumed feeding. In control broods, with no scaring device and uninterrupted parental feeding, seniors maintained negligible attacking throughout.[7]

I may have convinced you that behavioral ecologists get up to some weird antics to find out what makes chicks peck, but also, I hope, that lean rations prime these birds for escalation. In blue-footed boobies, black-legged kittiwakes, ospreys, and black guillemots, no matter how ingestion was experimentally reduced, this stimulated senior chicks to attack their siblings more vehemently, and this vehemence won seniors a greater share of the food brought by parents. In three of those birds, including guillemots, observations on the day after showed that the mechanism is reversible: aggression and food sharing fell back toward pre-deprivation levels. Only great white egret chicks failed to peck with greater conviction when their food consumption was experimentally reduced, but there, too, if more effective control of ingestion could be achieved, and its effects on individual chicks disentangled, I suspect hungrier individuals would be found to attack more. Indeed, suggestive correlations between food and aggression have already been reported for cattle egret chicks, one of the three herons claimed to peck at the same rate however much it eats.[1] Fine-grained analyses of populations in Oklahoma and central Japan documented increases in pecking as food consumption declined.[8,9] I suspect that underfeeding might universally elicit more intense attacking in bird species with aggressive chicks, but we cannot rule out the possibility of some species using different cues.

A postscript was added to the blue-foot food amount story when a Spanish graduate student named Miguel Rodríguez-Gironés joined our team. His advisor at Oxford University, Alex Kacelnik, informed me that Miguel's brilliance at statistics had other members of the Edward Grey Institute eating out of his hands, but what Miguel really hankered after, and needed, was not so much the rarefied intellectual ethos of Oxford as immersion in the natural history of a pristine field environment—like Isla Isabel. Or, as Alex put it, "Miguel wants to get some bird shit on him." Well, bird shit we certainly had and in spades. Indeed, that same year as I walked a trail with Cris and some volunteers one evening to visit the fishermen's camp, a male magnificent frigatebird roosting on a branch 1 m above me discharged, with perfect timing, what felt like half a liter of hot fetid guano onto me as I passed underneath, soaking my hair, head, and upper body. Cris, of course, was helpless with laughter. Mowgli had named such baptisms, usually courtesy of frigatebirds on the wing, "fragatazos." For a while I even suspected Mowgli was able to call fragatazos out of the heavens. Yes, Miguel should join us on the island!

Good-natured, energetic, and happy as a sandboy on the island, Miguel adapted immediately to camp routines: breakfast at 0630 h as a glorious golden sun rises from the sea by the Las Monas spires; dish-washing, bathing, and clothes-washing in the sea by the rock platform; arduous mornings with the boobies in the stifling heat and brilliant sunshine of the forest and beach on the colony's eastern flank; refreshingly breezy afternoons with the east-flank boobies shaded now as the sun sinks behind a ridge; and lunch and dinner at midday and after sunset, cooked on a gas stove by turn-taking teams of students and researchers. We eat at a long table under an awning, looking out over spires and islets, hearing the waves breaking on the beach, spouting through a blowhole, and sucking at the caves under the rock platform; and watching humpback whale mothers migrating northward with their calves.

Following our dismally failed attempts to prevent sibling aggression temporarily using tethers and custom-made waistcoats, Miguel came up with an ingenious way of experimentally testing the relative importance of chicks' hunger and size in determining, not just intensity of aggression, but dominance itself; using, of course, the statistical modeling *du jour*. For this purpose, during 3-hour trials he paired two junior chicks in the empty nest of a two-chick brood (vacated by temporary kidnapping), after previously controlling food ingestion in the juniors' home nests using neck tapes and hand-feeding. By this hand-feeding, each of the juniors was randomly assigned one of three hunger levels: underfed, normal, or overfed. And there was indeed an effect: the hungrier individual delivered more submission-eliciting pecks than the other, effectively dominating it. Miguel's elegant experiment (I have barely touched on its experimental design virtues) demonstrated more cleanly than our own the power of hunger to elicit sibling aggression and drive dominance inversion.[10]

Now, what would you expect of "obligate brood reducers" like the brown booby, whose elder chicks nearly always kill their younger sibs, and do so as soon as they are mature enough to mount an effective attack? Might food deprivation fire up the aggression of chicks that are automatically bent on siblicide anyway? This needed to be tested, so Mowgli and I carried out an experiment on Isla Larga, one of the Tres Marietas islands in Banderas Bay, 135 km south of Isla Isabel and much closer to the coast. While I languished at home in Mexico City, Mowgli and his team evaded rainy season storms by camping in an enormous auditorium of a cave with an impenetrable thicket of bromeliads at its main entrance. After dawn each day, as the blue-foots and brown boobies were setting off on the hunt, he and his stalwart assistants emerged from the dank gloom of the cave to insert pairs of kidnapped brown booby single-tons in abandoned nests as if they were pairs of brood mates. During the previous day, they had used a combination of neck-taping and hand-feeding to set the food consumption of the elder one in its natal nest at one of twelve percentages of normal food intake, from 0 percent through 110 percent. The 22 elder chicks were roughly 5 days older than their experimental brood mates and a week older than the normal siblicide age. On each pair's simultaneous insertion into an abandoned nest for a 30-minute trial, no adult was present, of course, but each chick suddenly had a nest mate to fight with.

And they fought like gladiators! The elder ones, in particular, pecked, bit, and pushed with gusto, and, critically, the less food they had eaten in the previous 24 hours, the higher were their rates of expulsion-pushing and their numbers of successful expulsions (expelled chicks were promptly reinserted). So, yes, in at least one obligate brood reducer, siblicidal aggression increases under food deprivation.[11]

Whether and how this flexibility is adaptive is puzzling because, anyway, every brown booby elder chick is going to urgently kill its sibling. That, at least, is how we have long understood their aggressiveness, but maybe that's not strictly correct. Possibly the experience of strong hunger functions as a sort of wake-up call for senior chicks whose failure to evict is giving their rival a dangerous respite. There may be more behavioral subtlety here than we have acknowledged. Might the brown booby be on the evolutionary cusp of transitioning from obligate to facultative brood reduction? Or could its food-sensitive aggression be an atavism, a behavioral leftover from a time when its ancestors were conditionally siblicidal like blue-foots? Just two of many unanswered questions.

The Brood Size Hypothesis

When the University of Oklahoma team reluctantly abandoned the idea that hunger makes herons more aggressive, they naturally sought a different mechanism for regulating sibling aggression, and they came up with the *brood size hypothesis*. Reasoning that current food ingestion may be an unreliable cue to food availability over the medium term because prey abundance can fluctuate greatly with the season and the weather, they argued that chicks might calibrate their aggression not to the availability of food but to the demand for it: the greater the number of chicks competing for what parents provide, the more aggressive each one needs to be. This wouldn't require an ability to count, just the ability to discern whether there are many or few sibs in the nest. Brood size is stable for lengthy periods, so it's a potentially useful predictor of how well fed a chick is going to be.

All very reasonable, and when the researchers removed either the C-chick or the A-chick from a brood of three great egret chicks for 3 days, the average per-dyad aggression between the remaining chicks declined, and then bounced back after the removed individual was reinstated.[12] Doubt remained though. In the case of C removals, a similar decline occurred in control broods, from which C was not removed, so no effect of brood size was shown and the hypothesis was not supported. In the case of A removals, average aggression per dyad may have declined because *the most dominant* chick was removed. Removal of A-chicks altered both the number of chicks and the dominance structure of the broods, and the reduction in *average* per-dyad aggression is hard to interpret.

When we tested the hypothesis using blue-foots, our approach focused on individuals rather than dyads, and we carried out our experiment on Isla Isabel in a year of high prey abundance and little chick mortality. We took it to be the essence of the brood size hypothesis that it predicts how *individuals* respond to the number of brood mates, so we removed C-chicks (always the most likely to die) and watched to see how this affected the aggression between A and B. To guard against the possibility of changes in aggression due to changes in food ingestion—remaining chicks might get more food—one experiment prevented chicks from swallowing any food during the day after removal, when aggression was scored, by taping both chicks' necks. And another experiment controlled for change in food ingestion by measuring the growth of the remaining brood mates during the 5 days after removal when aggression was scored. In both experiments, a comparison with control broods, from which C-chicks were not removed, showed that the removal of Cs did not affect the growth of the two remaining brood mates (probably because parents bring less food when they have fewer chicks). Therefore, any changes in the aggression of A or B could be attributed either to reducing the brood from three to two, or to modifying the brood's dominance structure, but not to change in food ingestion.

But there was no change in aggression. Neither of the two elder chicks attacked the other less frequently after C's disappearance, and control broods were similar. Therefore, in our simulation of the most common type of natural brood reduction— the death of the youngest chick—reduction of brood size did not encourage either survivor to attack less, contrary to the brood size hypothesis.[13] However, reaching satisfying conclusions about how chicks' aggressiveness varies with the number of competitors will require experiments that take into account chick ranks in the brood hierarchy, food ingestion, and parental adjustments of food provision.

The Feeding Method Hypothesis

Finally, we turn to the rise and fall of the most eye-catching of all hypotheses for the control of aggression among avian siblings. The history of the *feeding method hypothesis*—known also as the prey size hypothesis—shows how data, analyses, and interpretations can fall into place around a scientific idea that is appealing for its neatness and explanatory power, and then crumble under scrutiny and further testing. The idea is that the use of aggression against a sibling depends on *how* food is transferred from parents to chicks: if it goes directly from mouth to mouth, then elder chicks can control access to the parental mouth and the food issuing from it by intimidating, repelling, and outreaching their siblings. On the other hand, if parents simply dump food on the floor of the nest, that food is instantly available to all and cannot be aggressively defended by any chick. It must simply be gobbled down as fast as possible. Go for the food, not for your sibs! The hypothesis rests on the untested assumption that aggression is an efficient way of competing for food issuing from the mouth but not for food on the floor. Small prey are more likely to be passed from mouth to mouth, but large prey can be, too, for example, when raptors tear off pieces of prey to feed their chicks. Hence the two names for the hypothesis.

The feeding method hypothesis is broader in ambition than the food amount and brood size hypotheses. It attempts to explain which species of birds use sibling aggression to compete for food, how diet can determine which populations of a species use aggression, and how aggressiveness develops in individuals over the chick period. It offers an explanation for why Texas great white egret chicks that were fed small fish from mouth to mouth fought 18 times harder than neighboring great blue herons hurriedly pecking at large, predigested fish regurgitated onto the nest floor. Chicks of both species faced similar competition and similar rates of mortality and even inhabited the same bushes. The hypothesis purports to explain why great blue herons in Quebec that took small fishes directly from their parents' beaks fought each other eight times more than great blue herons in Texas that chiseled flesh off fish on the nest floor.[14,15]

These correlations are highly suggestive and convinced the researchers who spent long hours in blinds recording the data, but by themselves, they amount to only thin support for the hypothesis. A comparison of only two bird species falls well short of proving that sibling aggression in birds is associated with mouth-to-mouth feeding. For proof, a comparative study is required, but when Alejandro Gonzalez-Voyer made his statistical comparison of 69 species belonging to seven families of birds, including egrets and herons, boobies, eagles, hawks, anhingas, ibises, pelicans, and kingfishers, it provided no support at all for the feeding method hypothesis. On the contrary, it showed that chicks which feed directly from their parents' mouths are *less likely* to compete with each other aggressively.[16]

The contrast between great blue herons in Texas and Quebec suffers from the same sampling weakness. Two populations of a single species are scarcely sufficient for concluding that mouth-to-mouth feeding and aggression generally go together in avian populations, or indeed in great blue heron populations. Moreover, the obvious alternative explanation, that Quebec herons taking small fish from parents' mouths may have fought intensely because they were less well-fed than Texas herons, cannot be discounted because feeding, body weight, growth, and mortality of chicks in the two populations were not compared.

A dramatic field experiment in Texas seemed to nicely implicate mouth-to-mouth feeding as a cause of chick aggression. When 10 young great blue heron broods were reciprocally swapped into the nests of great white egrets, they fulfilled expectations by switching, over a 2-week period, from indirect to direct feeding from their hosts' mouths, and fought at a high rate.[14] Here again though, the cause of increased aggression could have been underfeeding rather than direct feeding, and in the absence of any measures of body weight, this alternative cannot be discounted. The egret parents were almost bound to underfeed the heron chicks because they are only one-third to one-half the size of heron parents, and their chicks are only half the size of heron chicks. To address this problem, the experimenters supplemented half of the heron broods by daily dumping fish equivalent to 33 percent of estimated normal consumption in their nests, but that amount seems meager and the supplement may not even have been consumed by the elder chicks.

Furthermore, the counterpart of the swap let the side down. Egret chicks that were fostered into heron nests cast doubt on the hypothesis by maintaining their normal intense aggression while feeding on large fish on the nest floor. Feeding like that was supposed to turn them into pacifists. These issues could be resolved by comparing the growth and aggression of heron chicks swapped into heron nests (the controls) with the growth and aggression of heron chicks swapped into egret nests *and fed well*, but pulling that off in a busy colony without provoking nest abandonment would be a tall order.

Remarkably, the developmental prediction of the feeding method hypothesis—that when growing chicks switch from feeding from the floor to feeding directly from the parent's mouth, they should simultaneously switch from pacific to aggressive competition—had not been tested by observations on any species.[17] The gradual switch in feeding method has been reported in many bird species and was convincingly documented in great white egrets and in the great blue herons of Quebec, but the accompanying increase in sib fighting was not sought by anyone until Dalila cozied up to the brown pelicans on the ridge above our camp on Isla Isabel and Alejandro Gonzalez-Voyer tethered his floating blinds in front of the cattle egret nests of La Mancha.

Over the first 30 days, brown pelican chicks in broods of three switched progressively and completely from picking fish off the nest floor to reaching, one at a time, into the parent's gigantic lower mandible or pouch. They pecked, bit, pushed, and called out threats, and every brood developed an obvious hierarchy of dominance-subordination. Many B- and C- chicks, which tended to be underweight, heavily beaten, and bloodied on the head and nape, died inside or outside the nest. In short, brown pelican broods of three developed classic dominance hierarchies and practiced siblicidal brood reduction. But chicks were no friends of the feeding method hypothesis: they didn't peck more frequently during mouth-to-mouth feeds than when picking up fish and mush from the floor, and they didn't peck more frequently as they got older and transitioned to direct feeds.[18]

The data on cattle egrets in the languid heat of La Mancha lagoon were more detailed and more damaging to the feeding method hypothesis. What Alejandro saw in the broods we rescued from the boa constrictor firmly rebutted the hypothesis's developmental prediction. During their second and third weeks of life, all chicks transitioned from grasping food indirectly from the floor to taking it directly from their parents' mouths, and in all broods, both chicks fought it out, particularly seniors, who of course tended to win and got more food than juniors. So far so good, but the relationship between feeding and aggression soundly contradicted the hypothesis's developmental prediction and two of its key assumptions. After the transition to mouth-to-mouth feeding was complete, seniors attacked nearly 30 times *less often* than during the early period of feeding from the floor. Nor was this surprising because the data clearly showed that aggression did not in general enable a chick to secure a greater share of food than mere begging, and it was no more effective for winning food during direct feeds than during indirect feeds.[19] Seniors tended to get more food than juniors in any particular feeding bout whether they used aggression or not.

Much of the cattle egret fighting may have been over dominance, just as occurs in blue-foot chicks. Attacking increased steeply over the first week and then dropped off after day 10, strongly suggesting a relaxation of aggression after successful subordination of junior. The existence of trained winning and losing in cattle egrets is highly plausible: seniors pecked juniors six times more often than vice versa; trained juniors signaled submission by crouching low with their neck stretched horizontally on the nest floor or hung over the edge of the nest; and, after the first pulse of violent battles, intermittent, mostly moderate attacking was the rule for the rest of the chick period. Differential training of the two siblings as dominant and subordinate, rather than recent pecking, could well be what allows dominants generally to get a larger share of food in a feeding bout.

The brown pelicans and cattle egrets have told us, quite eloquently, that they do indeed switch from ground-feeding to mouth-to-mouth feeding as they develop the ability to intercept food on its way down. Their competition for food is a first-come, first-served situation, so they have no choice but to grab food as soon as they can. However, they are no more inclined to use aggression for direct feeds than indirect feeds because aggression secures the same priority no matter how food is transferred. Whether the transfer is direct or indirect, dominant chicks get more of the food because dominance gives them an edge. The big deal is dominance, for the asymmetry it establishes, and that explains the early onset of sib fighting. Other avian species could be different, so the descriptive studies of these two species cannot by themselves sink the feeding method hypothesis, but they have holed it below the waterline.

Would that we could test the feeding method hypothesis on blue-foots! But that's out of the question. Blue-foot parents only ever feed chicks from mouth to mouth. In the first week after hatching, it's amazing that they can pull this off when the parental head dwarfs the hatchling and has to be held upside down to grant it access. On the other hand, with their closed eyes and thin, trembling necks, hatchling blue-foots look too feeble to help themselves from a dollop of fish mush on the floor. Anyway, it's a rare youngster that picks up accidentally dropped fish from the floor at any age. Young blue-foots ignore food on the ground, as do their parents. You can feed a blue-foot chick by carefully prizing open its beak and poking or dropping food into its gullet, but these are creatures that in the wild will only ever know two ways of feeding: engulfing food transferred to their mouths by parents, and plunge-diving for fleeing fish. That's their repertoire.

I believe that among conditionally siblicidal birds—the subjects of this chapter— trained winning and losing could be widespread learning processes that create and maintain competitive asymmetry in favor, usually, of elder brood members. Whether the sibling relationship is characterized by distinct aggressive and submissive pos- turing as in blue-foots, avoidance and separation as in oystercatchers, or more vig- orously contested dominance as in brown pelicans, in great white egrets and cattle egrets, trained losing and winning is what consigns some brood members, usually the youngest, to low rank. And low-ranking chicks get less food and are more likely to die. This learning has been demonstrated experimentally only in blue-foots, but there is every indication that trained winning and losing (including trained avoidance) are widespread among avian young that compete aggressively, and possibly among ag- gressive mammalian litter mates too. Chicks of the crested ibis may be an exception, although it's not inconceivable that the civility of their rotating dominance might fall away during famine to reveal a trained or size-based hierarchy.

There may have been some false starts, and we haven't finished yet, but some un- derstanding has been gained of the rules of thumb by which aggression in these birds is modulated in the service of dominance training, competition for food, and sibli- cide.[3,4] It may be too early to discard the idea that the feeding method could be influ- ential or that the aggressiveness of particular brood members may increase with the number of brood mates. These are unproven ideas that could be resolved by further tests that control for the effects of food ingestion on behavior, a factor that has often reared its head as an alternative explanation for the results of descriptive and experi- mental studies.

For now, the observations and experiments described in this chapter and earlier chapters entitle us to assert, with varying degrees of confidence, the following rules. First, elder chicks hurriedly increase their attacking as physical maturation progres- sively enables them to train their younger sibs before those sibs can turn the tables on them. It can take them a few days to develop the muscle power, but then they usu- ally seize the day. Second, elder chicks often slacken their attacking after subordinates are fully trained into submission. Third, at all ages, a dominant chick is inclined to treat more sternly any subordinate that is similar in age or size or is truculent, two situations that flag the risk of dominance inversion. And fourth is the all-important rule by which dominant chicks secure feeding priority and adjust the number of competing brood mates to the amount of food parents are bringing, in defense of their own inclusive fitness: as their rate of ingestion falls off, they intensify their aggression. This last rule is now well established, and it makes ecological sense.

The cast of all these family dramas usually includes not just the contending infants, but two parents. Parents finance the whole show by providing food to their chicks and potentially allocate the food among them, and their own fortunes stand to rise or fall with the dominant chick's siblicidal decisions. However, theory tells us that the fitness interests of parents and the dominant chick may often diverge. In some circumstances, parents and dominants should disagree over the siblicidal elimination of juniors, with parents siding with juniors. So far, for simplicity, we've deliberately left parents out of the picture, but they are the most powerful players in every family drama. The next chapter addresses the question of what parents get up to, and who wins *parent–offspring conflict*.

5

Are Parents Okay with Sibling Bullying?

After the trauma of its birth, one in ten fur seal pups born on the isolated, boulder-lined shore of Isla Fernandina in the Galapagos Islands discovers that it must compete for its mother's milk with a sibling that is 1 or 2 years older. Following conception of the second offspring, competition between the two sibs for maternal nutrients was mediated by the mother's physiology as the embryo burgeoned in her uterus while the juvenile suckled on her teats. But in the interval between the pup's birth and the juvenile's graduation as a self-supporting hunter, the undersized pup competes with its sibling for access to the mother's teats. This peculiar arrangement has advantages and disadvantages. Fur seal mothers who create a new pup before weaning the last one provide backup for when they lose an offspring, and they simultaneously position themselves to possibly raise two pups in the event of prey being abundant. However, their strategy burdens them with concurrent provisioning costs, sets two highly unequal offspring on a collision course, and risks sharpening mothers' conflict with their juveniles.

These ambitious mothers often find themselves without sufficient resources to satisfy the demands of both offspring. In a cold-water year with abundant fish and cephalopod prey, 95 percent of junior pups successfully survive the critical first 30 days of life, but in an El Niño year, 80 percent of them succumb to starvation. The parts played by family members in their deaths, observed by Fritz Trillmich over 9 years on Fernandina, can include lethal violence of elder sibs and resolute maternal defense of harassed and excluded pups. For, in refreshing contrast with avian parents, most of which are infuriatingly inscrutable when it comes to sibling violence, some fur seal mothers are proactive guardians of vulnerable juniors.

After its mother gives birth to a little sister or brother, a fur seal juvenile goes about its business as before, loafing, practicing hunting, and suckling eagerly on her teats. As far as we know, it takes what it needs, selfishly leaving junior to make do with the remainder of the milk. But half of the juveniles observed by Fritz took their selfishness to the next level: they aggressively denied their siblings access to the teats by biting them and chasing them away from the mother. Mothers often opposed these tactics by enforcing time-sharing at the teats. When a juvenile was merely usurping access to a teat, the mother would threaten it or conceal her teats by rolling onto her belly; but when a juvenile bit or chased a pup to gain priority, a mother would typically take it on, biting and tugging forcefully at the juvenile's thick hide, sometimes opening wounds as she did so. Some mothers dragged their pups out of harm's way. Through these interventions, mothers protected their pups, especially newborns, for

Blue-Footed Boobies. Hugh Drummond, Oxford University Press. © Oxford University Press 2023.
DOI: 10.1093/oso/9780197629840.003.0006

weeks at a time, often managing to raise both juvenile and pup to independence. But maternal love could not always be counted on. Some mothers turned their backs on newborn pups, abandoning them to die, to succor their juveniles.[1]

The Galapagos fur seal's siblicide and infanticide are loosely consistent with the expectations arising from Hamilton's inclusive fitness theory and O'Connor's model for avian brood reduction (Chapter 1). The fit with that model is no more than approximate because, among other differences, a pair of seal siblings belong to different generations and feed much of the time from different maternal sources (placenta versus teats). However, although food consumption was not measured, the family interactions of seals make for an arresting example of two of the model's main predictions: that *under moderate food shortage*, the senior offspring should attempt to kill junior while the parent thwarts it; while *under greater food shortage*, the senior offspring and the parent should jointly engineer junior's death.

In many avian species, including blue-footed boobies, cattle egrets, and brown pelicans, there's little room for doubt about the motivation of elder chicks to marginalize or kill their younger sibs, nor about the latter's will to survive. But what is the role of parents, the most powerful players in every avian family? In most birds, the parental role in younger chick deaths is notably enigmatic, in striking contrast with the up-front partisanship of fur seal mothers. This is why I opened this chapter talking about fur seal mothers. The example of their unambiguous behavior, showing either assertive parental opposition to siblicide or evident complicity in it, can help us get to grips with the more equivocal ministrations of avian parents and understand them as a different evolutionary outcome of the same parent–offspring conflict of interests. Fur seals help us grasp that parent–offspring conflict is a biological reality. In sexually reproducing species including *Homo sapiens*, parents and their offspring have truly discrepant interests, which can be channeled by natural selection into overt struggles or, as we shall see, some sort of accommodation.

This chapter will outline the basics of parent–offspring conflict theory and describe how the first wave of behavioral observations on boobies and other birds failed to confirm its predictions. It will go on to describe how second-generation theory modified expectations, allowing us to accommodate often enigmatic behavior in a more satisfying evolutionary framework. Finally, after describing our search for evidence that blue-foot mothers tweak the asymmetries between their chicks to facilitate or impede siblicide, I'll offer a speculative explanation of how cooperation over siblicide evolved in blue-foot parents and senior siblings despite their fundamental conflict of interests.

Parent–Offspring Conflict: What to Expect

Trivers's theory of parent–offspring conflict went mainstream long ago among behavioral ecologists, but when it was published in 1974, it stood out as a masterstroke of deductive reasoning that instantly offered radical insight into conflicts of interest among family members.[2] At a time when most evolutionary biologists were ignoring or silently mulling over Hamilton's earthshaking theory of inclusive fitness, Trivers grasped its wide-ranging implications and, in a paper full of unsettling ideas, explained how the dynamics of gene propagation in families can turn parents and off-spring against each other. The implications for human behavior were profound, and some social scientists and philosophers were scandalized. At a stroke, Trivers's pre-posterous ideas pulled the rug from under some of Sigmund Freud's best known and most exotic developmental hypotheses by substituting an evolutionary explanation for toddlers' tantrums and regression. He also made a provocative case for parents' dutiful socialization of their offspring unwittingly increasing parents' inclusive fitness to the disadvantage of those infants.

In general, the theory tells us that chicks and their parents should sometimes dis-agree over the allocation of food among brood mates and the occurrence of siblicide, and that each party should strive to impose its own preference on the other. As we saw in Chapter 1, because brood mates are 50 percent related to each other, each chick should value itself twice as highly as it values a brood mate and therefore should strive to bias parental provision in its own favor. Parents, on the other hand, are equally re-lated to all members of their brood, so they should prefer to invest equally in them. Parents and their offspring are at odds over the allocation of food among them, and in avian families, we expect jostling and tussles as each chick strives for extra portions and parents try to impose equality.

However, differences in age among offspring can affect their value to parents. At any moment, parents will normally do better by favoring the more senior one, and for two reasons. First, it has already given more proof of its viability than junior by surviving longer, and so represents a sounder investment. Second, raising senior to independence will cost less than raising junior because, being older, it needs less in-vestment from its parents to complete its development.[3] This is the situation of fur seal mothers with two dependent offspring. If current circumstances allow them to raise only one of them, it's a straightforward decision: they devote themselves to the more valuable juvenile and abandon the pup. In blue-foot families, the two brood mates are more equal. Senior chicks are 4 days older than juniors and dominate them, making seniors more valuable to parents. In this situation, parents value seniors more highly than juniors but don't overvalue them as much as they overvalue themselves. Parents and seniors should be at cross purposes but not as much as in fur seals.

The nub of the parent–offspring conflict in a blue-foot brood of two is that the senior chick should try to garner more than its fair share of investment from parents, while each parent should endeavor to reduce that bias. Competition among genes is what ultimately drives this conflict: offspring genes for selfishness ensure their own propagation by outcompeting genes for fairness, and parental genes for investing more equally among offspring ensure their own propagation by outcompeting genes for acceding to senior chicks' demands.

You might think, as did one theoretician who scrutinized Trivers's fitness mod-els, that the contradiction between being selfish when you are an elder daughter and egalitarian when you get to be a mother would somehow stymie the evolution of off-spring selfishness,[4] but population-genetic analyses have shown that this is not so.[5] And in humans as well as many other species conspicuous conflict between offspring and parents over allocation of resources among offspring is a simple fact of life. In fact, witnessing such behavioral conflict was what motivated Trivers's original search for a theoretical explanation. When he saw pigeon chicks outside his window besieging their two parents with noisy, exaggerated begging, he knew he wasn't witnessing ef-ficient communication and straightforward cooperation among family members.[6] Undoubtedly that was going on to some extent, but it couldn't be only that. If vul-nerable infants were expending so much energy and incurring such risk of attracting predators, Trivers felt it had to be because each individual was competing with its siblings to overcome its parents' egalitarian vocation and come out on top. Families of egalitarian parents catering to egalitarian offspring would surely be more furtive and waste less energy.

We won't stray from the theater of conflict over blue-foot feeding and siblicide, but I should at least mention that parent–offspring conflict of interests is far more ubiquitous than I have been suggesting. It permeates the family interactions of spe-cies that reproduce sexually and whose offspring have some behavioral or physio-logical means of influencing their parents' investment in them. For example, even in species where parents annually produce just a single offspring, this individual is in conflict with its caretaking parents over how much and how long they will nurture it. The selfish offspring wants more than its parents are prepared to give because it values itself more highly than the sibling its parents will produce next year, whereas the parents value the two more equally. This conflict of interests finds expression as weaning conflict, with juveniles imploring their parents for prolonged investment and parents repelling them ever more irritably, preferring to get themselves in shape for the next season. And so offspring evolve theatrical displays and feign illness to deceive parents and twist their arms, and parents, biting their lips, eventually refuse to indulge them.

And it's not only about the allocation of food. Depending on a species' character-istics and its ecology, parents and offspring can also disagree on a whole gamut of factors, including, in the case of some ants, the ratio of sons versus daughters to be produced by the colony. Yes, the queen and her daughters may not see eye to eye on the sex ratio of the offspring they are jointly producing![7] In bird species that live and reproduce in groups of adults, offspring can disagree with their parents on retaining fledged offspring as breeding members of the group, and sons can disagree with their widowed mothers over the latter's inclination to pair up with a new male.[8]

Parent–Offspring Conflict in Blue-Foots?

When I first sat down in the blue-foot colony with Alicia and Cecilia, the student founders of our research program, to watch the siblicide and parent–offspring con-flict predicted by O'Connor's brood reduction model, I was agog to see family dramas.

O'Connor's article was exemplary in its cautious theoretical reasoning and in its meticulous weighing of circumstances that could offset the need for battles among relatives over siblicide. No matter, my imagination was in gear so our behavioral sampling system was primed to score as-yet-unidentified parental tactics for either assisting or frustrating sibling aggression. I was half-expecting to see some blue-foot parents sparring with irate chicks over the prostrate, trembling figures of their younger sibs. To no avail though, because as far as we could tell both parents were totally indifferent to the fierce pecking and biting of their progeny, the abject subordination of juniors, and the unequal food consumption that went on daily in every young brood.

Taking our cue from the model, we had hoped to see parents colluding with siblicidal seniors in some nests and, by whatever means, rapping their knuckles in others. But to all appearances, parents and seniors were allies. We were looking out for subtle parental tactics for quelling aggression, such as separating, brooding, or distracting aggressive chicks; and such events did occur, but only accidentally, as parents adjusted their postures to maintain body temperature, salute their partners, or chase off intruding boobies and gulls. As far as we could tell, in every two-chick family, parents and the senior chick were on the same team.

Nor did more extensive observations turn up examples of patent conflict. After four more seasons on the island, our observations on the growth, mortality, and social behavior of the blue-foots in different ecological conditions—from abundance to dearth of prey—simply consolidated our perspective on parent–offspring interactions. So our first article on blue-foots concluded there was no evidence of behavioral conflict between parents and senior chicks over allocation of food or siblicidal elimination of juniors.[9] Rather, the data pointed to complex parent–offspring cooperation. First, parents establish competitive asymmetry between their two offspring by hatching their eggs 4 days apart. Taking advantage of that asymmetry, the senior chick promptly establishes dominance, and parents do not interfere. Dominance enables seniors to obtain more food and grow faster than juniors, to which parents accede. Finally, juniors that are allowed to fledge catch up in body size and weight before fledging, so we infer that juniors are allowed to survive only if the family food budget can ensure adequate growth.

Notably, seniors show moderation; they don't habitually abuse their power over juniors. Rather than taking all the fish they can handle and leaving their sibs to get by on the leftovers, they only take, however much food there is, a privileged percentage; a conclusion also reached by booby researcher Dave Anderson for the blue-foots in the Galápagos Islands. In years of bounty both sibs grow fast, seniors faster than juniors; in lean years both grow poorly, juniors more so than seniors.[9,10] Seniors' dominance is rarely contested, and they don't aggressively eliminate juniors unless their own body weight falls to 20–25 percent below normal. And when seniors intensify their potentially siblicidal aggression, parents do nothing obvious to interfere. Parents, in fact, seem aloof to the conflict of interests and aggression between their chicks. They rarely appear distressed by violent aggression, and they don't play favorites; they consistently transfer food to the most conspicuously begging chick, which more often than not is the senior one.

All in all, this looks like a conspiracy of convenience in which juniors are discarded whenever that will boost the fitness interests of both seniors and parents. Such collusion might be distasteful to humans, but it could well be favored by natural selection, which pays no heed to fairness and shapes each family member's strategy to promote that member's inclusive fitness. However, it's difficult to square conspiracy with O'Connor's prediction of *conflict between parents and seniors under moderate food shortage.* That key prediction was not confirmed by the behavior we observed in any of dozens of families in five ecologically variable seasons and hasn't been clearly confirmed in any of scores of families observed in the 35 years since then. Don't blue-foot parents and seniors *ever* disagree about killing juniors?

It wasn't only boobies. Colleagues looking for parent–offspring conflict over the survival of junior chicks in families of herons and raptors were coming up empty-handed, too,[11] and my own survey of field reports on reproduction in pelicans, boobies, cormorants, and anhingas concluded that parents have a hands-off policy, respecting the begging hierarchy of their chicks when allocating food among them.[12] In one or two species, there were signs of chicks begging extravagantly and even using bizarre tactics like feigning illness to extract extra food from parents,[13] but parents were not crossing swords with dominant chicks or obstructing siblicide in any blatant way.

Tussles Not Always Expected

Second-generation theory explains why the behavior we actually see in nature often looks more like parent–offspring harmony than parent–offspring conflict, and the reason is simple: *behavioral roles of family members coevolve to deal with their conflicts of interests.* Take the case of a singleton blue-foot chick and its two parents. Natural selection can act on singletons to make them beg more loudly and visibly, and thereby extract more food from parents, but it can also act on parents to devalue begging stimuli and allocate an amount of food more in line with the parents' own interests. In this situation, each party's behavior is a selective force on the evolution of the other party's behavior, begging and feeding coevolve, and genetic correlations arise between begging and feeding.[14]

On an evolutionary time scale, with luck, relative peace will settle as costly conflict behaviors are selected out.[15] The conflict of interests is fundamental and doesn't go away, but the two parties evolve to converge on a *solution.* The upshot can be that parents win the conflict, offspring win the conflict, or a compromise prevails.[16] And you usually cannot tell, at least not by simply watching families of your study species, which of those three solutions you are witnessing. The bottom line, from genetic modeling studies, is that fundamental conflicts of parent and offspring interests tend to result in evolutionary coadaptation, and the family interactions you end up with can vary greatly among species and even among individuals of particular species.[17,18]

Fig. 5.1 Imposition or communication? This blue-foot fledgling forcefully pesters its parent into finally providing more food than may suit the parent's interests. But maybe not! Maybe the parent is insisting that its offspring demonstrate its quality/potential or its neediness by begging convincingly before transferring a commensurate load of fish. Photos by Andrés Almitran.

All of which is disheartening for a simple-minded field biologist who wants to make a crisp test of whether their focal species measures up to theoretical expectations, because almost any family interactions observed can be accommodated only too readily into such sprawling model predictions. If anything goes, the theory is hardly being tested, and you are unlikely to know whether parents or offspring are winning unless you can pull off experimental analyses of fitness.

Parent–Offspring Conflict in Blue-Foots: A Second Look

Let's take a closer look at our focal species to see where behavioral coadaptation may have taken the fundamental conflict of interests between parents and senior chicks over the feeding and survival of junior chicks. The most solid information derives from our witnessing, during sustained observation of scores of nests, almost exclusively harmonious interactions between parents and seniors. Whether seniors allow juniors to beg freely or suppress them, and whether seniors peck juniors seldom or scare them from the nest with volleys of pecks, blue-foot parents confine themselves to standing by—they mind their own business. But if parents are not crossing swords with their dominant offspring, what might they be doing *behind the scenes* to more subtly orchestrate seniors' control over their siblings and its outcome?

It's mostly booby mothers, not fathers, that have the means to run the show, so it's mainly to them we must look for evidence of a covert egalitarian bias. Because mothers alone create and lay the eggs (helped by a risible contribution of sperm from the father), they have at their disposal several potential tools for influencing sibling conflict. Mothers could allocate nutrients and hormones differentially between successive eggs in a clutch, for example, to undermine seniors' dominance or buttress juniors against it. Mothers could hatch their eggs synchronously or several days apart to calibrate seniors' developmental head starts. Since daughters grow faster and end up larger than sons, mothers might also manipulate the order in which daughters and sons hatch, to facilitate or obstruct the establishment and maintenance of dominance. And because the genetic relatedness between offspring dictates the overlap in their fitness interests (relatedness is 50 percent between full sibs and only 25 percent between half-sibs) and consequently the level of altruism and selfishness between them, mothers could increase their chicks' relatedness and potentially reduce the conflict between them by remaining faithful to their male partners.[19] This would better align the interests of parents and seniors.

On Isla Isabel, we looked for evidence of blue-foot mothers deploying any of these four tools—egg contents, hatch interval, order of daughters and sons, and relatedness—to influence competition between members of the current brood and conflict between seniors and parents.

Maternal Control of Egg Contents

To assess what mothers might be achieving through egg size, yolk weight, and egg hormones, we framed our study as a contrast between the two-egg clutches of blue-foots and brown boobies on Isla Isabel. Our hypothesis was that blue-foot mothers should set up their clutches with moderate asymmetries between the chicks in order to favor the rapid and efficient establishment of dominance hierarchies, whereas brown booby mothers should set up their clutches with greater asymmetries to precipitate prompt siblicide.

To get the eggs for this test, every day for a few weeks we skulked around areas of the colonies where laying was still going on. Pairs with no clutch that were spending close to 6 hours a day on territory and repeatedly pointing their bills at candidate nest sites were on the edge of laying, particularly if the female was starting to crouch on the ground as if she had already laid. As soon as any female had laid her first egg—shiny, sky blue, and conspicuous on the forest floor when she stood up—we crawled up to her nest, substituted a chicken's egg for hers, and stole quickly away so that she could promptly resume incubation. Five days later, as soon as she laid her second egg, we removed it along with the chicken's egg, leaving the nest empty and the female free to start a new clutch if she was of a mind to do so. Through our intervention, the boobies suffered imposed egg loss and postponement of reproduction, but these mimicked losses to predation by gulls, which are common in the colony.

We stored the two eggs—first- and second-laid—in our gas-powered refrigerator, and then separated and froze their yolks and transported them to Hubert Schwabl, the German endocrinologist who discovered that avian mothers insert hormones into their eggs. In his laboratory at Washington State University, Hubert measured the concentrations of three hormones that could potentially affect the development of the chicks in ways that would affect their sibling struggles: 5a-dihydrotestosterone, testosterone, and androstenedione. The second of these was of special interest because testosterone circulating in the bloodstream boosts an adult bird's ability to rise to aggressive challenges. Maybe it could do the same for chicks.

Ever wondered how field researchers get their lab science done in the absence of laboratory facilities? Many of them work with the support of rather comfortable and well-connected field stations, but for "tree climbers" like us, it's actually a more complicated, iffy, and fraught business than you might think. At our campsite, we packed dozens of little glass jars of frozen yellow yolk into an ice chest, along with crushed ice brought to the island by the fishermen, and promptly shipped the ice chest to San Blas in the next panga, along with the fishermen's equipment and accumulated catch, and our own empty water bottles, gas tanks, and inorganic garbage.

In San Blas, we always repack our chests at the ice factory, and then one of us travels with the chest by local bus to Tepic, the state capital, by long-distance bus to Mexico City, and by taxi to their own house. It's the next leg of these journeys that carries the highest risk of warming and spoiling the samples. Someone from our lab repacks the chest with salted ice, and flies with it in the aircraft hold or dispatches it as air freight from Mexico City to our colleague's city in the United States. With it goes a wad of documents proving that all pertinent Mexican ministries have authorized the study, egg collection in the field, and exportation of samples to the United States, plus a document proving that the American authorities have approved the importation of the samples from Mexico; all documents obtained via much form-filling and visits to government offices over a few months. After that, to banish fantasies of our cargo thawing unnoticed on the floor of a U.S. Customs storeroom in the American airport, we sprinkle incense and chant prayers to the booby deities 24-7 until the customs officials at the port of entry approve admission of the samples and a waiting student hurries the cooler to our colleague's lab.

For transport of blood samples that need even lower temperatures, we fill a liquid nitrogen tank at the cattle insemination center in Tepic and ship it to the island in our truck and a panga. Weeks later we improvise a way of getting it back to Mexico City in the teeth of bus and airline companies' ever-changing regulations to restrict and ban shipment of liquid nitrogen.

For equipment and supplies, Cris every year does a formidable job of thinking in advance of *all* the stuff our camp will need over the 5-month season and shipping it out in the very first ship or pangas of the season. Then from Mexico City, the coast or the island itself, Cris coordinates transit of volunteers and delivery of additional supplies from Mexico City, Tepic, and San Blas using busline cargo services to the coast, then relying on the fishermen's extraordinary willingness to pick up fresh food and other supplies at the coast and deliver them to the wave-cut platform in front of our camp.

Drinking water arrives in 20- or 60-liter plastic drums. It sometimes used to look rather green or yellow, and in one of the early years gave me typhoid fever, but our camp has never run out. Only once in 40 years, when stormy seas prevented pangas from approaching our camp during several days, did I resort to dragging water bottles from panga to shore by swimming with them over the foaming coral reef. Of course, there are no shops on the island or within 50 km, and a favorite prank of the fishermen, when a novice from San Blas or Camichín joins their camp, is to send him across the island, via the crater lake and on through the forest "a la tiendita" (to the little shop) to buy provisions for their team. "Sigue derecho derecho! No puedes fallar!" (Keep going straight on! You can't miss it!)

Tropical fieldwork was ever thus and tree climbers, whether nationals or foreigners, mostly manage to ride the complications and crises in a state of amused anxiety. Naturally, they have to deal with some bureaucracy and, to coordinate with dilatory and skeptical officialdom, they cultivate patience and optimism, which comes naturally enough in the sultry heat of the Mexican tropics, with boat-tailed grackles prancing and screaming nearby, and frigatebirds and black vultures wheeling lazily in the sky. And for our team, any tropical tardiness is more than compensated by generous logistical support from naval and national park authorities, who are charming and willing and charge us nothing; and by the warmth and helpfulness of the fishermen and their families.

Our article on the boobies' eggs was entitled "Do mothers regulate facultative and obligate siblicide by differentially provisioning eggs with hormones?," and the answer, embedded in scientific jargon, was a no. Yolk hormone concentrations did not decline more steeply across the laying sequence in brown boobies than in blue-foots.[20] This suggested that parents do not set up brown booby broods for more rapid siblicide than blue-foot broods, not at least by spiking eggs with different concentrations of hormones. In both species, first and second eggs were similarly endowed with all three hormones, so we concluded that mothers don't tweak their offspring's aggressiveness by lacing eggs with hormones.

No joy in egg size either: at least under the demanding El Niño conditions females were facing in the year we got our samples, in neither species did mothers make first eggs heavier than second eggs. And to cap it all, comparisons of yolk mass actually contradicted our prediction: only in the blue-foot did first and second eggs differ, the latter containing 10 percent less yolk. It's tempting to think that blue-foot mothers give second eggs less yolk in order to handicap junior chicks, but there's an alternative explanation that can't be dismissed. Mothers that year may have been obliged to short-change second eggs simply because they were themselves underfed at the time of laying. The underfeeding explanation is quite believable because a mother booby's egg provisioning can vary with her recent food intake. After Nazca booby mothers in the Galápagos Islands swallowed fresh mullets tossed to them on their territories, they were more likely to lay a full clutch of two eggs, and their second eggs were heavier.[21]

With every successive test, there was less reason to suspect that blue-foot and brown booby mothers customize their eggs in any way to exercise control over offspring hostilities. But what about their hatch intervals?

Maternal Control of Hatching Interval

Surely the blue-foot's 4-day hatch interval, one of the longest in any species of bird, substantially affects aggressive competition between siblings? We had always assumed this was so but never formally tested it, although we'd shown that the relative size of two blue-foot chicks determines which of the two is more aggressive and becomes dominant.[22,23] Might two siblings' relative ages also affect the intensity of their pecking and submitting, the amount of food each one gets, how they each grow, and whether junior dies? These are matters of great concern to the chicks themselves and to their parents because they can potentially impact the inclusive fitness of all four.

Birds have to lay their eggs at least 1 day apart because it takes a day to produce an egg, and eggs cannot be accumulated in the female's body because flying around with so much ballast would be burdensome. However, mothers can control hatch intervals by prolonging laying intervals or delaying the onset of incubation, so in principle, they could adjust hatch intervals to their chicks' interests, or their own interests. Blue-foot mothers lay their two eggs 5 days apart and hatch them 4 days apart, implying some delay in the incubation of the first eggs. After laying the first eggs, they have to cover them to hide them from gulls, but they don't necessarily maintain them at incubation temperature. There's undoubtedly some scope for fine-tuning hatch intervals.

In a small sample, natural variation in hatch intervals didn't affect the probability that juniors would die, ages at death, or the growth of seniors and juniors that fledged.[24] But what about hatch intervals shorter or longer than those that occur in nature? What would be the consequences for chicks, and for their parents, if parents hatched their two chicks on the same day? Or 8 days apart? After approval of his undergraduate thesis on the territorial defense of blue-foot partners, Mowgli was rearing to go with a field experiment and delighted by the thought of working on this question. Completed for his masters' thesis, the experiment restored my faith in avian mothers' potential as puppet masters; it showed us that control of hatch interval is a maternal tool par excellence for modifying both offspring competition and parental investment in the brood.

Mowgli and his three volunteer assistants spent 3 months on the island. They identified clusters of up to four blue-foot nests that could be observed by a single person, and whose individual eggs or hatchlings could be swapped between nests to form broods with hatch intervals of 0, 4, or 8 days. Then two observers alternated every 2 hours watching each cluster for 12 hours a day during the chicks' first 3 weeks of cohabitation—the period when dominance relationships are established and sib aggression peaks. On the basis of our earlier studies, we anticipated, correctly, that these manipulations would induce changes in the aggressive behavior, growth, and mortality of chicks, but only within the range of variation arising from natural phenomena such as hatch failure, chick mortality, adoptions, and the crossing of growth curves when female chicks outgrow their elder brothers. Hatch intervals of 8 days occur naturally between successive eggs and whenever the second egg of a three-egg clutch fails to hatch.

You will not be surprised to learn that in control broods with an experimentally assigned 4-day hatch interval senior chicks established dominance by pecking as soon as they were able and steadily increasing the tempo, nor that juniors quickly learned to be consistently submissive. Subordinates, of course, grew more slowly than dominants and more of them died. What we were most interested in, however, was the contrast between those natural-interval broods and broods where both chicks hatched on the same day, which could tell us how things would work out if parents postponed incubation until the second egg was laid.

Same-age broods surprised us by forming dominance relationships routinely and rather quickly, showing that you don't need staggered hatching for this. And dominance-subordination consigned subordinates to slower growth and more frequent death by starvation, just as it did in natural-interval broods. Indeed, the growth and mortality of chicks differed little between same-aged broods and natural broods.

The important departure from natural-interval broods lay in the intensity of aggression and the amount of food consumed. In same-age broods, dominants pecked three times as frequently, and the brood consumed 20–30 percent more food because subordinates ate as much as dominants.

Our understanding of these results was that when the subordinate sibling is of an age/size that poses a threat of dominance inversion, the dominant sibling has to work harder to neutralize that threat. Whether all this extra aggression and eating ultimately affects the development of chicks we don't know, but the extra eating clearly affects parents: they must pay the energetic cost by stepping up their hunting effort and bringing more food. The take-home message is that same-age siblings themselves may be no better or worse off (unless their enhanced aggression incurs psychological costs), but their parents have to finance enhanced conflict by bringing more fish, an extra effort that could well impact their fitness the following year.

On the other hand, a hatch interval of 8 days was worse than a natural interval from every party's perspective, because junior chicks were more likely to die, probably prejudicing the inclusive fitness of parents, seniors, and juniors. What's more, pecking in broods with 8-day intervals was three to five times more intense than in natural-interval broods. Seniors seemed especially incensed at the sight of very small/young nest mates, a puzzling phenomenon that occurs also in natural three-chick broods reduced to just an A-chick and a C-chick.[22]

These results, along with similar results from experiments on other siblicidal birds[25–28] and non-siblicidal birds,[29] showed that the natural hatching interval confected by parents saves them from having to finance the elevated conflict that arises between evenly matched siblings and between terribly mismatched siblings, and probably saves all family members a fitness cost due to unnecessary deaths of juniors. Analysis of consequences over at least a year after fledging would be needed to discern longer-term fitness effects of manipulated hatch intervals, such as whether fledglings from same-aged broods are less likely to survive the challenging first year of life and breed, and indeed whether adults who cared for such broods pay a price in terms of their subsequent survival or breeding success.

Sadly, although the Isla Isabel blue-foots would have been ideal birds for such a follow-up study, we couldn't manage it because our booby marking system—three colored bands of PVC on an adult's legs—was failing. These spiral bands from Britain that curled around a bird's leg two and a half times wowed us when we started fitting them. They could be simply rotated onto or off the legs of a captured booby in a few seconds, and we could read a booby's three bands from any angle, at a considerable distance with binoculars and at least 20 m with the naked eye. But they let us down within days of fitting when their bearers returned from hunting trips with bands snarled around their toes, opened (we guessed) during plunge-diving by impact with the water or underwater paddling. Not to worry, in the first year Cecilia and Alicia had successfully used a powerful glue to reassemble severed rubber washers around boobies' legs, and those washers had lasted at least a year! So we recaptured all of our banded boobies and glued their spirals together. The glue held fast.

It took several years for the system to fail again, and this time the damage to our long-term research program was greater because we lost more historical information on known individuals than we could bear to think about. With the passage of time, seawater and the tropical sun took a toll on the colors and resistance of the PVC. As the colors of bleached bands converged, we increasingly confused red, orange, and brown, as well as green and blue; and when the bands started fracturing and dropping off legs, we were crestfallen. Our system was a shambles, so, apologizing profusely to repeatedly banded boobies, we discarded 9 years of records and started again, this time fitting each bird with a single stainless-steel band engraved with an individual number. Steel bands have to be fitted with special pliers, they're less readable at a distance and from unfortunate angles, and engraved numbers can get temporarily coated with guano, but the bands are durable—perfectly legible after 25 years of sea, sand, and sun, they outlast their wearers. If the Internet had existed in 1980, we would have discovered the steel bands at the outset, and not lost those 9 years!

Bands, by the way, are not just for revealing whether migrating birds fly from pole to pole or only from Alaska to Mexico. Combined with systematic monitoring and behavioral observation, they reveal and explain birds' lives. Without them you couldn't tell that Veronica, alias F271, is 15 years old and is back with Harry after breeding with Lance during three successive seasons; that sticking with Lance for 3 years earned her greater childcare equality; or that she's courting and copulating right now with a neighbor after her partner went off hunting. You couldn't know whether sibling bullying turns boobies into adult wimps, or whether growing up with an aggressive elder brother affects a female's survival, attractiveness, and reproductive success when she enters the breeding market. We wouldn't know that blue-foot males and females are both unfaithful to their partners, or that females are more secretive than males about their extra-pair liaisons.

Getting back to parent–offspring conflict, Mowgli's experiment shows, first that you don't need a 4-day hatch interval for chicks to rapidly form a dominance hierarchy, and second that a 4-day interval probably serves the interests of the whole family by discouraging contests over dominance and avoiding unnecessary mortality. Any shorter and parents would probably pay a fitness cost of financing their chicks' exacerbated conflict; any longer and parents would be penalized by both that cost and unnecessary starvation of juniors. And dominants and subordinates separated by a hatch interval of 4 days grow and survive as well during the nestling period as those hatched on the same day, while engaging in less aggression. So this is not a case of mothers winning parent–offspring conflict by imposing a hatch interval that suits themselves better than their senior offspring.

Sequence of Male and Female Eggs

Next up from the hypothetical maternal toolbox for masterminding dominance relations and siblicide is the laying sequence of male and female eggs. Provided blue-foot mothers are able to influence the sex of their eggs (which some other bird species manage), the large size of daughters is a characteristic they might be able to play with to increase parental control over siblicide. For example, by tending to lay male eggs in first place and female eggs in second place, and thereby diminishing the average size advantage of senior siblings, they might make it harder on average for seniors to kill juniors in circumstances where parents benefit from their survival.

However, Roxana Torres, a doctoral student at the Universidad Metropolitana in Mexico City, showed that blue-foot mothers probably don't use this tool to protect their junior chicks or, apparently, for any other purpose. By sexing 751 chicks in a La Niña year and following their survival through fledging, she showed that 56 percent of offspring were males, both at hatching and fledging, and that this same male bias was present in both seniors and juniors. Neither sex was more likely to appear in first place or second place in a brood. In good years at least, blue-foot mothers don't strategically manipulate the laying sequence of male and female eggs.[30]

Genetic Relatedness of Siblings

In theory, genetic relatedness of their chicks is something mothers could manipulate to reduce selfishness among their offspring. The idea here is that mothers could usefully maintain high relatedness (50 percent) between brood mates by chastely ensuring that both of them are sired by the same male. Similarly, they could ensure 50 percent relatedness between their offspring of successive reproductive seasons by repairing with the same male. By increasing the relatedness of simultaneous or successive offspring, females could reduce the discrepancy between the relative values they and their senior chick place on the two chicks. For example, they might less often find themselves maneuvering to prevent siblicide if seniors had a greater stake in their sibs' survival. In support of this conjecture, competitive begging is less intense in avian species where chicks are always full sibs and in barn swallow broods composed of closely related individuals.[31,32]

As we shall be seeing in Chapter 8, only a small minority of the many unfaithful female boobies on Isla Isabel produce offspring sired by extra-pair partners, and every year half of blue-foot mothers repair with their partner of the previous year. So, with a nudge, one might infer female restraint over granting paternity to extra-pair males and over changing sexual partners in the service of intergenerational harmony. Blue-foot females may indeed avoid creating offspring that are half-sibs in order to reduce sibling conflict and two hypothetical costs thereof—increased parental workload to finance sibling conflict and greater conflict between parents and senior chicks over investment in juniors. However, the conjecture has us skating on thin ice and, as we shall see in Chapter 7, alternative explanations for the within-pair genetic fidelity and repairing of this booby are more compelling and better supported.

Parent–Offspring Cooperation?

Do blue-footed boobies exemplify parent–offspring conflict? Our observations would certainly not sell Trivers's theory to a skeptic, but I am inclined to interpret the harmony we see between parents and senior chicks as cooperation evolved out of genetic conflict through coadaptation. Parents facilitate competitive inequality in their two chicks, not by lacing eggs with hormones but by hatching them 4 days apart, and they may fine-tune the inequality by providing seniors with an extra dose of yolk. Over the chick period, parents probably cap food provision to the brood at a level that safeguards their own longevity and future reproduction and they *leave the chicks to work out food allocation and possible siblicide between them*. Parents need do no more because they have already structured their brood in such a way that the senior chick can aggressively regulate food allocation in its own favor and eliminate junior if necessary while minimizing the cost of sibling conflict to parents. Chicks assume their roles—dominant and subordinate—and parents allow dominants to control food allocation and siblicide. Subordinates submit to this control under duress.

Despite their fundamental conflict of interests, parents and seniors are decidedly on the same team. The all-powerful parents consistently abstain from imposing food allocation on their own terms, interfering in sib hostilities, or killing juniors. In theory, the interests of parents and seniors coincide when the death of a junior will benefit both of them, and diverge (under moderate food shortage) when it will benefit seniors but not the parents. But rather than addressing the conflict of interests by struggling against each other, parents and seniors have apparently coevolved to charge senior with the stewardship of junior's growth and survival.

As for whose inclusive fitness interests are better served by this division of roles between parents and seniors, we don't know. Evolution of collusion between parents and seniors over junior's food allocation and survival could have been facilitated by the substantial coincidence of their inclusive fitness interests and by asymmetry in the information they possess.[33] Juniors are devalued by seniors because of siblings' 50 percent relatedness, but also by both parents and seniors because juniors' youth and subordination make them a poorer investment. In conditions of food scarcity, which selectively depresses the prospects of juniors, parents might devalue juniors nearly as much as seniors do. As for information, parents are probably less informed than seniors regarding the adequacy of food provision to the brood because each parent is largely in the dark about the other parent's food transfers, whereas chicks know about the transfers of both parents. And chicks know about their own body condition. If indeed the conflict of interests is minor and anyway seniors can assess it better than parents, parents might best promote their own interests by delegating decisions on food sharing and brood reduction to seniors. If so, then the evolved solution of conflict likely tilts in favor of the seniors' inclusive fitness rather than the parents'. Seniors probably win, but by a small margin.

No such harmonious solution has coevolved in Galapagos fur seal parents and their juvenile pups, but there the discrepancy between the mother's interests and the juvenile's interests is greater. Whereas mother fur seals are equally related (50 percent) to their two offspring, these are normally related to each other by only 25 percent because mothers rarely copulate with the same male in successive years.[1] Therefore juveniles ought to devalue their newborn half-sibs more than their mothers do, and this discrepancy could help explain why Fritz discovered fur seal mothers doing something booby mothers never do: defending juniors from their elder siblings' attacks.

Allow me to distract you with some anecdotes and a final thought about ghosts of ancestral behavior. In the spring of 2021, while I was drafting this chapter, Cris and Cheko (Sergio Ancona), a new member of the team who cut his teeth studying El Niño impacts on blue-foots, insisted on braving the covid pandemic to launch our annual long-term monitoring of the Isabel blue-foots. During their time in the colony, both of them happened to witness, independently, blue-foot parents quelling the sibling aggression of elder chicks. It was a La Niña year and flocks of blue-foots, brown boobies, and pelicans whirled and dived daily in front of our camp. The blue-foot colony was happily alive with broods of two or three spongy white chicks, begging, feeding, and flapping their stumpy wings into the winds. The fishing was good and many broods were attended by both parents because breeders had ample time on their wings for loitering on their territories.

Cris witnessed a parent thrust its bill deliberately downward between its junior offspring, frozen in a submissive posture, and the senior sib that was raining down blows on its cranium. Cheko witnessed a father hop down from a bough, walk half a meter to where his B-chick was 15 seconds into a bout of pecking at his C-chick, plant himself between them, and whistle. In both cases, the parental behavior looked purposeful and sibling aggression stopped immediately. I once saw something similar in another context. After I replaced the senior chick of an experimental nest with a different senior chick, the incoming senior pecked the resident junior and the father of the family promptly stilled it with an authoritative downward peck to its cranium. Wow! I gasped.

Were Cris's and Cheko's observations exceptional occurrences, under extraordinary ecological bounty, of a type of parental regulation of sibling conflict that in another era was commonplace but has been all but deleted from the repertoires of blue-foot parents and their offspring in the course of coevolution? Were they atavisms of parent–offspring conflict over suppression of siblings? I suspect that interactions between parents and seniors over the feeding and survival of juniors have vacillated over the decades or millennia between the collusion we're so familiar with and physically contested control. And I suspect blue-foot parents are fully capable of suppressing sibling aggression. Maybe in ecologically distinct eras, when the costs and benefits of their behavioral options were different, parents policed siblicide.

I have witnessed what I have taken to be atavisms before. My favorite was in snakes exposed to extreme hunger. The western terrestrial garter snakes of California's Sierra Nevada prey on tadpoles, fishes, and leeches in mountain streams and ponds, whereas snakes of the coastal lowlands feed only on terrestrial prey like slugs and rabbit kits, and avoid entering water. Might coastal populations have progressively lost the capacity for aquatic predation and the genes that underwrite it? In the laboratory, unfed, newborn snakes from the coast refused to either enter the water to hunt for live fishes just centimeters away or eat dead fishes left in their cages. But after 2 weeks without food, refusers reluctantly entered the water and successfully hunted fish using the aquatic repertoire of the mountain snakes. Although it long ago abandoned underwater hunting, the coastal population has retained the species' aquatic repertoire in its genome, and exceptional circumstances can provoke its expression.[34] What has evolved is the hunger threshold at which stimuli from fishes elicit aquatic predation—it has risen. I suspect that Isla Isabel blue-foots have likewise retained in their genomes behavioral repertoires by which ancestral adults and senior chicks contested the exclusion and killing of juniors when food was moderately scarce.

Finally, do brown booby parents and their senior offspring manifest parent–offspring conflict over the feeding and survival of juniors? Brown booby juniors are there only for insurance, so when both eggs hatch and two viable chicks are cohabiting in a nest, the junior one is living on borrowed time. Parents and the senior chick both benefit from junior's execution, which is better carried out by senior, lest parents kill the wrong one. By mistake, they could end up raising a sickly senior when they could have raised a vigorous junior. Siblicide demonstrates the killer's quality. However, the parents and the senior chick could well be in conflict over the age at which junior should die. By prolonging junior's survival for a week or two, parents may stand to benefit by extending their insurance cover against senior's developmental failure or loss to a predator. From the perspective of seniors, however, urgent dispatch of desperado juniors removes a mortal threat and is so advantageous that strong selection on seniors may have enabled them to impose it on reluctant parents. I suspect brown booby seniors have won the conflict with parents over when juniors must be killed and incidentally deprived parents of occasionally handy insurance coverage.

What are the developmental consequences for an infant bird, of being threatened and assaulted and forced to submit on every day of a months-long chick period? Sustained discomfort and pain, along with reduced rations and, for all we know, fear and humiliation could potentially affect a junior blue-foot chick for the rest of its days. We fully expect human infants to suffer long-term developmental impacts of deprivation, abuse, and cruelty throughout their adult lives, but maybe tales of woe and selective publication of alarming studies have made us overly pessimistic. After all, there's plenty of resilience out there in nature, too, and natural selection would be expected to prepare and equip infant animals for the typical challenges of their natural social environments. However, long-term impacts of sibling competition and conflict on development have gone virtually unstudied in any animal. Because the package of challenges endured by blue-foot juniors at the beaks of their siblings is extreme—involving chronic violence, starvation, and physiological as well as psychological stress—we expected them to teach us lessons about evolved resilience. So we compared the growth, survival, and behavior of seniors and juniors over their adult lifespans. The inspiring results of this decades-long research project are recounted in the next chapter along with some comparable data on human siblings.

6

Bullying and Lifelong Scars

Are Infants Vulnerable or Resilient?

Would you expect the multiple traumas of bullying to prejudice junior blue-foot chicks permanently? Are junior fledglings bound to be damaged goods? We are all familiar with the idea that stressful or harmful physical and social environments during infancy can set humans back for the rest of their lives. Because Sigmund Freud encouraged us to believe that experiences in infancy set patterns of behavior for life, and because psychologists have traced ever more causal arrows between early adversity and adulthood, we believe in the developmental potency of experiences during infancy, especially traumatic ones. But in boobies? They may have something important to tell us about the developmental impacts of sibling conflict on vertebrate infants, and an evolutionary account of boobies' behavioral development will surely be anchored more securely than Freud's speculative proposals.

From well before the time we finally managed to mark fledglings for life with steel rings, I was dying to probe the damaging effects of subordination by comparing how juniors and seniors fare in their adult lives. Over the years I have returned repeatedly to this issue because I am intrigued by the idea that competition between close relatives could lead them to do permanent harm to each other. We already know, of course, that seniors pile suffering on juniors and that juniors give in to this oppression, but we haven't considered their vulnerability to developmental impacts. Long-term damage shouldn't be automatic; evolution by natural selection ought to equip infants to deal physically and behaviorally with challenges that are predictable, including bullying by siblings. However, there are likely to be limits to how well an infant can be buffered against stresses, for example, because adequate defenses may cost more than its energy budget can afford, or because defenses may trade-off against other important traits like growth and capacity of the immune system.

Serious developmental damage is entirely to be expected when young animals meet an unnatural challenge invented recently by humans—like industrial pollution or habitat degradation—but we expect wild animals to handle *natural* challenges rather well and, generally, to be able to avoid, resist, or mitigate the long-term damage these may threaten. Yet in this century, dozens of experimental studies by biologists have shown that contrived, semi-natural stresses experienced by chicks can set them back in adulthood in numerous ways. The stresses most commonly administered are food shortage, increased stress hormone, repeated capture (fake predation?), and brood enlargement—giving chicks extra nest mates to compete with. And the list of

Blue-Footed Boobies. Hugh Drummond, Oxford University Press. © Oxford University Press 2023.
DOI: 10.1093/oso/9780197629840.003.0007

negative consequences is scary. In 2015, by our last count, 23 behavioral, reproductive, physiological, and morphological deficits of adulthood had been attributed to these experimental stresses in infancy.[1] This growing body of research has created the impression that chicks, and maybe infants of other vertebrate species, are highly sensitive to early conditions, and that unfavorable conditions readily disrupt their long-term development and damage their prospects for surviving and reproducing.

The truth may be somewhat different though. Experimenters have mostly tested chicks in captivity, where their options for dealing with stress are limited, and seldom attempted to make experimental stresses match the magnitude of natural stresses. Rather, they have generally opted for levels of stress likely to produce damage, because proven damage is what gets the research published. No one wants to come out of a lengthy experiment empty-handed. This understandable dynamic, which I exaggerate only slightly, may have contributed usefully to biological sciences by revealing developmental vulnerabilities and by linking them to looming environmental changes. However, it has also led to overestimation of chicks' developmental fragility and, I believe, underappreciation of their *resilience*, which for the most part remains to be probed experimentally in natural habitats using natural levels of stress.

In the early years of the blue-foot program, when I gave talks on sibling conflict in boobies at meetings of ethologists, psychologists, and behavioral ecologists, the most common question from my colleagues was "What are the long-term effects of subordination?" Colleagues were particularly eager to know whether the cohabitation of blue-foot chicks trains subordinates into adults who are mild-mannered losers and dominants into adults who are assertive, aggressive winners. One colleague claimed, tongue in cheek, "I would kill to get long-term data on that." And there were other questions. We also wanted to know how long such training effects last, whether former subordinates are less attractive to the other sex, and most important of all, whether they are in any way hampered in their reproductive capacities and outputs. It was easy to imagine that former juniors would be uncompetitive when it came to defending a territory, expelling an intruder, winning the favor of a potential partner, or even plunging for fish; and underperformance in any of those domains would likely impact a booby's fitness.

To achieve a balanced perspective on developmental vulnerabilities, one surefire research strategy is to describe systematically how stresses *naturally* encountered in infancy impact animals during the rest of their lives. Description doesn't give the leverage of experiments when it comes to tracing causal pathways between stresses and deficits, but it tells you whether developmental damage is there in nature and whether the damage correlates with the supposed stresses. Descriptive studies keep you on track in investigating natural phenomena rather than experimental artifacts, and you can often follow up with well-focused experiments rooted in the descriptive information. For descriptive research, the blue-footed boobies of Isla Isabel were ideal subjects because the markedly different roles of juniors and seniors in sibling conflict were well characterized (descriptively and experimentally) and because the lifelong attachment of fledglings to their natal colony made it possible to document and analyze complete lives. Over several decades our annual monitoring program, complemented periodically by physiological measurements and a field experiment, gradually revealed how sibling bullying can affect a vertebrate infant's psychological development and reproductive trajectory.

This chapter describes how, as the years rolled by and data accumulated, patterns emerged in the survival and reproductive performance of former subordinates and dominants over their lifetimes. It also describes a field experiment designed to compare their ability and willingness, in youth and old age, to muscle up in defense of their offspring. It concludes by looking at the evidence for long-term effects of sibling competition and conflict in *Homo sapiens*, a vertebrate species whose sibling impacts have been under scrutiny for over 100 years, yet are poorly known.

Junior Blue-Foots' Developmental Challenges

The three challenges posed to the junior blue-foot's development by sibling domination during infancy make up a potentially crippling package: aggressive subordination, partial starvation, and chronic corticosterone (stress hormone) exposure. The first of these involves both physical abuse and psychological subordination. From the time it first wobbles its little head to beg for mush until about 12 weeks later when it launches from the beach on its maiden flight, the junior chick is threatened and assaulted every day and possibly every night. The initial barrage peaks at about 3 weeks, as the aggressor grows in strength, coordination, and mobility, and then subsides after the victim's subjugation is consummated. Perpetual psychological subordination is then maintained by the daily, low-level aggression that junior experiences during the remainder of its 3-month chick period. Senior achieves this by instantly snuffing out any revival of the rebellious tendencies junior still harbors but seldom dares to express.

Starvation of juniors during the first several weeks consists of consuming 20 percent less food than seniors and growing more slowly. True, juniors catch up with their siblings in body size and weight by the time they fledge, but the experience of growing on limited rations during the earliest and most developmentally sensitive stage of life could have long-term consequences. And there's an additional risk: experiments have shown that growing faster to compensate for deficient growth early on can itself result in a variety of short-, medium-, and long-term deficits, including in the timing of clutch production and the speed of escape from predators.[2–4]

Then there's the chronic corticosterone exposure. An undergraduate medical student at the UNAM named Alejandra Nuñez de la Mora, supervised by me and John Wingfield, an English behavioral endocrinologist at the University of Washington, investigated this in the field. Ale discovered that the concentration of this hormone in a junior blue-foot's bloodstream during at least the first 3 weeks of life is roughly double what it is in a senior chick's bloodstream, due to partial starvation.[5] Animals secrete extra corticosterone to buffer themselves physiologically against stresses of different kinds, but buffering can come at a cost. Experimentally dosing up chicks has revealed that excessive corticosterone exposure can have deleterious effects on several traits, including brain morphology, metabolic rate, learning, and competitive ability.

My first experience of the damage that corticosterone can do, in fish by chance, left me shocked and ashamed. Before my doctoral studies, one day in my Mexico City apartment I carefully introduced some adult mouthbrooding cichlid fishes into a 1 m aquarium and watched with satisfaction as they explored their spacious new habitat and competed for space, nervously then boldly and happily I thought. To my delight, two of them eventually got together and produced dozens of millimetric fry that floated in a glittering cloud near the guarding parent's head. When disturbed, the fry fulfilled the promise of their name by crowding hurriedly into the parent's mouth and refuging there until the coast was clear.

But prior to this marvelous spectacle, which I had to crawl stealthily across the floor to see, there was a tragedy. I didn't realize it, but the space in the aquarium was insufficient for each adult mouthbrooder to acquire a territory. In consequence, the least competitive male was harassed daily by the others for intruding into their space and ended up corralled in a corner of the tank, head-up, gulping and quivering—terrified, by the look of him. Ethological writings on cichlids by Gerardus Baerends and Konrad Lorenz had not prepared me for this, and I naively expected that my little community of mouthbrooders would settle down. The next day though, the harassed male was floating belly-up, a victim of social stress and chronic corticosterone exposure (not to mention uninformed husbandry). Food, temperature, and water chemistry were all fine; he died of stress and chronic corticosterone. Could junior blue-foots resist the developmental harm of chronic corticosterone exposure? Might they have evolved developmental resilience to *all three challenges*: violent subordination, temporary starvation, and chronic corticosterone exposure?

After fledging, blue-foot juveniles have the opportunity to recover from sibling abuse, when the calvary ends and their sibs drop completely from their lives. During 4 years, on average, junior dedicates itself to learning how to survive independently in the ocean ecosystem; it takes that long. Life on its own, searching the ocean surface in all weathers for shoals of fish is demanding, but other plunge-diving birds often show where the shoals are, and social conflict over the ocean and on land is minimal. Aggressive experience during this period is probably confined to minor altercations with other boobies over access to perches on ledges and clifftops; there's never any fighting over fish, which are always either in the surrounding water or secure in another booby's mouth. A juvenile has to withstand and evade the piracy of magnificent frigatebirds, in the air and on the water, but it probably doesn't need to show any real mettle until it attempts to install itself in the colony as a breeder. This huge step involves a whole suite of novel activities that every booby first sees and hears and even practices in a make-believe way when it is a chick; for example, impressing candidate partners with courtship displays, standing its ground against mature and experienced strangers, and joining with partners in protracted pair-fights against belligerent neighbors. When it finally morphs into an adult with speckled white breast and splendid blue feet, and the time comes to engage, she or he takes to those activities like a duck to water. But do former juniors have what it takes to succeed?

Saved by Natural Selection?

On the bright side, evolution by natural selection has probably had thousands of generations in which to adapt blue-foot chicks to the trials and tribulations of cohabiting with a violent oppressor and emerging from it all in good shape. Of all the juniors that over the millennia were subjected to subordination, surely it was the most resilient individuals that went on to produce the most offspring? And if resilience itself is passed on genetically from parent to offspring—a reasonable supposition—then resilience has had abundant opportunity to evolve progressively in the population, generation by generation. This gradual microevolution is the main engine of Darwinian adaptation to challenges and, given enough generations, it can be transformational. For example, genes for peck-resistant skulls and eyes, genes for thriving after temporary starvation, and genes for bouncing back after being drubbed into abject submissiveness could all have cropped up by chance mutation and progressively spread to fixation in the population. Provided bearers of resilient genes produced more viable offspring over their lifetimes than bearers of less resilient alternative genes, microevolution could, over thousands of generations, cumulatively shape ever more resilient juniors. Juniors are not likely to be perfectly resilient though, because of trade-offs between design features like impact resistance and lightness of skulls. Therefore, we expect former juniors to manifest some flaws, and these could emerge during the juvenile period, young adulthood, middle age, or old age. Savvy evolution would probably consign their expression until late in life because, with normal luck, mortality would often ensure that they were never expressed at all.

What did I expect of junior blue-foot fledglings before I saw the data? Given the forbidding three challenges of the chick period, and considering my observation that even newborn, inexperienced animals like the snakes I used to study tend to be highly variable in their individual characteristics, I was ready for highly variable developmental damage. Specifically, I expected that (1) among fledglings, more juniors than seniors would die before graduating into the breeding population, and (2) compared to senior graduates, junior graduates would be less likely to survive from year to year, less attractive to potential sexual partners, and less able to prevail in territorial defense against an intruder. In consequence, junior fledglings would, on average, achieve lower fitness over their lifetimes than senior fledglings. However, many junior graduates would, one way or another, evade or overcome the developmental penalties of what is, after all, entirely natural and predictable adversity and perform as well as former seniors. Those were my naïve expectations for the first analyses of the long-term consequences of sibling bullying in a wild vertebrate. In the light of the experiments revealing the developmental vulnerability of avian chicks, I thought these were optimistic expectations. But I underestimated the blue-footed booby.

Fig. 6.1 Junior blue-foot chick approaching fledging. Its parents still feed it, but over a period of several weeks it will progressively shoulder that task, by learning to plunge-dive. Photo by Andrés Almitran.

Mortality and Graduation of Former Juniors

In view of the importance of our question, Cris and I threw the book at it! First, we made numerous comparisons of the lives of thousands of fledglings in our database, looking at a suite of variables, to reveal former juniors' deficits in survival or repro-duction, and identify the specific weaknesses behind their deficits. We also compared former juniors and seniors for their adult size, body condition, and immune responsiveness. When they became breeders, we compared their offspring's size, body condition, immune responsiveness, and success in graduating into the population of breeders. It was particularly important to include old age in some comparisons because great hardship during infancy might lead to accelerated senescence—a decline in functions with increased age. Finally, to answer my colleagues' favorite question, we tested aggressive mettle experimentally by observing whether former juniors are any less bold and vigorous than former seniors when it comes to repelling intruders from the vicinity of their offspring.

First off, we tackled the most important likely differences between former seniors and juniors by comparing the survival and graduation into the breeding population of 7,927 fledglings from 20 generations over a total span of 21 years, with the statistical assistance of Dani Oro, a Spanish population ecologist and seabird specialist from the Mediterranean Institute of Advanced Studies (IMEDEA). After Spain joined the European Union, evolutionary biology and behavioral ecology flourished there as young Spaniards, like Dani and Miguel Rodríguez-Gironés, interacted more with diverse European universities. Dani, who completed his advanced training in France, was part of a new generation of ecologists specialized in statistics. All of those fledglings were either seniors and juniors that grew up and fledged together or singletons from one-chick broods. To graduate, fledglings had to attempt to breed—that is, reappear in the colony at any age with a sexual partner, a nest, and a clutch. Life is dangerous and competitive in the wild, so in most species of birds, only a small minority of fledglings manage to pull that off. Graduation is a sterling test of a fledgling's quality and it's easy to measure in the Isla Isabel colony because dispersal to other islands for reproduction is rare.

My jaw dropped as I sat in Dani's laboratory in Mallorca and stared at the survival curves his computer had slaved all night to generate. Even with that huge sample, the analysis found no evidence that junior fledglings are, at any age, less likely than seniors or singletons to survive from one year to the next or to graduate into the breeding population. Whether they hatched in unfavorable El Niño years or favorable La Niña years, juniors showed no inferiority in these respects, and they were just as capable as seniors of staying alive through the stresses and strains of El Niño years. What's more, the lack of differences among juniors, seniors, and singletons clearly showed that just as growing up as a junior does not prejudice the ability to survive, nor does growing up as a dominant enhance that ability. Dani's rigorous analyses gifted us a stunning affirmation of resilience to the psychological, nutritional, and physiological impacts of sibling oppression in infancy on viability across the adult years, including viability into the depths of old age.[6] Dani was cool, he seemed unmoved by the results, but he'd never witnessed what junior chicks go through.

Nonetheless, there's an important rider to this result: during the chick period, junior blue-foots suffer more mortality (40 percent) than seniors (29 percent), and if death in that period falls selectively on duds (individuals of low quality), then the sample of junior fledglings included in our study was screened for quality more strictly than the sample of senior fledglings. Deletion of more duds from the junior sample than from the senior sample would represent a sampling bias that, if it exists, could have led us to underestimate the susceptibility of junior fledglings to mortality. There is no evidence that deaths in the chick period winnow out duds, and deaths of junior chicks seem to depend on their family food budgets rather than the juniors' individual quality, but we can't rule out sampling bias. Our article reporting Dani's analysis argued that any bias was probably of small magnitude ... but a gnawing doubt remains.

Reproduction of Former Juniors

Next in order of importance is the ability of former juniors to reproduce. To compare the reproductive prowess of nearly 1,200 fledglings that grew up as juniors and seniors, we analyzed its main components separately, for thoroughness and in the hope of turning up clues to specific failings. We looked at age of graduation because the highest quality individuals might manage to graduate earliest into the breeding population, although a strategy of postponing graduation to achieve greater longevity is also a possibility. The date of its first nest is another index of an individual's quality. Blue-foots strive to nest early in the season because early nesters enjoy the most benign conditions for raising a family: not only do they hatch more of their eggs and fledge more of their chicks, but they produce more fledglings and their fledglings are the most likely to graduate. The numbers of chicks and fledglings produced in the year of graduation are breeding outputs whose importance speaks for itself. High scores on these two variables speak to graduates' skills in hunting, competing for partners, competing for space with colony neighbors, repelling predators, and coordinating efficiently in incubation and brood care with a partner while sustaining a schedule of productive hunting. Finally, and most conclusively, we compared total chicks and fledglings produced across the first 5 years of life and, with a smaller sample, the first 10 years of life, as well as laying dates over those periods.

The above battery of statistical tests, controlling for sex and year of birth, turned up not a single difference between former juniors and seniors on any of the listed variables, except that juniors tended to lay 7 days earlier than seniors, implying the superiority of juniors! As noted for the mortality analysis, we cannot discount the possibility that greater mortality of juniors in the chick phase of every generation left us with fewer duds in the junior sample than in the senior sample. Even so, I find it truly remarkable that fledglings who endured protracted hunger, beatings, psychological subordination, and chronic corticosterone as infants showed absolutely no deficits in the important variables we measured when they launched their breeding careers. This implies that former subordinates, both females and males, are just as healthy and strong, just as skillful and smart, and just as attractive to the other sex as former dominants.[7]

But that reading of the data could be wrong. Might it be that maintaining dominance over a sibling incurs such high costs in terms of energy or time, that the dominance-subordination relationship imposes similar costs on both parties? We might be failing to detect juniors' deficits in reproduction due to subordination because they are matched by deficits of seniors due to possible costs of establishing and maintaining dominance. Keeping up all that attacking costs energy and may reduce opportunities for feeding or resting. We always think of dominance as a privileged situation, but maintaining dominance can be costly to male primates,[8] and it could conceivably be costly to chicks. We should take into consideration though, that annual survival up to age 21 years was consistently as good in juniors and seniors as it was in singletons, implying that no long-term price is paid by either of the parties that grow up in a dominance-subordination relationship.[7]

Immunity, Old Age, and Intergenerational Effects

A bird's body, physiology, and behavior are so complex that there's literally no limit to the number of questions that could be posed in the search for former juniors' deficits. We could go on searching forever and never satisfy all the skeptics. So, after much fruitless scrutiny, at some point, we should be willing to accept that juniors are *roughly equivalent to former dominants* and garland them as champions of developmental resilience. Before calling it a day though, we tackled a final clutch of questions that took us into morphological, physiological, and intergenerational terrain, and probed into old age. In the first part, we asked whether female juniors graduate into breeders with deficits in their body size (skeletal measures), body weight, or immune system and whether their own infants show such deficits in the chick period. In the second part, we accumulated monitoring data over another 10 years before asking our database whether juniors' reproductive success declines in old age more than does seniors'. And we asked whether the fledglings produced by juniors are up to standard—fully capable of becoming adult breeders. Mightn't juniors produce substandard offspring unlikely to ever become breeders, as observed in captive songbirds that grew up in the midst of intense nest mate competition due to experimental brood enlargement?[9,10] To our delight, two of the findings from this clutch of studies made for an unprecedented natural history nugget!

The first part was the thesis project of a meticulous undergraduate called Momito (and named María Cristina Carmona-Isunza) who was undaunted by the prospect of spending 3 months on a waterless island making fine-grained measures of well over 100 breeding females. We had ringed them all 3 to 8 years earlier as junior, senior, or singleton fledglings. Momito and her crew of volunteer assistants hand-captured and subdued each frightened and unspeakably angry mother at the nest where she was caring for a young brood. Sitting or standing about 6 m from her nest, they weighed and measured her, and injected into the skin of her wing a plant lectin that would provoke her cellular immune system to inflame the tissue. Using a digital micrometer, Momito measured skin thickness to the nearest one-thousandth of a millimeter before injection and again 24 hours later. The 1-day increase in thickness is an index of immune system capacity. After measuring each mother's body and her inflammation, Momito would gather her in her arms with her neck extended for flight (the booby, that is), allow her a few seconds to take stock of her location, and launch her from about 6 m directly toward her nest. This rarely fails. The booby flutters down to its nest in a little cloud of dust and down feathers, seizes control, and turns to vigorously threaten its captor. If it's too frightened or disoriented for that, it makes a half-kilometer loop over the ocean before landing in its territory, after inspecting it from the air. It always returns and quickly attends to its brood by covering the chicks if they are small or standing beside them if they are large.

Momito made the same set of measurements on every mother's eldest chick at an age of either 5 days or 20 days, and also took one-twentieth of a milliliter of blood from the legs of all chicks in her sample at 15 days, for sexing them after amplifying the genes in our laboratory. To know how well they're growing, you need to know whether they are female or male.

Compared to former seniors and singletons, former juniors, although full-sized in skeletal dimensions, were roughly 8 percent underweight at 2 years of age, and the older mothers in the sample showed that they take 3–4 years to gradually achieve normal body weight. This weight deficit of these mothers was a significant finding: it could be the subordinate booby's Achilles heel, and it might easily have repercussions on reproduction. And yet we already knew that junior boobies nest just as early and produce just as many chicks and fledglings as seniors during the first 10 years of life. How then do former juniors prevent their 8 percent weight deficit from impacting their reproductive success during the 6 years that they are laboriously building body mass? We would answer that question 10 years later in the second part.[11]

In contrast, the cellular immune responses of mothers that were former juniors, and of those mothers' own young chicks (which grew normally), were fully up to standard. This is only one component of a complex immune system, so we should not overinterpret the data, but the inflammation of former juniors was no smaller than that of former seniors and singletons, and the same was true for their respective offspring, hinting that the three stresses of subordination in infancy may have no impact on a female's immunity in adult life, nor any intergenerational impact on the immunity of her offspring.

The 10 years of additional records of fledgling production by close to 1,300 males and females carried our monitoring of former seniors and juniors into the depths of old age, specifically from peak production of fledglings at age 10–12 years through continuous senescent decline until age 16 years. And juniors weathered decline in their faculties just as well as seniors, if not better! Among males, juniors produced just as many fledglings as seniors, and among females, juniors actually produced slightly more fledglings than seniors. This was a stellar repudiation of any who doubted that aging juniors would eventually be unable to hide the damage from their beleaguered infancy.[12]

But there *was* a deficit, a palpable deficit. Only it lay elsewhere: although junior fledglings were on all fours with seniors when it came to immunity, surviving, graduating into the breeding population, and successfully breeding well into old age, it turned out that the offspring they fledge at the very start of their reproductive careers tend to be duds. Fledglings produced by 2- or 3-year-old juniors of both sexes had only a 20 percent chance of ever becoming breeders, compared to a 40 percent chance for blue-foot fledglings in general.[12] The 8 percent body weight deficit that Momito had discovered 10 years earlier in young mothers who were former juniors does indeed have a consequence: while underweight, juniors produce substandard fledglings, presumably because their energy budget is not sufficient to care for chicks adequately. I expect that when their reserves are low, they can't go the extra mile on behalf of their offspring without putting their own survival or welfare at risk. To address this deficit, natural selection has a cunning plan though. Blue-foots seldom breed when they are only 2 or 3 years old, and those that do produce few fledglings; the average junior that graduates into the breeding population produces only 0.18 fledglings during its first 3 years of life. Hence, expression of the developmental damage that subordination does to juniors is, first, passed to the next generation and, second, consigned to an age where it probably has a negligible impact on former juniors' fitness. Not a noble strategy, but brilliant, and all achieved by unseeing natural selection!

Can Juniors Stand Up for Themselves?

It's highly unlikely that younger siblings who are beaten into submissiveness during the first 3 weeks of life and remain submissive over the next few months of cohabitation with their oppressors will ever be able to match the assertiveness and aggressiveness of their sibs. Right? After all, their training in losing is so effective that even when an oppressing sib is experimentally doctored to become smaller than junior, and junior responds by going on the offensive to invert dominance, most juniors buckle and capitulate as soon as they meet resistance. They yield; they can't help themselves (Chapter 3).[13] So should we assume that juniors manage to have reproductive careers as successful as those of seniors by handling competitive situations unaggressively, through alternative tactics? Well maybe, but maybe not. What was needed first was a formal experimental test of juniors' defense of their most precious possession—their current brood—in adulthood.

And for this, an ingenious undergraduate from the Universidad de Morelia— Oscar Sánchez-Macouzet—arranged for a fake booby to intrude into the territories of former juniors and seniors that were caring for their own chicks 5–13 years after fledging. Each of them had fledged alongside a single sibling, so it had been totally immersed in either subordination or dominance throughout its first few months of life. When Oscar arrived surreptitiously at each one's nest, it was alone on guard;

solitary parents defend their broods with all they've got, even against an advancing human, whose blood they are quite willing to spill. The intruder was "a stand-up, full-frontal, life-sized, fiberboard-mounted color photograph of a male adult intruder in an alert standing posture presented at a distance (90 cm from the subject) and location that mimicked a natural intrusion." Like those fiberboard, life-sized, smiling Kodak sales-lady photographs standing outside photography shops, but less alluring and more imposing. You may be skeptical, but it often doesn't take much to fool birds; a static bunch of songbird feathers can be enough to elicit aggressive mobbing or attempts at copulation.

Oscar sneaked into each booby's territory, fixed the intruder photograph to the ground, and crept to his tripod-mounted video camera, where he waited 5 minutes before tugging a cord to unveil the intruder. He filmed for 4 minutes. All 79 tested boobies took the intrusion very seriously, responding almost immediately with aggressive displays, including yes-headshaking and "menacing," some of them escalating to the point of wing-flailing, approaching the photo, and pecking at it. But no matter how the data were computed, there was no hint that the aggression scores of former juniors and seniors differed.[14] Even the oldest juniors of both sexes were every bit as fierce as seniors in defense of their offspring.

This experiment, incidentally, cast doubt on the hypothesis of siblings as disposable sparring partners, proposed by a South African field ornithologist, Rob Simmons.[15] He suggested that black eagles, which raise only one fledgling, may well lay their second, doomed egg *so that*, among other benefits, the elder chick can permanently bolster its aggressive prowess by attacking and killing a rival. Senior booby chicks seem not to bolster their aggressive prowess in later life by dominating their sibs.

There's an outside chance that a different test of aggressive prowess could reveal a relative weakness in former juniors, but I doubt it. Their proven ability to survive and produce viable offspring over their adult lifetimes demonstrates competence in the physical and social domains and qualifies them as fully equivalent to former seniors except regarding the viability of offspring produced in the first 3 years. This latter is the mote in the former junior's eye, the cost of second rank—an ounce of underperformance tucked away where it can do little harm.

Importantly, descriptive observations of blue-foot chicks enduring and recovering from a set of natural stresses revealed resilience that is a far cry from the experimentalists' demonstrations of contrived stresses disrupting long-term development and damaging infants' prospects for surviving and reproducing. As far as we can tell, after age 3 years, former juniors can go toe to toe with senior fledglings on immunocompetence, physical strength and confidence, sexual attractiveness, and probably fishing skills. This testifies to the ability of natural selection to craft resilience to sustained trauma in the most formative stage of life and raises the possibility that similar resilience to sibling bullying is commonplace in animals. But what does it imply for a species as cognitively and socially complex as *Homo sapiens*?

Resilience in *Homo sapiens*

Notwithstanding the blue-foot's ability to shrug off the developmental impacts of sibling dominance, I expected that, with their years-long psychological development, emotional complexity, and complicated social lives, many human adults would carry emotional scars from ill treatment by siblings in their infancy and youth.

As we shall see in Chapter 9, human infant siblings are competitive and aggressive and tend to establish asymmetrical relationships. A large minority of youngsters experience sibling bullying, defined as repeated unwanted aggressive behavior that intends to inflict harm either physically, psychologically, or socially and involves an imbalance of power.[16] The elder sibling usually prevails in the toddler and preschool years using physical coercion.[17] However, after age 4, even when altercations are frequent, actual violence progressively diminishes and becomes more sporadic, although it can continue well into adolescence. During adolescence, younger sibs emerge from their subordination and relationships become increasingly egalitarian.[18,19] One might expect that the first decade or so of two siblings' asymmetrical cohabitation is when the younger one experiences a level of oppression that could have a long-term developmental impact.

However, elder sibs have two additional tools of oppression that are potentially more damaging than violence and increasingly adopted during adolescence. They use psychological aggression to threaten, taunt, and denigrate, and also relational aggression, maneuvering socially to isolate, humiliate, and dominate. The scale of this tripartite aggression—physical, psychological, and relational—is understudied and hard to appreciate, but in a recent sample of thousands of 12-year-olds in southwest England, *on a weekly basis*, 31 percent of them suffered physical violence, 41 percent were called names, and 31 percent were made fun of by a sibling, among other abuses.[20]

Might these painful and humiliating experiences debilitate humans for the rest of their lives? Or at least into adulthood? I suspect the childhood chant "Sticks and stones may break my bones, but words can never hurt me" is the opposite of the truth, but how long does the hurt last? And where does it show? In the abusive relationships described by family therapist Jonathan Caspi,[21] younger siblings are overpowered and diminished, and the hostile language they experience can stay with them for the rest of their lives. However, the dyads that reach the attention of family therapists like Caspi are not broadly representative of the general population and, in truth, we're in the dark regarding the proportion of younger siblings that suffer prolonged trauma. We don't know whether the long-term negative effects of sibling aggression tend to be substantial, trivial, or nonexistent. But we know a little about sibling domination's medical consequences in late childhood and early adulthood and more about the effects of birth order on personality across the lifespan.

Sibling domination's damaging consequences for youths in the United States and Britain are not trivial. They include low self-esteem, loneliness, anxiety, peer difficulties, chronic interpersonal stress, depression, eating disorders, antisocial behavior, and delinquency. How far and in what form these effects persist into adulthood have been little investigated, but findings of a recent study of 24-year-olds that carefully separated out the effects of peer bullying (which often accompanies and is propitiated by sib bullying) were alarming. Men and women who suffered sibling bullying during childhood were more likely to be diagnosed with depression and also to experience suicidal ideation and suicidal self-harm.[20] Apparently, sibling hostilities can lead to a variety of mental health problems that have some potential for persisting into adulthood. However, we need to be cautious not only about persistence but also about prevalence and causality. We don't know about prevalence, and we cannot discount the possibility that sibling conflict and later mental health problems are both caused by an unidentified underlying condition, rather than sibling conflict affecting health.

But we do know something about effects of sibling interactions on personality. Possible effects of sibling relationships on personality and temperament in adulthood have been suspected and studied industriously for over a century. The *birth order* literature uses statistical techniques to identify systematic differences in the personalities of first-borns, second-borns, and later-borns and attributes the enduring differences to asymmetric social interactions between different-aged siblings, or to differential treatment by parents. It's a contentious field of study. Many fascinating effects have been discovered and then discarded after criticism because important variables such as age, sibship size, and socioeconomic status were not properly taken into account.[22,23] Other studies have failed to find effects,[24] or they have been criticized for reporting spurious effects on the basis of unrepresentative sampling, dubious statistics, or highly questionable measures of personality. For example, if the researcher gets their data by asking one sibling in each sibship to rate themself and their sibling, they're effectively inviting them to bias results in accordance with their own preconceptions about siblings.[25,26] However, a rigorous and statistically sophisticated study by a team led by Julia Rohrer of the Max Planck Institute for Human Development,[26] based on 17,000 British, German, and American women and men, recently yielded a compelling finding.[27]

Using self-evaluations based on standardized questionnaires composed of numerous items, the subjects scored themselves on each of the "Big Five" major personality traits that were identified in the 1980s and have maintained mainstream consensus among psychometricians and personality psychologists since then. Each trait represents a cluster of personality and temperamental facets grouped together under a broad name because they are correlated. The total scope of the Big Five seems appealingly comprehensive, so my intuition is that if sibling bullying in infancy in any way changes the bully's or victim's personality or temperament in adulthood, then the changes are likely to be captured somewhere among the five traits—Imagination, Conscientiousness, Extraversion, Agreeableness, and Emotional stability—and their numerous facets.

To give you an idea, scores on the *Extraversion* scale locate each individual somewhere between two extremes. Introverts are low-energy, socially unengaged individuals who tend to be quiet, low-key, and deliberate, not because of shyness or depression but because they are less dependent on socializing than others. Generally reserved and unassertive, they seek less stimulation than extroverts and are more comfortable alone than extroverts are. Extraverts are their high-energy, enthusiastic, all-action opposites, and they tend to be assertive and dominating. At the low extreme of the *Emotional stability* scale, neurotics habitually experience negative emotions, such as anger, anxiety, and depression, they readily understand situations as threatening, and they often perceive minor frustrations as hopelessly difficult. They are vulnerable to feeling anxious and pessimistic about their work, and they are moody. In contrast, those with high scores on this trait are typically calm, emotionally stable, and untroubled by negative emotions. They tend to be confident and resilient.

There's no doubt that these traits develop in us under the influence of our social interactions with others. Fully half of the individual variation in a population is induced by lived experience (rather than genetic inheritance), so one would suspect that a decade or more of cohabitation with a sibling might nudge younger siblings toward greater introversion and neuroticism, and elder siblings in the other direction. Even in the face of great variation among dyadic relationships, including those in which the younger sibling is dominant, and assuming that most nudges are of small magnitude, I expected Rohrer's sample of many thousands of adults 18–98 years old to reveal some consistent personality differences among siblings of different birth order.

But there were none. For none of the Big Five traits was there a clear sign of a systematic difference between siblings of different birth order. This unambiguous result demonstrates that human personality development is robust to the documented developmental challenges posed by hostile, manipulative, and overweening siblings. But couldn't it be that there *are* systematic differences between people of different birth order in adulthood, only they happen to lie in specific psychological variables distinct from the broad personality traits? The Rohrer team rigorously tested this possibility, too, using samples of 6,500–10,500 German women and men. Fifty-one years old on average, they had been interviewed to assess their life satisfaction, locus of control (feeling of agency), interpersonal trust, reciprocity, risk-taking, patience, impulsivity, and political orientation. All of which drew more blanks: scores on all eight variables were uninfluenced by birth order.[28] Similarly, a slightly less sensitive analysis of 377,000 American high school students failed to find any substantial evidence of birth order effects on personality at the end of childhood,[29] when effects of cohabiting with siblings might be expected to be at their strongest.

The notable impacts we saw of bullying on mental and emotional health in childhood and adolescence may be too uncommon, or fade too rapidly, to generate systematic personality differences in adulthood among younger and older siblings. True, before closing the file on the long-term effects of siblings, it would be prudent to explore additional psychological variables handpicked for their well-established relevance to sibling relationships. But thus far at least, the evidence suggests that human psychological development is less vulnerable to perturbation by siblings than many of us suppose or have been led to believe.

Resilience in Vertebrate Infants

Vertebrate infants may often be more resilient to stresses and challenges than biologists suspect. The researcher's perpetual search for "effects"—impacts of environmental variables on animals—likely drives them not only to pose unrealistic experimental challenges to animals but also to emphasize or preferentially publish studies that link stresses to deficits. An example of the latter was provided by observations in the natural habitat of that great workhorse of avian research, the great tit, the one-time poster child for negative effects of privation in infancy on performance in adulthood. Far more attention was garnered by the original study showing that underfed female fledglings lay few eggs when they graduate than by later studies showing that, curiously, the laying deficit doesn't impact their brood size, the growth of their chicks, or their reproductive success.[1] The two findings are equally interesting, but the first one—an effect—won more attention than the second one—no effect.

We certainly need to know about the vulnerabilities of infants to the environments they encounter, but it's also fitting to give animals credit for their resilience and take heart in it. Take the case of the blue-foot chicks that have the great misfortune to hatch in an El Niño year. On Isla Isabel, parents struggle to spend enough time attending their broods, chicks grow slowly, more of them die, and those that fledge are 15 percent underweight on average. This is a serious ecological constraint: each additional degree (°C) of ocean temperature in the natal year means a reduction of 50 percent in fledglings' probability of graduating into the breeding population. But then the graduates from warm-water generations breed just as successfully as graduates from cool-water generations, although differently. In a 10-year database study for his doctoral thesis, Cheko found that warm-water graduates survive as well, live as long, and produce just as many fledglings as cool-water graduates, and achieve this equivalence by following a different strategy: they graduate into the breeding population at an *earlier* age and take more frequent sabbaticals (non-breeding years).[30,31] Selective elimination of dud chicks and fledglings by El Niño, if it happens, might be part of the explanation for this equivalence, but we shouldn't lose sight of the fact that infants nurtured in extreme ecological austerity can nonetheless blossom into fully functional and productive adults. That's part of the story, too, and you may have heard it first from blue-foot juniors.

Indeed, close to half of the studies based on descriptive monitoring of wild bird populations over many years have uncovered admirable resilience of chicks of one or both sexes to environmental challenges that overshadow their development. As agriculture expanded into the forest habitats of the Mauritius kestrel, chicks suffered underfeeding, and young adults suffered greater mortality, but by increasing their early-life reproductive investment, females from agriculturally degraded areas ended up producing just as many offspring over their lifetimes as those in forested areas. Male great tits that grow up and fledge under poor conditions live *longer* and produce *more* offspring, and male tawny owls that have to put up with low prey density in their natal year suffer no deficit in their annual survival or reproductive success through at least age 12 years and do not age any faster.[1]

Most dear to my own heart, however, is the resilience of younger siblings of the blue-footed booby and *Homo sapiens* to oppression by more powerful siblings. As adults, they manifestly are not damaged goods. In both species, I suspect that the ancestral juniors whose development was least easily derailed by overweening sibs were those that prospered in spite of adversity, left more offspring, and contributed genes for resilience to the next generation. Understood in this way, the developmental robustness of blue-foots and humans reflects evolved adaptation to highly predictable challenges of family life. And after the moral shock of Hamilton's and Trivers's postulation of fundamental conflicts of interests between relatives, it's strangely gratifying to conceive of that robustness as part of a coevolved solution to the conflict between relatives who benefit indirectly from each other's good fortune.

The next chapter is the first of three that are dedicated to a different domain of family cooperation and conflict: partnerships between adult females and males. Like humans, blue-footed boobies mate monogamously, and every reproductive event requires the extended commitment of two individuals to an extraordinarily demanding project. It involves evaluating, choosing, and winning over a partner in the colony mating market, bonding with that partner, agreeing on a nest site, and coordinating joint contributions to incubating eggs, feeding chicks, and defending the pair's territory 24 hours a day for several months. This would be quite a task even for two birds whose fitness interests are completely convergent, but they're not. Females and males differ substantially in body size and capacity to provision chicks and, critically, their fitness interests overlap only partially because, in principle, each can benefit by exploiting the other.

7
Happy Marriages with Blue Feet

Social Monogamy

Of course, boobies don't get married. No pair of boobies makes explicit vows before witnesses, and no authority gives blessings, issues certificates, or enforces promises. Like humans though, boobies form reproductive partnerships. The tongue-in-cheek reference to marriage encourages readers to compare booby and human partnerships and ask themselves whether any similarities in behavior might imply similar causes, functions, and mechanisms. Through this exercise, we may better understand blue-foots, and they can help us gain perspective on our nature and our place in the animal world. It takes humans centuries to assimilate conceptual revolutions like realizing that Earth is not the center of the universe, and most humans are still wrestling with the implications of Darwin's shocking discovery that *Homo sapiens* is one among millions of evolved animal species. Specifically, we are a bipedal, manually dexterous, tool-using, cognitively endowed hominine ape that commonly breeds in socially monogamous partnerships resembling those of many birds. Moderating our fond belief in human exceptionalism and understanding our evolved reproductive behavior for what it is, is a work in progress that can only benefit from thoughtful, open-minded comparisons with species whose lives are similar. For this, blue-foots are a particularly handy model.

The similarities between blue-foot and human reproductive partnerships will be discussed more fully in Chapter 9. For now, it's worth noting that blue-foots evolved to reproduce in monogamous pairs and that this is the most common pattern in human societies, although the extended family often helps with childcare. In both species, pairs of partners invest jointly and intensively in the care of highly dependent offspring for several months (boobies) or years (humans) until they become independent. Boobies move on to their next reproductive event a year or two later, while successive events of humans tend to be more widely spaced but overlap in time because children mature so slowly. In both species, partnerships generally arise through a period of courtship that bonds the two breeders and commits them to reproduce together, and deviations from this standard pattern are common, including extrapair relationships and copulations (sexual relationships with additional mates), mate-guarding, and partner-switching (deserting one partner for another).

Our focus here though is on boobies. This chapter will present a cameo of the blue-footed booby's socially monogamous reproduction. Centered on the relationship itself, it will focus on how two essentially selfish individuals whose interests do not

Blue-Footed Boobies. Hugh Drummond, Oxford University Press. © Oxford University Press 2023.
DOI: 10.1093/oso/9780197629840.003.0008

entirely overlap collaborate closely to achieve together what neither could manage alone. This enterprise involves evaluating and choosing a partner, adjusting to each other's qualities and dispositions, arriving at joint decisions, and coordinating relative contributions to the care of offspring. The chapter will also outline the improved coordination and reproductive gains that blue-foots achieve when they take collaboration to the next level by prolonging their partnerships over more than one reproductive event. Chapter 8 will modify our perspective by shining a bright light on the amoral sexual infidelity, deception, and countermeasures that are rife in booby partnerships, so we can appreciate the complexity and strategic cunning of boobies' deviations from strict social monogamy. Here, I won't belabor, and will seldom point out, specific similarities between human and booby partnerships—we'll be tripping over them anyway; reflection on similarities and differences between the two species will be the subject of the book's last chapter.

Before we get down to the nitty-gritty of how blue-foots construct and manage their monogamous relationships, let's be clear about the ambitious mission of every partnership and the built-in tension between the two partners' interests. Over a 6-month period, while daily spending enough time hunting to maintain their bodies and finance their immense investment in reproduction, blue-foots must acquire a quality territory; bond with a quality partner; lay a clutch of eggs and incubate them for 6 or 7 weeks; defend territory, clutch, and brood against neighbors, intruders, and predators; and feed chicks round the clock until well after they outweigh their parents at age 10 weeks. To achieve all of this in the midst of a hostile colony and ever-changing ocean, continuous collaboration with a partner is essential. No booby could possibly raise a clutch and brood by itself, and the partners that reproduce most successfully are surely the ones best able to coordinate their contributions.

However, there's a fly in the boobies' collaborative ointment: as far as each partner is concerned, besides being its bosom buddy, the other is an indispensable resource that should, when convenient, be exploited. For example, the amount of work, wear, and risk each partner needs to shoulder depends on how much the other shoulders, leading to a fundamental sexual conflict of interests.[1] Just as there is a conflict of interests between siblings and another between parents and offspring, so is there a conflict between male and female partners. Over the millennia, males that managed to unfairly shift workload onto their partners will have produced more offspring over their lifetimes than males that play fair, enabling their unfair genes to progressively replace fair genes throughout the population. Likewise, females should have evolved to exploit males. This pitting of the two collaborating sexes against each other in every domain of contention—including but not limited to incubation, brooding, and provisioning—should result in the evolution of stable solutions, in which either the male wins, the female wins, or compromise is achieved. Victory by fait accompli occurs habitually in some bird species and opportunistically in others when one sex— usually but not always the male—deserts after the brood is established, leaving his partner holding the baby while he dedicates himself to his next family. Mowgli found that magnificent frigatebird males do this on Isla Isabel when their single chick is just a few months old. They suddenly desert, leaving their female partners to provide all care during the last 9–12 months, while the males prepare themselves to reproduce with a different partner.[2] More often though, the evolved outcomes of fundamental conflicts of interests between male and female birds over workload and other issues look superficially like harmonious agreements, and we are left guessing which partner's interests have prevailed.

Keep in mind this vision of booby partnerships as finely honed collaborations based on mutual dependence but potentially affected by conflicts of interests, as we examine how males and females evaluate each other, form partnerships, switch partners, make decisions together, and divvy up their workload. The Isla Isabel colony is no Garden of Eden, and the separate evolution of males and females for selfish personal advantage, smooth collaboration, and mutual exploitation has resulted in intriguing inequalities and interactions that demand ecological and evolutionary explanations. I'll finish the chapter by suggesting that males provide far less food to chicks than females because, historically, females shot themselves in the foot by forcing males to become smaller!

Choosing a Partner

Blue-footed boobies owe much of their charm, not just to their surprising blue feet (booby species with red feet and yellow feet have fewer followers), but also to their lovable courtship antics, and to the background buzz in their colonies of scores of bonding pairs, along with a sprinkling of unattached females and males, grunting and whistling as they intently strut their stuff. In my tent, I often awaken chuckling as dawn excites a groundswell of ritualized displaying around the campsite. Just as I'm easing into consciousness, there they all go again: all through the forest boobies are tirelessly posturing and vocalizing, switching each other on, and loudly announcing the colony's urgent imperatives to court and copulate, go hunting and push back those insufferable neighbors. Despite the amusing garb and moves, it all looks terribly serious, and of course it is, because the colony is a hive of competitive adults doing their utmost to impress, cajole, and outdo each other—uncomprehendingly, I believe—in the service of every individual's lifetime fitness.

In the majority of birds and mammals, males are larger and more elaborately decked out in colors, ornaments, and weapons than females, a difference mostly attributed to *sexual selection*: competition among males to win and fertilize females. We're all familiar with the massive thoraxes of silverback gorillas, the imposing racks of antlers on bellowing stags, the exquisite and absurdly large tails of strutting and strumming peacocks. To win the favor of females, males need all that bulk and gear to intimidate and defeat their rivals or to directly impress the females themselves, so they invest heavily—in energy, time, and risk—in gaining access to females, and only lightly in actually caring for offspring. This can pay off in spades for an impressive male when he manages to sire the offspring of several females, but tends to leave other aspiring fathers out in the cold and often lands females with all or most of the childcare.

Boobies don't fit this stereotype at all! Just look at them: blue-foot females' brilliant feet, fetching displays, and resonant, intimidating voices are different but right up there with those of blue-foot males. And that's to be expected in a species where it takes two full-time caring parents to raise a brood and, consequently, *both* sexes stand to gain by choosing a high-quality partner who will make a weighty contribution to the offspring's development.[3] Since male and female blue-foots are both more likely to be chosen if they can convince members of the other sex that they are worthy partners, sexual selection—a major subcategory of natural selection, identified by Darwin—has equipped both sexes with the means to convince. According to theory, male and female boobies with impressive displays get preferentially chosen as partners, and they tend to produce more or higher quality offspring, who themselves are likely to bear the genes for impressive displays and the genes for being impressed by them. Through sexual selection, the genes for displaying and for responding to displays propagate and become established in the population, and individuals of both sexes give priority to pairing with the highest quality partner that will have them.

In theory, a high-quality booby partner would be one with good genes to pass on to chicks or one that provides several months of good parental care to them, and an ideal partner would cover both of these bases. So a male or female who hooks up with such a partner can expect to produce more or better-quality offspring, and maybe avoid depleting itself and its reproductive prospects for the following season. Partner choice is key to success, so at the start of each season, both sexes invest heavily in displaying impressively and scrutinizing the displays of candidate partners. First, males stake out territories in the colony and defend them against neighbors and intruders, while females meander through the colony on foot interviewing potential partners by exchanging courtship displays with them. These interactions may last just seconds or minutes, but when there's mutual interest, courtship goes on for days or weeks; if the courters begin early in the season, they can be at it for a couple of months!

Finally, it's the female who sets a seal on the partnership by laying a clutch, and then you know the two of them are bonded for the whole 6-month project. It comes as a lovely surprise. For a day or so she has been fooling our monitors by crouching on her territory as if she had laid an egg, but when she stands up, there's nothing there. Then one day, a distant monitor spots her standing upright with an egg on the ground between her feet, sky-blue and so pristine it's almost glowing. The male is nowhere to be seen, but in an hour or two he'll return from the ocean, and a few minutes later she'll allow him to incubate, so she can take her turn fishing.

Both sexes show off, bond, and stimulate each other's sexual readiness using the repertoire of 10 terrestrial and aerial displays and other minor elements that have won over so many naturalists.[4] For the iconic "sky-pointing" display, which signals sexual arousal, a standing booby partially extends both wings while rotating them forward at the shoulder until the deep-brown, upper-wing surfaces face the other booby; stretches its neck vertically upward with the beak pointed skyward; and slowly pulsates its crooked wings back and forth while vocalizing gently and seductively. A "parading" booby walks or stands on the spot while deliberately raising its feet high to display the blue webs. Having walked on land in flippers myself (not recommended!), I know that wide stiff webs force anybody into a high-lifting, widely spaced gait, but even knowing that, blue-foot parading looks ostentatious and is more exaggerated than their usual waddle. Bouts of solo, reciprocal, and mutual sky-pointing and parading can go on for minutes, the synchrony increasing as each bout progresses, the solemnity of it all never faltering. The two partners' gentle bill-touching, bending necks over each other, and mutual grooming are suggestive of tenderness. Their greetings, on the other hand, suggest passion. Returning from the hunt, a male or female greets its partner with excited "yes-headshaking" as it lands on the shared territory, and intermittently thereafter as it catches the partner's eye. Their frenzied nodding as they rush at each other, whistling and grunting as they lean forward and rattle their bills together, speaks and spells joy: to all appearances, they're pleased as punch to be together again.

Fig. 7.1 Female blue-foot sky-points at male during a sequence of reciprocal courtship. He's attentive to her posture, her protracted grunts, and probably the color of her feet.
Drawing by Jaime Zaldivar-Rae; based on photo copyright Pablo Cervantes.

Fig. 7.2 Simultaneous sky-pointing by male (left) and female (right) blue-foots.
Copyright Pablo Cervantes.

Close to the very start of courtship, before an incipient pair of blue-foots coordinate in bouts of displaying on their territory, they can sometimes be seen "flighting." The two of them repeatedly take off, fly around the zone of the territory, and land together. For the male this can be a culmination of the period when he first establishes control of the territory through a combination of solitary flighting and aerial "saluting" with his feet (see below) and, of course, aggressive on-site displaying to male neighbors.[4]

What do we guess, then, about their tender feelings? On the basis of his extensive, sensitive, and thoughtful observations on a flock of greylag geese, the Austrian co-founder of ethology, Konrad Lorenz, described animals cackling passionately, falling in love, and finding their love reciprocated.[5] Impressed by similarities in greylag pair-bonding and human pair-bonding, and undaunted by the impossibility of entering into the actual experience of other species, Lorenz was even more incautious about inferring animal emotions than Charles Darwin. If he had witnessed booby courtship and bonding—so similar in important ways to what he described in geese and cichlid fishes—he would surely have inferred love, or at least deep affection, in blue-foot partnerships, too. There's a key difference though: whereas greylags' bonds persist through the nonreproductive season and last for life, boobies' bonds expire when their fledglings depart (but can be renewed the following year). My own guess is that the rules of thumb and extensive rituals involved in establishing and maintaining a bond are backed up by emotions and that these motivational mechanisms, whatever they actually feel like, are the blue-foot equivalent of human romantic love.

Fig. 7.3 More blue-foot courtship displays: parading by male (left), bill-touching by male and female, and neck-bending by male.
Photos by Hugh Drummond.

When copulation eventually comes, usually during a bout of sky-pointing and parading, it's a serious business demanding gymnastic poise lest the male topple from the female's back. She braces herself in a standing posture with wings folded as he steps precariously onto her back, facing forward, and balancing by flitting his wings and treading her feathers. Thus steadied, he curls his tail end under hers to silently press his cloaca against hers for a second or so, all it takes for his ejaculate to pass. After promptly stepping down, he often sky-points, in satisfaction it seems, and the two partners may continue displaying for a while. Maybe they're just reinforcing the bond, but there could be more to it. He might be persuading her to put extra nutrients in the next egg or to give priority to his sperm, in the event of her reproductive tract containing ejaculates of a rival. There's no question of copulations ever being forced, as occurs in mallards; not only are female boobies bigger and stronger than males, but the dicey gymnastics during ejaculate transfer depends on collaboration.

Animal courtship displays demonstrate the displayer's sterling qualities—health, immune status, and physical prowess—and vouch for his or her potential contribution to the offspring: good genes and valuable parental care. But it's probably not just about information. The repetitious and insistent nature of courtship displays implies that boobies coax each other into committing to the partnership and investing generously in parental care.[6–8] Striking postures, parasite-free plumage, and loud vocalizations all potentially inform and persuade and probably prime for reproduction, but we strongly suspected that the most eloquent organs of a female or male blue-foot are its feet, with their variable hues. They are not ordinary feet. The webs are highly vascularized because boobies, lacking a brood patch (patch of bare skin on the belly), incubate with their webs, wrapping these over the eggs or settling the eggs on top of them. If their hues reflect an individual's health and other qualities in a way that cannot be faked, webs could be exactly what's needed for evaluating candidate partners and making decisions about bonding with them and investing in offspring.

Probing the significance of web color fell to Alberto Velando, a researcher at Spain's Universidad de Vigo whom I met at a behavior conference in Granada, Spain, and Roxana Torres, who investigated the sex ratio of booby offspring for her doctoral thesis and is currently on our own faculty. One of this productive team's crucial early findings was that a male's feet do indeed encode information on his current status that should be of more than passing interest to females. After noticing that male feet faded during a 2-day storm, the next year they hand-captured males who were actively courting females in the colony and isolated them in cages on the forest floor for 2 days. Some they hand-fed 200 g of fresh bonito fish every day, others they fed 1 mg of carotenoids (pigments with antioxidant and immunostimulant function), and still others got the premium package, bonito plus carotenoids, while the remainder got the short straw—they simply fasted.

During the first 2 days, the hue of the fasting males' feet changed as caging took its toll, shifting from (presumably attractive) light turquoise toward a dull darkish blue. This darkening was less marked in males supplemented with fish or carotenoids and still less so in the lucky few who got both. The causal link between nutrition and

Fig. 7.4 Blue-foot copulation. Teetering on her back by briefly bracing his tail on the ground and levering his beak against hers, the male inseminates her with a cloacal kiss, and then touches her beak affectionately with his before dismounting and parading contentedly. Oddly, in this pair, the male is larger than the female.
Drawing by Jaime Zaldivar-Rae; photos by Hugh Drummond.

web hue was nicely confirmed when the fasting males recovered their light turquoise after being fed. Moreover, a link between hue and an important index of health was shown when males with the most handsome pale turquoise feet showed the highest cell-mediated immune responses.[9] In sum, the foot color of male boobies varies over the short term with nutrition and testifies to their current physical and immunological condition. Waved under the beak of every female interviewer, a parading male's webs exhibit real-time information on his quality as a sire and his aptitude as a caretaker of chicks.

Male feet also encode information on another two timescales. Using a spectrometer, Alberto and Rox analyzed the web colors of 84 courting males we had banded at fledging, up to 16 years earlier. Our database contained their precise ages and their complete reproductive CVs, allowing tests of how age and reproductive effort affect web color. What would you expect? Maybe color improves gradually with maturation and then declines after age 11 (when breeding success peaks), in synchrony with the rise and fall of a male's ability to produce fledglings? Or color might steadily decline over the whole reproductive lifespan, as age and effort progressively take their toll on the body. And what about sabbaticals? It wasn't known whether male blue-foots skip the occasional reproductive season because they fail to secure a partner and a territory, or because they are spent and choose to take a year of rest and relaxation. An evaluation of web color after sabbaticals might give us an idea.

It turned out that the handsomeness (green chroma) of male feet declines linearly over the lifetime, well into old age.[10] It's hard to know whether age itself or reproductive effort—the wear and tear of pairing and raising broods—is responsible for the deterioration because age and effort are tightly correlated. Either way though, a male's age/experience can be read from his webs, and such estimates could be of considerable interest to females. For example, young females might do best by choosing experienced partners, if these are better carers, whereas females of a certain age might do better by teaming up with youthful males, whose genes could be in better shape (more on this in Chapter 8).[11] And if, as in some avian species, female boobies prefer their extra mates to be old, some females could even prefer to pair with young males to reduce the risk of their partners getting lured away. Whatever it's used for, age information is there on every male's feet.

As for sabbaticals, it turned out that indulging in one of these raises the attractiveness of a male's webs to celebrity level among his own age class during the next year's courting and bonding.[10] Precisely how this enhancement benefits males or their partners we don't know, but it's plausible that skipped breeding seasons are partially or amply compensated by increased reproductive potential the next year, or by an increase in longevity. From the point of view of unattached females, it could be very handy to hook up with males who've just taken a year off if they can shoulder extra brood care.

Finally, as if nutrition, age, and sabbatical recovery weren't enough information to encode in those famous webs, it also turned out that their color gives an accurate clue to the reproductive benefits a male can provide to his chicks. This, too, is crucial.

For a female it's not enough that a male's feet signal great reproductive potential; she needs to "know" she can count on him to come up with the goods: his beauty needs to be tied to benefits for the two partners' chicks (or for the female herself). Alberto and Rox tested this by swapping chicks between pairs of contemporaneous broods just after hatching and then weighing the chicks at age 15 days. The swap would reveal whether the growth of a chick is correlated with the foot color of its foster father or of its biological father. A yes to either of those questions would prove that females can improve the growth of their young chicks by pairing with colorful males; and whereas a correlation with the foster father's webs would suggest that colorful males are better at *caring for* chicks, a correlation with the biological father's webs would suggest that colorful males provide their offspring with *better genes*.

Studies of adopted human infants have provided similar opportunities to compare the influence of parents on their children's development through their childcare practices versus through their genes. There, numerous analyses based on quantitative genetics have posed a serious challenge to orthodox opinion by demonstrating that in their growth and some psychological measures children are very similar to their biological parents and hardly similar at all to their foster parents.[12] The staggering implication is that in ordinary families (with no adoption) resemblances between parents and their children in body size, cognitive skills, temperament, agreeableness (altruism, caring, kindness), and grit are due to genetic inheritance rather than child-rearing practices. The exceptions are measurable effects of foster parents on intelligence and school achievement during childhood, but these disappear during adolescence except for effects on performance in the humanities, which persist on a modest scale through the university years. Amazingly, although parental care is undeniably of great importance in humans, the different ways different human couples raise their kids have a negligible effect on their physical growth or the aforementioned psychological characteristics. On the other hand, biological inheritance tends to account for roughly half of the variation among individuals in all of those variables.

Boobies are different, however, because parental genes and brood care are both important. Any behavioral ecologist would expect care-providing adults to substantially affect the growth of their chicks, and it turned out that, for fathers at least, that is true. Growth of the swapped chicks was positively correlated with both fathers' foot colors, but foster fathers were three times as influential as biological fathers![13] So booby females could improve the growth and probably the reproductive prospects of their offspring by choosing partners whose feet are pale turquoise because those males bring to chicks both high-quality care (plenty of food?) and good genes.[14,15]

So male webs reveal useful information on nutrition, health, age, and sabbatical experience, and pairing with a male who flashes pale turquoise can boost the growth of a female's offspring. So far so good, but a third crucial question must be answered: Do females actually base any of their reproductive decisions on the web color of males? For all we know, sensory or other limitations have prevented natural selection on females capitalizing on what looks like a handy technique for increasing reproductive success—bonding selectively with turquoise-footed males. Note that even if females tend to end up with turquoise-footed males, they might get there by ignoring web hue and basing their choice on any characteristic correlated with foot color (e.g., out-of-this-world sky-pointing). Answering the question confidently requires experimentally manipulating color.

Alberto and Rox didn't feel technically qualified to enhance the beauty of male feet, but they reckoned they could sabotage it. To this end, they daubed the webs of courting males with water-resistant blue make-up, making them resemble the dull webs of undernourished males. Then they released the daubed males where they had captured them and watched as they continued courting with their partners. As expected, daubing degraded male attractiveness; the males themselves were no less eager about courting, but their partners courted them less intensely and copulated with them less often.[16] Female blue-foots do indeed base their responses to males at least partly on web hue, but whether they inherit their specific preference for turquoise webs or acquire it by interacting with different-colored males remains to be tested. Preferences we are tempted to call innate, congenital, or instinctive often develop in the course of an animal's social or other interactions; they are instinctive and learned.

Guess how females responded when the male's webs were daubed blue on the day *after* his female laid her first egg! With an egg in the nest, it could be too costly to abort the breeding attempt, and too late to satisfactorily switch to a different partner, but what can a mother do when her partner's health or vitality suddenly collapses *after* starting a family together? Should she soldier on regardless? Or cut her losses? Reduced investment in the next egg was the females' answer; they laid undersized second eggs.[17] It's hard to know how short-changing the second egg could actually benefit females (apart from saving on the cost of egg formation), but maybe they were abruptly winding down an unpromising breeding attempt, all the better to breed successfully next year.

How the information on a male booby's nutrition, carotenoids, immune status, age, and reproductive experience is *combined* in his feet we don't know. It's conceivable that his scores on all those factors are summed in the hues of his webs, as were the scores on fish ingestion and carotenoid ingestion. Or maybe discrete information on each of the variables is more useful to females than pooled scores. And don't forget that feet provide only a part of the information that males furnish. Females might benefit by interpreting foot color in conjunction with the information contained in a male's voice, the motor coordination of his displays, and the quality of his territory. Some sort of integrated evaluation of his health, experience, nutritional condition, commitment, and real estate holding would be ideal, but whether boobies (and other animals) can manage such integration is a mystery.

Similar considerations surely apply to the feet of females, too. Female web hue is a sexually selected social signal,[18] and it deserves just as much if not more attention because, as we shall see, mothers not only make the eggs, they are the main providers to chicks.

Switching Partners

Given the catalog of qualities that a booby would ideally consider when choosing among potential partners, coupled with its information-processing constraints, you might expect to see some ambivalence, vacillation, and mistakes over partner choice. This is especially likely to be so in the busy pairing market of a colony, where unattached males are staked out on clustered territories, lone females are meandering through, members of bonded pairs are slyly courting with neighbors, and the social composition of each neighborhood fluctuates each day as its temporary and permanent members alternate on their own schedules between wooing and hours-long fishing trips. The highest quality candidate partners are few, and most unattached boobies will have to be satisfied with the best candidate that will accept them. But what should they do if the partnership goes poorly or the array of candidates in the neighborhood changes?

Evidence for great instability in some boobies' partner choice and bonding was incidentally documented by two graduate students in the course of their independent theses on blue-foot infidelity. Marcela Osorio-Beristain pulled off a record three successive theses on the Isla Isabel blue-foots before becoming an ecology researcher at the Universidad Autónoma de Morelos, and Diana Pérez-Staples completed her master's thesis on blue-foots before morphing into an insect pest specialist at the Universidad Veracruzana. In the forest beside our campsite, their modus operandi was decidedly ethological. In busy neighborhoods of unattached and bonding blue-foots, they and their volunteer assistants alternated 2-hour shifts in stuffy cloth blinds, systematically scoring all comings and goings, courtship, copulations, and altercations. Critically, they knew who was who because, by that point, most of the local boobies were banded and their numbers could be read through binoculars. They worked from dawn to dusk for weeks at a stretch, bearing systematic witness to behavioral interactions that in other species have to be inferred from pairing outcomes and molecular paternity analyses because they can rarely be observed.

Although there were many stories of straightforward loyalty—seamless bonding progressing to clutch and brood care—6 percent of the pairs they watched experienced the turmoil of partner-switching before the female started laying; that is, desertion followed by pairing with a different individual.[19] Around our camp dining table, these entertaining and revealing tales are known as "bobonovelas," after the local name for TV soap operas (telenovelas).

Before recounting Marce and Diana's bobonovelas, let's consider some of the panoply of extant theoretical proposals for why natural selection might favor monogamous birds switching partners.[20–22] Adaptive switching could happen because a male or female discovers that their partner is of poor quality or finds that the two of them are incompatible and likely to coordinate poorly on the demanding project they are about to undertake. Rather than making do, the switcher deserts its unpromising partner and keeps its options open, in the expectation of finding an upgrade.

Alternatively, even when a partner is of satisfactory quality and coordination is going well, a better-quality candidate shows up in the neighborhood, so the booby, male or female, opportunistically and adaptively switches to the newcomer. More Machiavellian and potentially more profitable, a male bird might desert his partner after fertilizing her, leaving her to raise his progeny with another (unsuspecting?) male while he pairs with a new female and raises a second brood.

In any given species, the same few conditional switching strategies could coexist along with loyalty in the repertoires of every male or every female, each strategy being preferred, and, if possible, adopted, only in the particular circumstances where it is expected to outperform all of the others. Unsurprisingly, none of the proposed switching strategies is terribly romantic or noble, but we can readily imagine how individuals who used them would end up producing more, or better quality, fledglings than individuals who are inflexibly loyal to their bond. Evolution of the strategies is therefore credible. All of them theoretically increase the fitness of the male or female that implements them, without regard to the interests of the deserted partner, which ordinarily are irrelevant to the deserter. Just think about it: which strategy would outcompete the other, selfish maximizing of personal fitness through switching or self-sacrificing loyalty to the partner? Other things being equal, the latter strategy loses because it restricts the propagation of the genes supporting it into the next generation: genes for loyalty find themselves in fewer offspring bodies than genes for selfishness.

Marce and Diana witnessed five males initiate switches by deserting their partners at a time when they were courting and collaborating with those females in the defense of each pair's joint territory. Three of the males had hitherto been faithful but switched shortly after observing their partners court and copulate with an extra male. Some of those copulations with extra males occurred during the 5 days before laying, at a time when the females were almost certainly fertile, so the copulations threatened the paternity of the switching males. All three males deserted their partner and territory before switching to a novel partner, and every one of those desertions was preceded by an additional adverse event that cast doubt on the original pair's prospect of success—loss of freshly laid eggs to predatory gulls during incubation change-overs. Love lives are complex! But the simplest interpretation may be that faced with the infidelity of their partners and the risk of raising another male's offspring (an absolute no-no!), the three males switched to new partners. This situation may have been aggravated by egg loss, or the breakdown of the relationship may have precipitated the careless loss of eggs.

The other two males made their switches in quite different circumstances. Both deserted their current partner in order to pair with their current extra female. In the first case, the male's partner had been faithful but he progressively committed to his extra female by spending more time courting with her on her nearby territory and consummated the switch before his first partner laid any eggs. The other case was similar but complicated by simultaneous infidelity of the male's partner and some paternity-protecting infanticide. That's right, this was a bonded pair in which both parties simultaneously had extra partners, and the male destroyed all three eggs laid by his partner before finally decamping to the nearby territory of his extra female.

In both cases—the two males who switched from their partners to their ongoing extra partners—the apparent trigger for desertion and switching was not precisely the female partner's infidelity but the onset of laying by the male's extra partner. So I can't help suspecting that each male held on to both of his females until the extra one satisfactorily laid a clutch in whose paternity he had confidence; only then could he switch without risk of ending up without a family.

And, call me a misogynist, but I can't help suspecting that the two extra females in these last two cases had "cunning" game plans. Could it be that a female who has no partner sometimes resorts to establishing her own territory (not a common tactic) next to the territory of a paired male, lures him into an extra-pair relationship, and then lays in her own territory *before his partner lays*, to induce him to switch? Not intentionally, for I suspect the requisite understanding and planning are beyond a booby, but female and male boobies may be equipped with naturally selected behavioral urges—rules of thumb—that usher them, step by step, into unwittingly poaching other boobies' partners.

Marce and Diana also witnessed desertion and switching by two females. The first of them suddenly disappeared from her neighborhood after 5 days of steady courtship, copulation, and territorial defense with her partner, leaving him on the territory, where he continued courting and copulating with other females. The second of them deserted the same male twice, in quite different circumstances. The first time, when she was bonded to him and copulating regularly with both him and an extra male, she just took off, leaving them both behind, and wasn't seen for 38 days. When she reappeared in the neighborhood, she was accompanied by a new partner and they defended a territory together, but the new partner got chased away by her original partner, with whom she re-paired. Occupying and defending their original territory, these two copulated daily during 16 days, and she also copulated with two unattached males when her partner was absent. During this period, this femme fatale (courted, in addition, by several unattached males) seemed ambivalent about her bond, attacking females who courted her favorite unattached male but helping her partner expel him. Finally, she and that unattached male disappeared permanently from the neighborhood on the same day, leaving her partner behind. The common pattern in these three female switches is the female deserting her partner and territory in order to establish elsewhere with an existing extra male, seek a new male elsewhere, or maybe postpone laying until she is ready. Did the females actually switch? With incomplete narratives and only three cases, female desertion and switching remain a most intriguing mystery. (Remember these females when we get to female hunter-gatherers with multiple partners in Chapter 9.)

And the boobies deserted by the seven switchers? None of them remained on their own. All five deserted females paired anew, the four unfaithful ones taking only a couple of days to pair with their extra-pair males, and the faithful one switching gradually to a new male as she was progressively abandoned. The two deserted males bonded with new partners, too, one of them taking just 2 days to get started and the twice-deserted male taking 32 days the first time and just a couple of days the second time.[19] There's plenty of action in the colony pairing market.

These bobonovelas resonate with us because we can so easily map onto their narratives the human emotions, temptations, and entanglements that populate our songs, movies, and literature: romantic love, unrequited passion, jealousy, and anger, not to mention bonding, duplicity, rejection, betrayal, and desertion. Only here, the protagonists are animals that may well lack imagination or intentions and, although some of their behavior looks familiar, we can only guess at the feelings that may accompany it.

Sticking with the Same Partner

At the start of each reproductive season, blue-foots decide whether to bond again with last year's partner or move on to a new partner, assuming both of them survived through the winter in good enough shape to breed. Two ex-partners should be able to locate each other in the colony because every year boobies prospect for mates in the vicinity of their own first nests. Indeed, roughly half of each season's pairs re-pair in the next season, and pair-bonds are sustained over as many as nine consecutive seasons, presumably because there's an advantage in sticking with the same partner. Many species of bird re-pair, and the most credible explanation is that individuals who have bred together coordinate better and enjoy greater reproductive success. Familiarity through working together on a demanding project might enable them to produce more fledglings, fledge them earlier in the season, or end the season in better shape. But how do you test the familiarity effect when individual breeding experience, breeding experience together, and age are all correlated? If you confirm that partners with more experience together breed more successfully, their success could be due, not to experience together per se, but to being older birds or more experienced birds. After his experimental study of whether subordinate chicks grow up to be less than brave parents (Chapter 6), Oscar Sánchez-Macouzet took on this question for his doctoral thesis by analyzing a special sample of adults from our long-term monitoring.

He found 752 males and females in the database who were making the fourth breeding attempt of their lives, but who differed as to whether this event was the first, second, third, or fourth consecutive event with the current partner. Controlling statistically for age and for the calendar year of the attempt, Oscar's statistical analyses revealed incremental benefits of breeding with the same individual in several successive seasons. First, the longest-bonded partners laid their clutches a full 5 weeks earlier in the season, a big plus for any breeder because chicks that fledge early are more likely to develop into reproductive adults. Second, the longest-bonded partners hatched a greater proportion of their eggs and consequently produced more fledglings. Both results testified to a positive effect of repeat breeding on breeding success.[23]

But maybe the individuals that re-pair most often are the highest quality boobies, and their high quality rather than their long bonds explains their breeding success? Not so! Detailed scrutiny of the boobies' CVs in the database, starting with their hatch dates and size at fledging, turned up no evidence of them being superior boobies, so we could conclude that repeating with the same partner was what made them so successful. When partners meet up on the colony at the start of a new season, familiarity is what catalyzes their efficiency. Familiarity allows them to collapse mate choice and bonding—all that displaying and mutual evaluation—into a shorter period. And familiarity eases the strains of turn-taking during incubation.

There is, however, a drawback to repeating with the same partner, at least for males. As we'll see in the next chapter, many paired females are unfaithful to their partners in the run-up to laying, and some males end up raising the offspring of male rivals, to the serious detriment of their own fitness. Might males whose partners know them well get duped more often? After getting to know her partner's vulnerability to deceit or discovering his shortcomings as a partner or father, a female might be more able or more motivated to have an alternative male sire her offspring. If so, repeating with the same female increases a male's risk of spending 4 or 5 months raising another male's offspring. This may sound unlikely, but a sample of 384 families from a single season revealed that the risk increases from 9 percent on two partners' first pairing to 22 percent on the second, and it declines on subsequent pairings.[24] The mechanism is speculative, but, to judge from a single season, males run the greatest risk of raising an offspring that isn't theirs the second time they breed with the same female. I doubt that this cost of re-pairing outweighs the benefit of producing more offspring and fledging them 5 weeks earlier, but that's a calculation for another day. Oddly, males that raised other males' offspring were no more likely to divorce afterward than males that raised their own. Maybe duped males never find out.

With so much to be gained by sticking with the same partner, you'd think the average bond duration in the Isla Isabel colony would be longer than 1.7 years, but separation may be unavoidable when a partner dies or takes a sabbatical, and nearly so when two partners return to the colony on different dates. Waiting for a partner to show up could be contraindicated because time is of the essence. Late clutches are less likely to produce fledglings, late fledglings are less likely to become breeders, and absent partners may be dead. It's probably better to find a new partner and get on with it.

Joint Decisions

For a newly bonded pair, few decisions are more momentous than where, within their territory of several square meters, to lay the clutch. Not only do bothersome boobies transit past the territory or reside on its borders, but the understory is littered with rocks, tree trunks, and branches and harbors a community of irksome land crabs, iguanas, and ticks, as well as prowling snakes. Guano-drenching frigatebirds have favorite perches in the canopy above, and gaps in the canopy allow patches of sunlight to creep across the leaf litter and gulls to scan that litter for vulnerable clutches. Predation, scavenging, parasitic infections, and temperature regulation are all at stake down there, as well as intrusions and infanticidal attacks by neighbors. No wonder decision-making can drag on for weeks! But whose decision is it? Casual observations of other bird species have suggested that the female and the male settle on a site through a process involving individual evaluations and communication, but the decision-making process itself, and the roles and balance of power between the partners, are obscure.

Who better to decipher this process in blue-foots than Judy Stamps, a renowned behavioral ecologist at the University of California at Davis who grew up in the ethological culture of painstaking observation of wild and captive animals? I met Judy at my first Animal Behavior Society annual conference and was truly taken aback by her intellectual effervescence. She is more curious, busy, and quick than any whiptail lizard, has one of the sharpest and most wide-ranging analytical minds in behavioral ecology, and has distinguished herself by meticulous observations and experiments that systematically laid bare how juvenile anole lizards on Caribbean islands resolve conflicts over territorial boundaries. And I didn't need to invite her twice! Judy was delighted to camp on a tropical island and grapple with joint decision-making of boobies; and she brought to the task not only Californian camping accessories and chocolates but a promising conceptual framework. Coupled with the right observations, this framework might reveal how nest site–choosing partners collaborate or contend with one another. Elaborated by social psychologists,[25,26] the dual concern model *successfully predicts strategies used by human dyads to make joint decisions over the use of space and other resources.* No one had applied it to any other animal, but in Judy's judgment, it could illuminate the decision-making of a marine bird.

According to the model, blue-foots are a shoo-in for using *collaborative tactics* to choose a nest site because a bonded female and male have *broadly overlapping interests*—fledging a few viable chicks, basically—along with a few discrepant interests. For example, at different potential nest sites, the two partners may be differentially favored or prejudiced by the opportunities for sneaky liaisons with opposite-sex neighbors. In this situation, the model predicts compromise solutions to discrepant preferences rather than win-lose solutions, and these are expected to be achieved by extensive communication of personal preferences. To this end, both parties should use exploratory tactics. The first tactic is "feeling out"—signaling strong preference for an option only after confirming its acceptability to the other party. Both parties should retain a power of veto—if one expresses dislike for an option, the other takes it off the table. And if both can veto, then disagreement over initial options should be followed up by "expanding the pie"—identifying and considering additional options.

In contrast, when two parties don't have broadly overlapping interests, the dual concern model predicts that each party should use *contentious tactics* that increase the probability of the pair selecting an option that favors one party without regard to, or at the expense of, the other. Thus, two options are contemplated: two parties can work it out together, through communication, or alternatively one can just impose its will on the other. Which of these would you expect to see in a pair of bonded boobies, bearing in mind that both are essentially selfish and females are bigger and stronger than males?

In 1997, Judy's first season on the island, the analysis of contentious tactics—for example, aggression, escalation, and submission—was the daily bread of many be-havioral ecologists, for whom conflict over resources is a central issue of animal behavior, while collaborative tactics were more of a mystery and likely to be more subtle. Identifying and distinguishing between the different decision-making tactics used by animals *that cannot speak* is quite a trick, but Judy was not intimidated by the task. With the enthusiastic assistance of two thesis students, biologist Carmen Pérez and psychologist Miriam Calderon, she plunged in.

As the team settled into long hours watching the signaling of dozens of blue-foot pairs near the edge of the forest, we were on the edges of our seats. Would boobies, undistinguished for any cognitive abilities, prove to have diplomatic skills? The dual concern model is not mathematical and generates only qualitative rather than quan-titative predictions, but in the light of Judy's preliminary observations and identifi-cation of relevant behaviors, we had a sporting chance of success. Statistical analyses of behavioral frequencies, contingencies, and contexts would hopefully allow us to infer whether blue-foots have or haven't evolved to make joint decisions by announ-cing their preferences, weighing the preferences of their partners, and resolving dif-ferences by satisfying their partners' concerns. Not what you expect of brutes, but bonded pairs of birds who coordinate closely for months at a time fit the bill and might be up to it. There would be no accounting of benefits and no fitness analyses, but we would at least discover whether the tactics used by booby partners are con-sistent with a well-established diplomatic approach to decision-making.

A key early discovery was that a booby expresses its preference for a potential nest site by "nest-pointing," a previously unreported form of communication. The booby points its beak downward at the ground with the tip moving laterally and vertically at a height of roughly 15 cm, often while grunting or whistling contentedly. Two part-ners will point, singly or together, at up to nine potential sites in their territory over many days, often alternating between sites. Tellingly, neither of them points when alone on the territory or after the first egg is laid: nest-pointing is for communication about where to lay. The average pair investigated 3.5 potential nest sites in their terri-tory before finally laying in one of them, and they nearly always chose the one where they had pointed most frequently; the two of them pointing there with increasing frequency as laying approached, especially the female. We don't know what, if any-thing, boobies think when they are pointing. I imagine they simply feel positive about the site, and that they enjoy removing debris from it with their beaks, or seeing it look cleaner; I doubt that they actually imagine a clutch there, although experienced breeders might have that advanced cognitive capacity. How I'd love to get inside a booby's mind! Anyway, pointing indicates a positive urge or inclination toward a can-didate site.

Fig. 7.5 Nest-pointing female and male partners simultaneously signal their agreement on the acceptability of a candidate site.
Drawing by Jaime Zaldivar-Rae; photo by Hugh Drummond.

Scrutiny of the boobies' behavior for collaborative or contentious tactics was enabled by establishing a criterion for when two partners agree on the merits of a site. The lag in days between pointing at the first site chosen by either partner and pointing there by the other partner was taken to be a measure of the initial agreement on the merits of that site, being only one-quarter as long for sites that were finally chosen. It turned out that pairs who disagreed about the merits of the first site they considered (the lag was long) were much more likely to expand the pie by investigating additional sites. Moreover, there was a marked tendency for individuals to point maximally at a particular site only after their partner had begun to point there, consistent with feeling out. And there was evidence that both partners have the power to veto the other's preferred site: although many pairs had sites approved (pointed at) by only one of the two partners, virtually none of those pairs laid in such a site.

As for clearly contentious behaviors, the team detected none. Boobies did not, for example, interfere with each other's nest-pointing, and there was no evidence that boobies use escalation, a contentious tactic, to prevail in dispute: there were no shouting matches with each partner pointing maximally at its own preferred site at the same time. Despite having no language (communication using symbols and syntax), the boobies used collaborative tactics similar to those used by humans to resolve differences between dyads with broadly overlapping interests. And their *attitudes*? Endearingly, they maintained their dignity and courtship display rituals throughout the pre-laying period, appeared almost gracious as they stated their inclinations, and attended to the inclinations of their partners. No crockery was ever thrown!

What about sexual equality? Did the male rule the roost in any sense, despite his small size? At the outset, an egalitarian spirit was evident. The sites that got the eggs were just as likely to have been proposed in the first place by the male as by the female. However, some female bias arose over the course of the negotiation: in every pair where more than one site had been at least cursorily approved by both, it was the site most popular at the end with the female, and not necessarily the male, that won. This isn't surprising; it's the female who lays the eggs.

There is a curious anecdote suggesting that by the time laying must begin, the male is on board with the female's final decision even if he personally prefers another site. In one pair where each partner preferred a different site right up to the time when laying was imminent, the female inexplicably disappeared from the territory, never to return. (Did she die out at sea?) The male remained on their territory, alone. Animals can be so highly motivated to do something that they'll do it even if the requisite props or circumstances are unavailable, and that male placed a little rock in the center of one of the two finalist sites and incubated it there. Nothing extraordinary about that because our monitors occasionally discover clutchless boobies incubating little rocks, not to mention an empty plastic film capsule. Once, even a panicking hermit crab! Significantly though, this male placed his rock and incubated it in the site preferred, not by him, but by his partner. I didn't say it out loud to Judy, but the word "gentleman" popped into my mind.

Female versus Male Care of Offspring

Sit and watch any blue-foot colony one day (recommended!), and the boobies will leave you thinking that between laying and fledging, female and male share all of their duties roughly equally. But how could that be if the two birds differ so much in size? Males are one-third lighter than females and correspondingly smaller in body length, wing length, and beak size; although their tails, the rudders and brakes of their aerial agility, are just as long as females' tails.[27] Body size affects pretty much everything an animal does, so it would be intriguing if male blue-foots can contribute similarly to the care of offspring despite their wimpy size. To weigh the roles of males and females in their collaboration and grasp the nettle of how those different roles evolved, first we needed to measure their respective contributions after territory establishment by males and laying by females. We'll compare their contributions to round-the-clock incubation and brooding, and the amount and types of food they bring to the brood and finish that accounting by pointing up the male's curious inflexibility over his unimpressive provisioning effort. Then we'll be ready to tackle the vexed issue of how it could possibly benefit males to be so small.

After watching blue-foots with his wife during 4 months on Hood Island in the Galapagos, Bryan Nelson floated the idea that male and female blue-foots spend roughly equal time incubating eggs and brooding chicks except during the first month of the chicks' lives when males are the main food providers and females do most of the brooding.[4] The idea was that very young chicks need small and frequent food transfers, and males are better equipped than females to provide them because their small size and relatively long tails facilitate plunge-diving in the shallow waters near the colony. Only he can make the frequent local fishing trips needed to provision small chicks, and only she, with greater cargo capacity, can make the long foraging trips needed to satisfy big chicks during their last 4 months of growth. Economically driven division of labor between the sexes is an appealing idea, and it offers an ecological explanation for why males evolved to be small. Despite slim empirical support, the idea is plausible and it has lingered. But would an idea based on preliminary observations in the Galapagos Islands hold up on Isla Isabel?

When they launched our program on Isla Isabel, Alicia and Cecilia sampled dozens of nests containing eggs or chicks on nine 24-hour days spread across 3 months, noting every hour on the hour which parent was incubating or brooding. Disappointingly, there was no evidence of a division of labor during the day, the night, or the whole 24 hours, at least not during either incubation or the first 2 months of the chick period. On Isla Isabel at least, females and males spend similar amounts of time attending their offspring, even in the chicks' first weeks of life.

Defense of those offspring is another matter. Successful reproduction depends on maintaining continuous control of the family territory for at least 4 months, so both parents, acting alone or together, attack and chase away predators and intruding boobies. Boobies are the lesser danger, but expansionist neighbors try to force back territorial borders and sometimes destroy unguarded eggs or kill chicks that wander across borders. I expected that females, being more formidable than males, would defend more willingly and effectively, but over the years, my casual observations on couples facing off with neighbors at mutual borders have not confirmed this. Each couple stands side by side, fencing with the couple in front by repeatedly jabbing at them and yes-headshaking into their faces with open or closed beaks, their mandibles vibrating so fast they blur. It's frenzied aggression, ritualized but potentially damaging, too; sequences of bouts can go on for an hour or so, and occur day after day, testing strength and perseverance. Some boobies have an empty socket where an eye once looked out. Surprisingly though, males seem to stand their ground and gain ground just as well as females.

And to cap that, two formal quantifications of behavior proved that blue-foot males are *more aggressive* than females. First, when Mowgli and another undergraduate watched 86 clutches and broods for nearly 1,000 hours during the severe El Niño event of 1983, males attacked intruders earlier in their approach and 50 percent more frequently than females.[28] Second, 24 years later when Oscar planted his standup photograph of a booby in the territories of scores of boobies with young chicks (Chapter 6), his meticulous behavioral measures revealed that through at least the start of old age, males threaten and attack intruders with twice the determination of females. In the absence of any observations, descriptive or experimental, showing that

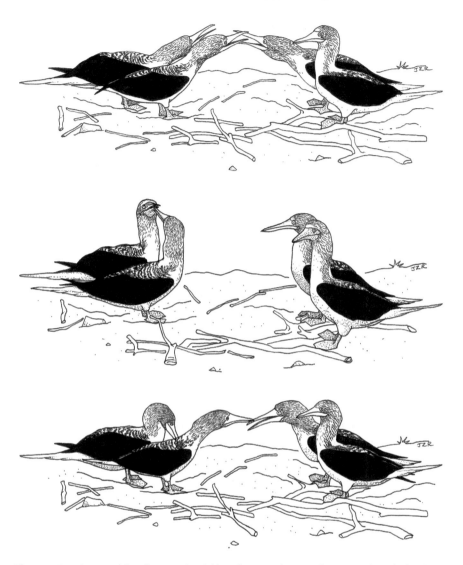

Fig. 7.6 Tension at a blue-foot territorial border. For about an hour, two bonded pairs face off tensely at their mutual border, straining for territorial gain by separately or simultaneously jabbing and yes-headshaking into their neighbors' faces, occasionally reassuring each other with mutual beak-touching.
Drawings by Jaime Zaldivar-Rae; based on photos by Hugh Drummond.

females defend more impressively or effectively than males, the weight of evidence is that male blue-foots defend against intruders more aggressively than females. Maybe they have to, to compensate for small size.

When it comes to repelling *predators*, there was no credible evidence that the sexes differ, but no one had looked either. On Isla Isabel blue-foots defend against gulls, iguanas, milk snakes, and biologists; they use vocal threats and yes-headshakes and, at close quarters, jabbing, pecking, and biting, as well as seizing and tossing. During monitoring, I believe I've been intimidated by females more often than by males, and it is females that bite harder and threaten so loudly and insistently that I'm sometimes tempted to back off, just to get some peace. Loud threatening calls can wear you down. One female, in particular, amazed me on my fourth visit to her nest by incorporating a detour in her aggressive defense; she ran in a wide arc around the forked stick I held out to keep her at bay and pounced on my bare legs. Where did she learn that trick? Then she startled me on my sixth visit with a fluttering leap *over* the stick as she growled and lunged at my chest.

That's just an anecdote, though; to conclude that females are more aggressive than males takes data. What's more, a counter-anecdote spoke to a surreal, verging on saintly, level of restraint in another nesting female. We routinely disarm the boobies we capture by folding an arm around their wings and feet, to constrain wing-beating and forearm scratching, and we always—but always—firmly grasp the top of the booby's neck behind the cranium, to prevent the head swiveling and biting. We hold the head at waist height so that no sudden escape convulsion can release the bird's head and beak within biting distance of anyone's face. Then—and don't ask me how this came about because I'm still dumbfounded—there I was one day in the colony training Tania del Rio, a gentle biomedical student, to take blood samples from boobies' wings, when it dawned on me that a big female was cradled *totally unrestrained* in Tania's arms, her long neck sweeping tensely but freely from side to side just centimeters from Tania's smiling, unconcerned face. I all but swallowed my clipboard. The female's passivity was unbelievable because seized boobies instantly struggle and bite for all they're worth, growling with the effort. I was paralyzed for about 10 seconds (probably just a couple, really) before I could blurt out a hollow-voiced warning to Tania and secure the female's head and neck. But there you are: a fearsome female snatched from her nest by the world's most dangerous predator can be too discombobulated to defend herself.

Eventually, our question was answered in an undergraduate thesis. When Santiago Ortega tested how the intensity of an incubating or brooding booby's defense against a predator varies with its age, a comparison of females and males was a welcome spinoff.[29] In the trials, Santi ably played the part of the predator, walking slowly toward each solitary parent on its nest while a volunteer followed discretely in his tracks, recording the parent's reaction. Defense against Santi by the 150 sampled adults varied from headlong panicky flight when he was still several meters away (low intensity), to pecking at him on his arrival at the nest (high intensity). As expected, the intensity increased steadily with maturation and experience through age 10 or 11 years, and then fell off progressively to zero by age 25 years. Old age gets to boobies, too. To our surprise, however, sex made no difference to defense against a predator at any age: males were no less ready to flee, no more willing to defend, and no more violent in their defense than females.

It was in relation to feeding of chicks that our expectations of female superiority were highest, although we were mindful of Nelson's suggestion that males' agility might give them an edge during chicks' first month. To make this comparison, we watched broods of two chicks continuously from dawn to dusk, noting the time and frequency of all parental food transfers and weighing both chicks at the start and finish of every day, every hour on the hour, and just after each feeding bout. Nine broods were watched daily from hatching through 10 days, and 21 broods with a spread of ages between 12 and 35 days were watched on a single day. This sampling enabled us to estimate rates of weight loss between meals and the amount of food transferred by each parent to each chick at every meal. It laid bare the poverty of provisioning by blue-foot fathers. Whether measured in grams of fish or frequency of transfers, mothers and fathers provisioned roughly equally during the chicks' first week of life and highly unequally thereafter, with females providing two to three times as much food as males overall. Unless these proportions change late in the chick growth period, which seems unlikely, mothers are the main providers and, on Isla Isabel at least, there is no chick age when males provide more food than females.[30]

Small body size is a reasonable explanation for males' decidedly lean contribution to the family economy. Males very likely work as hard as females, but they generally supply less food to the brood because their limited cargo capacity obliges them to make shorter fishing trips, bring smaller quotas, and in all likelihood provision less efficiently. Might males compensate for quantity with quality, providing more nutritious food to chicks than females provide? We have a good idea about that because Cheko, in an outburst of analytical fervor one day, resolved to tackle the dietary samples accumulated in the first 5 years of the blue-foot program. It may sound unsavory, but field biologists often store and get a lot of analytical mileage out of materials excreted from animals' bodies. Fishes regurgitated by a minority of frightened parents on the approach of monitors were always going to be a valuable (if not marketable) asset one day, so Alicia and Cecilia established a tradition of dutifully preserving the partially digested specimens in alcohol. Years later, student volunteers extracted the precious booty from the 312 plastic bags where it was stored and measured all the fish. They got no further, but they had done the nasty part; then Cheko rose to the challenge of delving into the nutrition literature and comparing the regurgitations of males and females.

Engraulid fishes (mostly Pacific anchovies) and clupeid fishes (mostly Pacific thread herrings) prevailed in the samples, and although fishes of these two families have similar protein contents, clupeids offer more to chicks because, weight for weight, they contain 16 percent more calories.[31] But it was females, not males, that brought a higher proportion of clupeids to their broods in all oceanographic conditions; for example, respectively, 24 percent versus 0 percent in an El Niño year, and 42 percent versus 23 percent in a La Niña year. Large body size, it seems, enables female blue-foots to provide their chicks with more and better-quality food, whether the ocean is warm or cool.

Small body size probably impedes provisioning by males, but might it enable them to step in and save the day during food crises? To buffer against such contingencies as injuries, illness, underperforming partners, or storms, it would be handy to have some flexibly managed reserve capacity. The two sexes could differ in capacity, or they could work to different rules of thumb, and Nelson suggested that there could be circumstances in which males' small size permits them to outperform females at provisioning. Maybe when the going gets tough, it's the males who get going? It was a long shot, but worth looking at in our largely fruitless quest for relative male strengths.

Alberto addressed this question through experiments on Isla Lobos, a Pacific island more than 4,000 km to the southeast of Isla Isabel. Just 19 km off the coast of Peru, this colony of 12,000 blue-foots may be the world's largest. At nests with two hatchlings, Alberto handicapped either the mother or the father by clipping 2.5 cm from the tips of its flight feathers, a manipulation that increases the energetic cost of flight by 5 percent and is reversed naturally at the next annual molt. The desired effect was underfeeding of the brood: a sudden food crisis demanding extra feeding effort by one or both parents. To find out which parent would compensate by increasing its hunting effort, 6 weeks after clipping he compared both parents' body weights with those of control parents at nests where neither parent was handicapped. The sex that lost more weight would be the one that volunteered to compensate for the shortfall in provisioning.

Whether it was the mother or the father that got clipped, it was the mother who lost weight.[32] Mothers responded by working harder, and this intervention resolved the crisis by reducing the chicks' growth deficit.[33,34–36] In flagrant contrast, whoever got clipped, blue-foot fathers lost no weight, presumably because they couldn't or wouldn't sacrifice themselves to save their chicks' lives or development. So much for counting on males when the going gets tough! Might this result have been an artifact of temporary mutilation rather than underfeeding itself? Probably not, because when Alberto contrived underfeeding using a different technique—adding a chick to two-chick broods—the results were similar: mothers but not fathers lost weight, as did chicks.

When underfeeding of chicks is an issue, a blue-foot mother compensates by flexibly increasing her provisioning effort, but a blue-foot father works rigidly to rule, leaving his partner and his offspring to suffer the cost. There's a revealing exception though. When Alberto *reduced* brood size by removing one of two hatchlings, fathers gained weight while mothers stayed the same, implying that when for any reason food demand falls off, while mothers continue working as before, fathers flexibly reduce their effort. Fathers can show flexibility when it's in their own interest! I suspect small body size not only makes males less capable provisioners; it may also make them more vulnerable to starvation, limiting their scope for occasionally working overtime.

So far, our balance sheet of female and male relative contributions to investment in offspring is decidedly lopsided. Mothers and fathers participate similarly in incubation, brooding, and repulsion of predators, but mothers provide food in far greater quantity and better average quality than fathers, and only mothers are prepared to resolve provisioning shortfalls by working overtime. On Isla Isabel at least, fathers outperform mothers in relation to only one duty; they repel more vigorously boobies that intrude on their territories. I expect that males, being small, can only match females' repulsion of intruders by working harder than they do.

Why Then Are Males So Small?

The elephant in our room is the conundrum of small body size in males. Why did such an apparently disabling characteristic evolve? In some hawks, males are small because it is fathers who provision the offspring (while mothers brood) and small individuals of those species are more adept at capturing swift prey. In blue-foots though, there is little division of labor and small body size seriously limits the amount of food males can provide to their chicks, apparently without enhancing any of their contributions to the family, except possibly expulsion of intruding boobies. Other things being equal, we should expect the largest males to bring the most food and father the most or best-fed fledglings, and, provided body size is heritable (an uncontroversial assumption), genes for large male size ought to be progressively accumulating in the population. Males ought to be larger than they are, and getting larger by the generation unless some powerful force is constraining that evolution.

The question is, which selective force could be strong enough to limit males to a body size that cripples their hunting capacity and their ability to provision offspring? It's no small matter that fathers provide to chicks only one-third to one-half as much food as mothers. Just think of all the junior chicks that every year are doomed to elimination by hungry siblings, but which would fledge alongside those siblings if only fathers could match the provisioning of mothers! Well, the most plausible answer is *sexual selection*. It took me years to get there, but I was finally persuaded by a provocative hypothesis: the selective force that has famously burdened male gorillas, stags, and peacocks with oversized bodies, weaponry, and trappings that hamper their normal activities has also shrunk blue-foot males into compact, underperforming fathers who can barely feed a brood.

In the most familiar vertebrate examples, sexual selection outfits males with bodies and displays that give them access to females by intimidating or defeating rivals, or by impressing females with their magnificence. However, Jehl and Murray proposed that in a peculiar minority of avian species males get chosen by females,[37] not so much for their might or belligerence as for their aerial acrobatics.[38] These males demonstrate their health and athletic prowess, and in some species testify simultaneously to their ability to hunt and the quality of the genes they will pass to offspring, by executing spectacular aerobatic displays, including tricky and energetically demanding maneuvers. But for males of some species there can be a catch: aerodynamic models and flight energetics tell us that displaying longer or more skillfully is facilitated by small size, so females who choose the most aerobatically proficient males for courting, pairing, or mating ipso facto select for small male size.

Confirmatory evidence for males evolving small size through sexual selection comes from field observations on male dunlins (a species of wader) showing that the smaller the individual the faster his rate of aerial displays and the more hovering flights he performs.[39] But the real clincher came at the turn of the century with comparative studies of scores of species showing that among shorebirds, alcids, and gulls, as well as marine birds generally, male use of aerobatic displays goes hand in hand with small body size relative to females.[40,41]

No link has yet been sought between the body size and aerobatic prowess of blue-footed boobies. Designed mainly for flying fast and straight over vast, obstacle-free stretches of horizontal water, and plunging into shoals of fish, blue-foots are aerobatically inept to the point of suffering dangerous, and sometimes fatal, collisions with trees when they fly through the garlic pear forest. On the other hand, Nelson was impressed by the athleticism of males performing dramatic aerial salute displays early in the breeding season, before their partners lay. Significantly, this behavior is frequent in males but rare in females. In Nelson's admiring account, "males fly in low to their territories, whistling loudly. At the last moment the tautly spread feet are flung up into the air, their soles at right angles or more to the ground, and held there until it looks as though the bird couldn't possibly recover in time to pitch down, but he does so with a flourish. Usually the feet are held in front of the underparts, but sometimes stuck right out to the sides.... The salute places the largest possible area of ultramarine web against white underparts and by jerking the feet makes them flash conspicuously.... Since this elaborate landing has nothing to do with the aerodynamic requirements of the situation ... it is undoubtedly ritualized behavior derived from ordinary landing. Its extreme development in the male is related to his unusual lightness and agility."[4]

When Nelson wrote that, the sexual selection explanation for male aerobatic agility and small size in shorebirds had not been proposed, and he suspected that saluting is primarily a territorial display directed at other males. When I reread his description of saluting after discovering Jehl and Murray's sexual selection hypothesis, I was intrigued but not persuaded. At the time, it seemed preposterous that sexual selection would shrink males down to a size where they can scarcely put food on the table for their offspring. Surely females would be selected to prefer males who can at least feed their offspring properly! How could capricious sexual selection for a mere display triumph so stupidly over natural selection for the ability to bring food to chicks? Then those comparative studies at the turn of the century twisted my arm; I understood that when female birds choose males based on aerial agility, males evolve to be smaller, and I swallowed the idea that this happened in blue-foots *despite male shrinkage surely prejudicing the reproductive success of both partners.*

Female choice is the filter that sexually mature blue-foot males have to get past in order to produce any offspring, so female preference for the most aerobatic males places downward selective pressure on male size and, it seems, has driven that evolution to the point where males are now too small to contribute more than a substantial complement to females' provisioning of chicks. There surely is selection for large male

Fig. 7.7 Saluting display of male blue-foot alighting on his territory.
Copyright Pablo Cervantes.

body size during the brood-rearing phase of reproduction, and conceivably during courtship, when larger males might have greater control over copulations, but this has been successfully counteracted by stronger selection for small size at an earlier phase when females choose their partners. A peculiar balance has been struck in one of those bizarre evolutionary outcomes of sexual selection that we would not be discussing if an intelligent designer were in charge of creation!

Egalitarian Asymmetry

Every blue-foot pair comprises a female whose body size was naturally selected to shoulder the burden of feeding up to three chicks, and a male sexually selected for agility and (just because it's correlated) also for smallness, who consequently can bring only a fraction of the food brought by his partner. The upshot is a dramatic departure from the more symmetrical parental contributions we might otherwise expect from two partners who need each other to sustain 24-hour care of offspring for months at a time.

Disharmony and inequality in the female–male relationship could potentially have arisen from another quarter: sexual conflict of interests. As mentioned earlier, conflict is expected over the two sexes' relative contributions to parental care because, in theory, either can benefit by selfishly inducing or obliging the other to shoulder more than its fair share. I actually expected that the larger sex should come out on top in sexual conflict because females are more formidable. What's more, they're expected to have a negotiating advantage in partnership formation and division of labor because they are less abundant than males,[42,43,44] which makes them a limiting resource.

However, the collaborative relationship between females and males does not align well with this expectation; it is more egalitarian. First, it's notorious that attraction and choice of mates follow a similar course in both sexes. Females and males display similarly to demonstrate their quality, even if females skip the saluting part, spending as long as 2 months evaluating, choosing, and bonding with a partner. Prominent in the displaying of both sexes is the flaunting of webs that provide reliable information on recent nutrition, immunological condition, age, and recent reproductive activity in males, and quite likely also in females.[18] Moreover, both sexes sometimes exercise the option of switching to a different partner. Second, females and males enjoy similar status in decision-making. Discrepant preferences among potential nest sites are resolved, not by contention or imposition, but by a collaborative pantomime in which preferences are expressed and weighed by both partners until an agreement is reached. There's no bullying by females. Indeed, the spirit of this nonlinguistic dialog is egalitarian and diplomatic, although females can't help having the last word because it's they who designate the site by laying the eggs.

Female and male boobies are boxed into an asymmetrical partnership by the quirk of sexual selection that downsized males, and they both make a good fist of it by sharing responsibility for all aspects of parental care roughly equally. True, there's a necessary division of labor at the very start of the season when the male stakes out a territory and the female lays the clutch, but thereafter the two parents apply themselves with similar dedication, if not equal effectiveness, to care of the clutch or brood. Males bring the brood considerably less food, with slightly lower protein content, only because small body size constrains their hunting, and, oddly, males defend against intruders more vigorously than females. The willingness of females but not males to shoulder extra provisioning when parental supply to the brood wanes I put down to the greater provisioning capacity of the larger sex, although it might also reflect male victory in sexual conflict over relative contributions.

Although the relative contributions of females and males to reproduction are distorted by sexual selection, their relationship includes no obvious friction, aggression, or coercion. Essentially, it's an asymmetrical but egalitarian relationship between a female and a male who differ markedly in size and strength. Both sexes dedicate themselves equally to care of offspring because, in contrast with magnificent frigatebirds, it would be counterproductive for either blue-foot partner to leave the other holding the baby just to get a head start on its next attempt at reproduction. A deserter would surely lose more fitness through prejudice to the current brood than it would gain by making a head start on the next. I have been describing the two sexes' contributions to breeding as "fair," but let's not overinterpret. The fairness arises from the evolution of behavior that selfishly promotes each sex's own fitness, not from the evolution of moral concern, which boobies almost certainly lack. Blue-foots nicely illustrate how fair, and conceivably *loving*, relationships can evolve in amoral and probably uncomprehending beasts when the care of highly dependent offspring by two partners is indispensable, and how sexual selection can make such relationships lopsided.

But this rosy overview of the nature of blue-foot social monogamy is deliberately incomplete. Sometimes there are three or more participants in a booby partnership. We turn in the next chapter to the fascinating and perplexing issue of blue-foot sexual infidelity.

8
Cheating, Infanticide, and Egg-Dumping

In the mid-1960s, when I was studying law in Bristol, it was well known that the great majority of bird species reproduce monogamously, with two partners laboring mightily to feed, defend, and fledge a whole brood of chicks. Exceptions aside, the female and the male were thought to share feeding duties rather equally and were models of parental devotion to their sons and daughters. Compared with human families, with their affairs, illegitimate children, and paternal desertion and abandonment, birds offered exemplary models of family loyalty and harmony. Why would it be otherwise in nature, in beasts untroubled by original sin and moral frailty? There were some exceptions, of course, and Peter Marler,[1] a distinguished ethologist, had reported male chaffinches sneaking away from their own incubating partners, foraging silently and inconspicuously through the undergrowth, and copulating with the fertile paired females they encountered. But few ethologists gave such observations any importance; in an imperfect world, some behavioral aberrations and pathology are bound to occur and should not be overinterpreted.

Then the myth of widespread faithful monogamy in birds was shattered by a series of developments in evolutionary theory and molecular biology as the Hamiltonian revolution gathered momentum. At the time I moved on to teaching English and exploring Mexican landscapes, Robert Trivers,[2] then a precocious doctoral student at Harvard University, published a paper with a clutch of seminal ideas on the evolution of strategies for sexual reproduction. Among these, he argued that natural selection on males should often favor the evolution of a *mixed reproductive strategy*, partnering monogamously with one female while applying some time and energy to fertilizing other females on the side. Likewise, selection should favor males and females engaging in strategic desertion and cuckoldry whenever these tactics favored their own fitness, and outmaneuvering partners who used such tactics to their own detriment. Why wouldn't selection smile on cynical, self-serving strategies that boost the actor's individual fitness? It's not about high-minded fairness; it's about the differential success of genes at getting into the bodies of the next generation, and genes for self-serving behavior are more likely to be winners than genes for fairness or generosity.

By the time I completed my research on reptiles in the United States and Panama and returned to Mexico as a behavioral ecologist in 1980, the concept of sexual conflict had taken hold, and many colleagues were persuaded that whenever male and female interests diverge, for example, in relation to mating, fertilization, or division of labor in parental care, selection on each sex would favor it pursuing its own selfish interests. Inevitably, behavioral conflict between partners would follow and

Blue-Footed Boobies. Hugh Drummond, Oxford University Press. © Oxford University Press 2023.
DOI: 10.1093/oso/9780197629840.003.0009

eventually be solved by coevolution (Chapter 5), leading to victory by one sex or some sort of compromise.[3,4] Then in the 1980s molecular biologists invented genetic fingerprinting, a methodology that empowered field researchers to identify offspring that are not the progeny of their putative father and sometimes put the finger on the biological father, by analyzing blood samples. Now the truth about avian monogamy could come out!

These developments motivated and enabled behavioral ecologists to probe the sex lives of wild birds, and they uncovered, over a period of decades, a wealth of previously unsuspected reproductive tactics. This was a sea change, from a vision of partners reproducing in harmonious lockstep to the benefit of the species, to dissection of the strategies whereby individuals interact both collaboratively and selfishly with others to enhance their own fitness and, more fundamentally, propagate the genes that induce the strategies. Scores of my colleagues were involved, and their discoveries transformed our understanding of animal mating systems.[5,6] Belief in harmonious partnerships of two adults pursuing identical objectives now seems quaint after we have strained for decades to understand the tactics used by each partner as it connives (unwittingly, we assume) to promote its own selfish interests in the context of a collaboration.

Incidence of Female and Male Infidelity

On Isla Isabel, in 1991, we could finally divert attention from the conflict between blue-foot chicks to the conflict between their parents when our study areas housed enough banded adults for observers to know which booby was which. We decided to observe sexual interactions in the colony's densest and busiest neighborhoods, where there was likely to be more competition among males and among females, and more infidelity. After a season of volunteer monitoring, an undergraduate thesis on philopatry (fidelity of breeders to their natal nest sites), and my class on avian reproduction, Marcela Osorio-Beristain was ripe for such a study and itching to get back to the island. Marce wanted to see in the flesh whether freshly bonded males and females are strictly loyal partners or engage in extra-pair liaisons, and what if anything the two sexes do to counter their partners' infidelity. For years we had confined our attention to how adults interact with their warring chicks; now we'd get into their love lives.

The second prong of this new line of research, paternity analyses, would be carried out by molecular biologists. Only they can work out which chicks are not the progeny of their supposed fathers or mothers and identify the true fathers of offspring produced by unfaithful females. But the boobies didn't yield their secrets easily. A Canadian colleague who had successfully analyzed paternity in other bird species, and an American colleague whose early successes at paternity analysis had persuaded him to set up a commercial molecular lab, were both unable to tease the desired information from the blood samples we sent to their labs. Blue-foot genomes weren't sufficiently variable for their techniques to be informative, possibly because there is limited gene flow between their small, widely separated populations on marine islands.

This was a blow and held us up for years, but out of the blue, my plans were rescued by another friend I had known since my first animal behavior conferences. Patty Gowaty, formerly a behavioral ecologist at the University of California in Los Angeles, is a passionate all-out researcher, and a feminist firebrand on a mission to reveal the proactive and controlling roles of females in sexual reproduction. In 2009, Patty buttonholed me at a conference because she was fired up to analyze paternity in an insular population of vertebrates. She had her eye on the Isla Isabel boobies because she had visited our camp many years earlier and was impressed by the small size and isolation of the population. A population in a jar, she called it. Blue-foots, she imagined, would be ideal for testing whether female animals selectively mate with males whose immunity genes are different from their own, in order to fortify their offspring's immune systems. And she had hired an ingenious postdoctoral student, Brant Faircloth, who would do his best to come up with a technique for assaying variability in immunity genes.

It was not to be. Blue-foot immunity genes did not submit to Brant's coaxing and that project was shelved, but a nice spin-off of his work was the invention of a technique for analyzing booby paternity. This was a breakthrough: that technique allowed his collaborator, Stacey Lance, a molecular biologist at the University of Georgia, to decipher the paternity encoded in blood samples from 478 booby families in the eastern plot of our main study area. These were furnished by Alejandra Ramos, a student who joined our group to do a doctoral thesis after monitoring California condors in the mountains of northern Baja California. Ale quickly warmed to the paternity project, picked up molecular techniques in Patty's lab, and led the team of volunteers who extracted the blood samples during a 4-month stay in our camp. On dark, moonless nights, when the colony would be less spooked by snatch-and-puncture biologists, the team worked furtively among the boobies in the forest understory—identifying, capturing and subduing, measuring, drawing blood, and gently releasing.

But let's start with the boobies' behavior. For her master's thesis, Marce observed, in addition to many part-banded pairs, 13 newly formed fully banded pairs in two forest neighborhoods where the terrain was open enough for an observer in a blind to see all courtship and copulation in a 15 m radius. Over a 3-month period, she and a small team of volunteer voyeurs watched their assigned neighborhoods during all daylight hours, taking turns in pairs and using binoculars when useful. Boobies mostly move slowly and deliberately in the colony, even when copulating, so our observers could systematically score all behavior of interest, even picking up on details like female cloacas turning pink a few days before laying, and sometimes extruding drops of ejaculate after copulation. A second team, led by Diana Perez, later expanded that behavioral sample by observing a further 18 pairs in another season.

At the start of bonding, the 13 monogamous pairs of partners were together on their territories for about 1 hour a day, standing around, defending the borders of their territory, and intermittently courting and nest-pointing.[7] Then over a period of up to 7 weeks, they progressively increased this joint presence to about 6 hours per day by the time the female was ready to lay. Throughout those weeks, they copulated occasionally, progressing from one copulation every 3 days to nearly two per day by the end. In addition, all paired females and males lingered alone on their joint territories for roughly an hour a day over the whole pre-laying period.

They were all unfaithful. Every male and every female occasionally courted with opposite-sex neighbors, and half of the individuals of each sex consummated their relationships with those extra partners by copulating with them. The seven promiscuous males each copulated with one, two, or three monogamously paired female neighbors, averaging a total of five extra-pair copulations per male, the most unfaithful male notching up twenty extra-pair copulations. Of the seven promiscuous females, six had one extra-pair partner and two had a couple, and they averaged seven extra-pair copulations each, the most unfaithful female managing twenty-one. Willing participation by females in this promiscuity was undeniable, and not only because females are larger and postural coordination is a must for achieving a "cloacal kiss" (brief conjunction of cloacas). All of the females' extra-pair copulations occurred in the course of reciprocal courtship with regular extra-pair partners; these were relationships sustained over days or weeks of repeated interactions, not fleeting encounters.

Fig. 8.1 Blue-foot liaison. After walking rightward 4 m from his territory, a paired male (left) touches beaks affectionately with his extra-pair partner on her territory when her partner is away fishing, before courting and copulating with her in the absence of their respective partners. The pair in the foreground simply continue incubating and grooming.

Drawing by Jaime Zaldivar-Rae; based on photo by Hugh Drummond.

Observations spread over a period of 22 years yielded a more complete picture of the sociology of blue-foot infidelity. By 2012 we knew that in dense neighborhoods, 53–61 percent of males and 33–53 percent of females copulate with habitual extra partners, and that these can include not only paired neighbors but also unpaired birds that linger in a neighborhood or defend territories there, as well as foraging males, whose provenance is still uncertain. Six foraging males of unknown reproductive status but who were not territory holders there or anywhere within sight were seen to approach resident females and copulate with them after an exchange of courtship displays.[8] Whether female boobies also foray on foot through the colony for sexual encounters, we don't know.

But were copulations with neighbors for an exclusively social function, or was it on the cards that they'd result in fertilization? Sex is not always about reproduction, and bonobos, our close chimpanzee-like relatives, are notorious for easing social tensions by copulating. And if fertilization by extra partners is an issue, which male is more likely to win the "sperm competition" that goes on between competing ejaculates in the female's reproductive tract—the female's partner or her extra male?[9] Timing of copulations can be critical. Since boobies lay eggs at 5-day intervals, a female is likely to be most fertile in the 5 days before each egg is laid, and the last ejaculate to enter her during that period may be best placed to fertilize. So it was surely no accident that, although copulations by both sexes with social partners *and with extra partners* increased progressively in frequency as laying approached, copulations with social partners were even more concentrated in the last 5 days than were copulations with neighbors, and that every promiscuous female's last two copulations were with her social partner. This timing would be expected to put the sperm of extra males at a disadvantage, and indeed the paternity tests showed that only 11 percent of broods contained an extra-pair chick.[10] Assuming that the promiscuous females were fully in control of paternity, only a minority of them were inclined to favor their extra-pair partners with paternity.

We concluded that infidelity is commonplace in both sexes of socially monogamous pairs, and that whatever social functions female infidelity may serve, it can result in sperm competition and sometimes jeopardizes the paternity of the social partner. So what do social partners do about it?

Countermeasures to Infidelity

Guarding

Trivers suggested that male birds should protect their paternity by guarding their partners, for the obvious reason that genes for protecting paternity ought to outcompete genes for indifference to a partner's infidelity.[2] And observations followed theory. Once Trivers's suggestion was out there, researchers began to suspect that males who accompanied partners that were nest-building and feeding were actually mate-guarding, they noticed that accompanying males drove other males away, and they added ever more species to the list of suspected mate-guarders. In some cases, there was doubt whether guarding benefitted the male by protecting his paternity, or the female by liberating her from male harassment, but the concepts of infidelity and mate-guarding transitioned from daring proposals to everyone's idea of standard avian equipment. The paradigm had changed.

In our blue-foots though, unambiguous mate-guarding was uncommon. Some males were provoked by their partner's courtship or copulation with a neighbor to disrupt it by approaching quickly on foot or launching a wing-beating assault, but more often than not males seemed indifferent. Three of the first thirteen females we watched copulated promiscuously a total of eight times within 1 to 5 m of their partners without provoking approach, aggression, or any hint of an inclination to interfere. I suspect booby males use aggression only sparingly to discourage infidelity because, although it instantaneously prevents insemination, it does not dampen the extra-pair relationship, which resumes and can be pursued when the males are not around. Or maybe males with insecure bonds are leery of provoking desertion.

In theory, there are other options for winning sperm competition. One of them is for the male to strategically adjust his own schedule of copulations. We tested for this, but males who witnessed a partner's copulation with a rival did not promptly respond with forced or consensual "retaliatory copulations" and, compared to males with faithful partners, males with promiscuous partners were no more likely to copulate with them on return from hunting trips or during the 5-day fertile period before laying.

Mere Presence

However, the male blue-foot's bow is not without a string. He may achieve little by disrupting his partner's extra-pair copulations, but his mere presence on their territory is a powerful disincentive to promiscuity. Females' rates of extra-pair copulations dropped six-fold when their partners were on the pair's territory, and during the 5 days before laying not a single female copulated with her extra male in the presence of her partner. Although there are blatant exceptions, females avoid copulating with extra males when their own partners are around. And this rule of thumb for copulatory discretion may be even stricter than it appears because when we only considered periods when their extra-pair partners were present in the neighborhood, promiscuous females copulated with them *35 times more frequently* if their own partner was absent. So, yes, some female blue-foots have the gall to copulate with an extra male right in front of their partner (who probably will not interfere), but as a general rule, the mere presence of the male partner is a compelling deterrent.

But why? Could it be that females are normally discreet about infidelity because males might retaliate in a way that hurts female interests? Later in the game, when the partners are caring for eggs or chicks, males may have ways of limiting the delayed costs of their partners' infidelity to them. Whether during clutch and brood care males mitigate the costs they incur from their partners' infidelity was a question we would eventually tackle experimentally, by duping males into "believing" their partners had escaped supervision and possibly been unfaithful.

But first, given that females are so sensitive to male scrutiny, and allow it to severely curtail their infidelity, we looked for signs that females sidestep scrutiny by modifying their schedules of territorial attendance, and signs that males outmaneuver them by modifying their own schedules. Hunting schedules must respect the constraints of the solar cycle, weather, and ocean conditions, but boobies can even plunge-dive at night when convenient, so there must be some room for evasive scheduling. Territorial attendance of the 18 pairs in the second year of the project tentatively confirmed this suspicion. In comparison with faithful females, promiscuous females partially shifted their hours of attendance from late morning to early and late afternoon, and their partners shifted their hours of attendance similarly.[11] These coordinated shifts of the two sexes could be variously interpreted, but they are at least consistent with promiscuous females modifying their schedules to escape supervision, and partners modifying *their schedules* similarly to maintain supervision.

Infanticide

When they have grounds for suspecting that their partner has been inseminated by another male, what remedy might males apply during clutch and brood care to reduce the impact on their fitness? Raising offspring that are not yours is a serious misuse of resources and bound to incur a fitness penalty, but what can cuckolded males do about it if they are unable to recognize eggs or chicks sired by another male? Males of no bird species have been shown to recognize eggs sired by another male. Well, they could abandon the nesting attempt and either start again with a new partner as soon as possible or wait till next season, but the cost to their fitness would be high. A less radical and costly solution would be merely to reduce their investment in the current clutch/brood and apply that saved energy to their next brood, in which, with normal luck, all offspring would be theirs. Genes that incline males to reduce investment in offspring when their paternity is in doubt ought to end up in the genomes of more offspring than genes that incline them to ignore the risk of paternity loss or drastically abandon a whole clutch. It was worth testing whether suspicious blue-foot males flexibly adjust their investment in this way.

Marce and a new team of volunteers tackled this question for her doctoral thesis (her third thesis on Isabel blue-foots!). To determine whether males who are uncertain of their paternity invest less in their offspring, they captured 17 experimental males on their territories shortly after dawn when their partners were just 4 days from laying (i.e., in their fertile periods, signaled by pink cloacas), detained them in cloth-covered cages elsewhere in the colony for 11 hours, and then released them at their nests before sunset. Inability to supervise their partners during one day would, we hoped, undermine their confidence in paternity. Experimental males would have less confidence than 17 control males who were captured and caged a couple of weeks before their partners laid, long before their fertile periods. Less confident males ought to invest less in the offspring, so we predicted that experimental males would skimp on clutch incubation and defend their clutches less resolutely against approaching predators (two biologists walking in single file).

The kidnapped males stood and slept calmly on the forest floor of their cages, apparently relaxed and not attempting to escape. When released at twilight by lifting the cage, they promptly flew out to sea; during our 30-minute watch, they did not return to their nests. At dawn the next day though, they were all on their territories. In the cages, they suffered food deprivation as they presumably would during stormy weather, but all appeared unharmed and resumed their relationships with their partners.

We did not have high hopes because similar experiments on songbirds had failed to confirm the researchers' predictions,[12–14] and we feared that kidnapping might just upset and disorient the males rather than creating the sensation that they hadn't been able to monitor their partners. To our relief though, the males kidnapped when their partners were most fertile emphatically confirmed our expectation of reduced investment, although not by fulfilling either of our predictions. Marce was shocked by their robust response to diminished confidence in paternity: their rule of thumb was to destroy the female's first egg, at the first opportunity. Roughly 2 hours after laying, when females stood up from their precious, freshly laid first eggs for the first incubation change-over, seven of the seventeen experimental males stepped forward as if to assume their paternal responsibility, and then either beak-rolled the egg until it was well outside the nest, or grasped the egg in their beaks, carried it a few meters, and tossed it. The two female partners who were standing right by the nest at this moment did nothing to prevent the infanticidal expulsion, recover the eggs or scold their partners. The seven expelled eggs were predated by gulls, pecked by neighbors, or collected by biologists. Needless to say, none of the control males expelled eggs; all simply incubated them, as did the remaining experimental males. As for incubation and defense of remaining eggs, the subject of our predictions, experimental males cared for their clutches with the same devotion as control males; there was no strategic skimping.[15]

You may suspect that destruction by males of eggs in their own nests is an experimental artifact, pathological behavior induced by capture and confinement a few days before laying, but we've also witnessed this behavior in unmanipulated nests. It's part of a male booby's reproductive repertoire. In one season, after two paired females were watched courting in their fertile periods with extra males in their partners' presence, and copulating with them in their partners' absence, both male partners destroyed their partners' first eggs at the first incubation change-over, one by piercing the egg with its beak, the other by rolling it out of the nest. Male boobies do indeed destroy eggs of dubious paternity, and they do it after their partners have been unfaithful in their fertile periods.

Fig. 8.2 Egg infanticide. At the start of his first incubation shift, a male who was experimentally kidnapped a few days ago during his partner's fertile period seizes and tosses the first egg rather than risk caring for a rival male's progeny.
Drawing by Jaime Zaldivar-Rae; based on photos by Marcela Osorio.

The seven experimental males that expelled the first-laid egg cared for the second and third eggs as assiduously as all the other males, implying that the switch that was flicked after not monitoring their partners in their fertile periods flicked back after the first egg. So their infanticidal rolling and tossing targeted only the egg most likely to be sired by a rival male. In the 5-day intervals between the laying of successive eggs, two booby partners normally copulate with each other but not with extra mates, so the second egg is likely to be sired by the male partner, and he does well to care for it. Therefore, I'd put my money on first-egg infanticide being not only a better-targeted strategy but also a more profitable strategy than reducing investment in the whole clutch, because it does no harm to the male's own offspring. Our original predictions may have underestimated the unseeing wisdom of male blue-foots.

The surprising results of that experiment vindicated our earlier inference that female boobies generally avoid copulating *in flagrante delicto* with their extra partners because their partners may respond by doing something that affects the females' fitness. That response turns out to be infanticide and its function is not, as far as we know, to punish unfaithful females but to save males from investing their hard-won resources in the offspring of other males. However, the response has also favored the evolution of strategic sexual restraint in females: they are discreet in their infidelity

Fig. 8.3 Laying is imminent, and attendance of the male (left) at the nest site is maximal, maybe because the risk of female infidelity is greatest when she is fertile. Photo by Hugh Drummond.

because discovery can cost them a chick or, as we saw in the previous chapter, provoke a partner to desert. But why then do some females copulate with extra partners *in flagrante*? Are they making a point? That they have options for deserting and pairing with a rival? I'd love to know.

Mere Female Presence, Too

The pre-laying *infidelity of males* is a quite different matter because it cannot prejudice the fitness of their partners by lumbering them with eggs and chicks of other females. What is it to a female if her partner has sired offspring elsewhere in the colony? He's hardly going to invest in them. However, a male's infidelity could prejudice a female partner's fitness in other ways. If dalliance develops into desertion and switching to the extra female, then the male's first partner suffers a major setback, particularly if she has already invested much time and energy in bonding with him. Starting again implies absorbing the time and energy costs of bonding twice, and results in nesting later in the season, which dims chicks' prospects for fledging and graduating into the breeding colony. So male infidelity ought to be a matter of concern for female partners, and indeed they do respond to it. Females courted their partners twice as frequently after witnessing them courting another female, but this reaffirming of the bond may be their only means of preempting desertion.

Nonetheless, when their partners are present on territory, males cut their rates of copulating with extra females by half. This reduction is modest compared to females' six-fold reduction in the equivalent circumstance, and it may be nothing more than a side effect of males being occupied courting their partners, but it could imply that female partners have some leverage over unfaithful males. Females do sometimes desert males and switch, in circumstances that are unclear (Chapter 7), but could be disinclined to do so because the cost of desertion is likely to exceed the benefit it could bring them.

Ecological Influences on Infidelity

Habitat Structure

After leaving high school and working first in the offices of a manufacturing company in London's northeastern suburbs, and next in a government office in central London, a friend once suggested, in the vernacular, that in pair-bonded humans infidelity is a prevalent motive but is context-dependent. As he saw it then, in the mid-1960s, many wives and husbands were motivated to have affairs with other specific individuals, but few of them did so because the physical and social environment constrained opportunity. Humans do their copulating in private, and privacy is not readily available. The idea shocked and intrigued me. It was just a few years later that Trivers pointed out why infidelity should be expected in avian and human males, [2] and about 20 years afterward that "effects of habitat structure and population density on extra-pair interactions" became a recognized but under-researched topic in avian mating systems research.[16] Influence of habitat structure was a rather intractable question though. For example, although open habitats seem to discourage female infidelity because they make it easier for males to monitor and guard their partners, female restraint there could just as easily be attributed to food scarcity and predation risk.[17,18] Researchers can't be sure whether infidelity is brought on by concealment or allowed by freedom from hunger and danger.

Blue-foots are a model species for addressing this question because, from what we had seen, all of their sexual behavior goes on in the colony, a setting where there is no food for boobies and no predation on adults. Using Ale's paternity data on whole families at hundreds of precisely mapped nests in our 8,000 m^2 plot, we would be able to test for effects of *habitat structure*—ground cover, obstacles to movement, visual access—and *proximity to sexually active boobies* on the probability of chicks being sired by an extra male. That's right, those data would tell us how the *social structure* of a neighborhood affects extra-pair paternity and, by inference, infidelity. They would also tell us how far paired males foray to sire chicks at other nests, and they would reveal whether habitat structure around a male's own nest affects his opportunity to sire chicks elsewhere. The study promised a bonanza of highly original data that would tell us whether and precisely how my friend's intuitions about Londoners apply to blue-foots and, by extension, other birds. Ale would be too busy obtaining blood samples to monitor sexual behavior, which requires observation by more researchers than we were willing to house in our camp, so behavior would be creatively inferred from the distribution of extra-pair chicks in the study plot (inevitably leaving us with some puzzles and doubts).

At the end of the season, after the boobies and their fledglings had migrated from their nesting habitat in the colony to the ledges and crests of the island's cliffs, Ale and a tireless volunteer measured habitat variables at over 150 nest sites in the study plot. Around each site, they counted obstacles to walking boobies (boulders, trees, and fallen trunks) and estimated percentages of rock cover and grass cover. They also crawled around the forest floor taking sightings, to compute an index of 360-degree visual access from each site to distant boobies, at booby eye-level.

To our satisfaction, it was obvious just glancing at a map of the study plot that nests with extra-pair chicks were distributed unevenly, and statistical analysis revealed two factors correlated with that distribution. Along the east-west axis, moving inland from the beach and wave-cut platform to the forest interior, there was no pattern, but from south to north, the probability of a nest containing an extra-pair chick increased progressively, culminating in highest incidence in the area around our campsite. This was not because the biologists were egging the females on, but probably because, as we found out, the density of obstacles around nests decreases from south to north. On the other hand, the presence of extra-pair chicks was in no way related to ground cover or inter-booby visual access.[16]

No causal arrows can be confidently traced because the statistical analyses revealed only associations and we were loath to test causality by experimentally modifying the boobies' habitat. We knew for a fact, though, that in the study plot the further north a female's territory, the fewer obstacles to booby locomotion in its immediate vicinity, and the more likely she was to have a chick sired by an extra male. We inferred that obstacles like boulders and tree trunks complicate boobies' access to each other, making approaching and displaying awkward and hazardous to the point where the benefits of dalliance, whatever they are, are outweighed by the costs of injury and expenditure of time and energy. We concluded, tentatively, that high obstacle density in the vicinity of a female's nest deters sexual hookups with foraging males.

Another, quite different, causal pathway seemed seductively plausible and gave us nagging doubt. Might it be all about the quality of the boobies rather than the structure of the habitat? The highest quality boobies might be best able to compete for territories in the most obstacle-free habitat to the north, and their high quality could also make them more desirable as extra partners. This idea makes perfect sense, so we needed to test it. And test it we could, because we already had the data; we had up-to-date CVs of most boobies in the study plot in our database, and Ale's team had weighed and measured all of them when they sampled their blood. Results of the tests were crystal clear. There was no hint of any association between location on the north-south axis and any index of quality: body size, body condition, age, or previous reproductive success. One alternative explanation down ... but one more still to go!

The last alternative explanation we could test had to do with the insidious inhabitants of the study plot's boulders and tree trunks. On the island, snakes are nonvenomous; scorpions are rare and keep to themselves; spiders, though diverse, abundant, and omnipresent, bother nobody; the mosquitos blown sporadically from the coast don't survive long enough to harass anybody; and the resident mosquitos that reproduce in tree-fork pools after rain are too few and too ephemeral to matter. But the invisible mites that burrow into biologists' softest skin, and the soft-bodied argasid ticks that infest boobies' webs and their chicks' axillas, are bloodsuckers to be reckoned with. Larval ticks lodge themselves for up to 3 days where chick down is sparse, and adult ticks swarm onto the webs of breeding boobies for leisurely nocturnal meals. About 150 species of marine birds are known to be parasitized by ticks, and heavy infestations can drive them to abandon clutches and broods. So boobies ought to minimize their exposure to ticks, and that could mean giving a wide berth to boulders and other objects that harbor them. Risk of being parasitized might even discourage dalliance-seeking males from foraging south to find willing females, accounting for the scarcity of extra-pair chicks in southern nests. And indeed, when Ale counted larval ticks in the axillas of young chicks at 50 nests, she found that infestation increases southward.[19]

In sum, the correlations with habitat variables told us that extra-pair paternity, and presumably female infidelity, increase as the number of boulders and other obstacles declines, possibly because foraying males prefer to search for receptive females in terrain where access is easy or where the risk of parasitism is low. There are, as far as I know, no comparable data on Londoners in urban habitats.

Neighbors

It would not surprise anyone who has lingered in a blue-foot colony amid all the sexual and aggressive antics among partners and neighbors, and the discreet passage of solitary individuals prospecting for partners or commuting between territory and shore, that the probability of chicks being sired by an extra male also varied with the social structure of the colony. The more nests with sexually active males—males whose partners have not yet laid—within 10 m of each fertile female's nest, the *lower* the probability of her producing an extra-pair chick; and rarely did a female nesting *within a cluster* of such pairs produce an extra-pair chick. Paradoxically, the nearby presence of sexually active males puts a damper on female infidelity. How could that be?

Males foraying for receptive females undoubtedly meet resistance in crowded neighborhoods, where most walkable space is in someone's territory and territorial boobies of both sexes reliably attack and repel intruders. So, if a receptive female's territory lies in the midst of couples without clutches and therefore free to approach and repel intruders, I suspect it can be challenging for foraying males to access her. However, the data also indicated an opposing tendency: at neighborhood densities *below* one nest per 10 m^2, the more nests with sexually active males around each female's nest, the *higher* the probability of her producing an extra-pair chick. This, too, makes sense, because medium concentrations of females may offer more opportunity for foraying males than lean concentrations. Why foray into places where there are hardly any females to check out and court? If these inferences are correct, then the sweet point is one nest per 10 m^2; at that density, candidate females abound, but there aren't so many territory-defending residents around that a male can't get to them.

The paternity analyses also put the finger on 17 extra-pair sires, all of them paired males with nests and broods in the study plot—of course, because only such males were blood-sampled and genotyped. Surprisingly, the sires' own nests were located an average of 44 m, and as far as 100 m, from their extra females' own nests. We had expected most of them to be neighbors, but this was an exceptionally benign year in which nearly all chicks fledged and roughly half of the male extra-pair partners were forayers. However, the poor representation of male neighbors among extra-pair sires—the closest one was 14 m from the female he fertilized—could imply that forayers have some undetermined advantage over neighbors in sperm competition.

How and why would a forayer bias in sperm competition come about? Females, big-bodied and plainly in control of copulation, may well have the means to favor one inseminator over another. In theory, they could do this by dictating the timing of the two males' inseminations in relation to the females' fertile period, or by "cryptic choice"—selectively ejecting ejaculates after copulating (as witnessed by Marce) or prioritizing them through internal control of sperm storage and use.[5,20,21] But what might females gain by favoring distant males over neighbors? It's hard to imagine how this preference could benefit females personally, and much easier to imagine a genetic benefit for their offspring. The blue-foots on Isla Isabel mostly nest only 10–30 m from where they hatched and continue nesting in that vicinity for the rest of their lives,[22] so there's a chance that a female's offspring by a more distant male will be less inbred and hence more viable than her offspring by a neighbor.[23]

That was not the only puzzle to arise from the paternity analyses. Ale also found that a paired male's chance of siring a chick in another pair's nest declined with crowding in his home neighborhood, similar to what we saw for paired females. We can explain this tantalizing result if we assume that paired female blue-foots foray in the colony for copulations with extra partners rather as males do and, like males, find crowded neighborhoods unmanageable because of all the territorial defense there. We haven't seen female blue-foots foraying, but females of other avian species do it.[24] How male and female blue-foots go about foraying, and whether it can involve flying over distant neighborhoods or nocturnal excursions, are unanswered questions we'd dearly love to answer.

El Niño

The last environmental influence on infidelity we put to the test was El Niño Southern Oscillation, probably the greatest ecological challenge to the blue-foot's reproduction throughout its geographic range. Are liaisons and foraying a luxury activity that boobies indulge in when benign cool-water conditions cut them some slack, and renounce when warm waters deplete fish stocks and force them to manage on a tight budget? Is infidelity a luxury they can't afford in years of privation?

To tackle this and other questions about female infidelity, we could count on Lynna Kiere, a newly recruited doctoral student from West Virginia. In a historic tour de force, Lynna and her teams of volunteers watched well over 200 bonded pairs in the course of four breeding seasons—2010, 2011, 2012, and 2013—happily including an El Niño year, a La Niña year, and two intermediate years. They watched each pair for 40 hours over 4 or 5 days, continuously recording each adult's presence on territory and its interactions with its partner and any extra partners. Over the same dates, the satellites, buoys, and ships of the USA's National Oceanic and Atmospheric Administration measured the surface temperature of a 12,000 km^2 patch of ocean centered 56 km southwest of Isla Isabel. Downloaded gratis, these data allowed Lynna to probe for correlations between water temperature and blue-foot social behavior. There would be little doubt about the direction of causality: warm waters make prey fish scarce, reducing the boobies' access to food and obliging them to modify their priorities among feeding, breeding, and self-maintenance.

Lynna found that in both warm-water years and warm spells within years females spend less time on territory, copulate less frequently with their partners, and are less likely to lay a clutch, implying that food scarcity depresses reproductive activity generally. And warm-water depression extends to their infidelity: they don't give up on it; indeed, they join in extra-pair courtship and copulation with their accustomed gusto, but they indulge less often.[25]

We concluded that blue-foots deal with food shortage by cutting back on reproductive effort, whether with partners or extra partners, in order to devote their resources to maintaining their health and vitality. Although capable of going ahead with reproduction, underfed females often decline to do so, playing the long game, preferring to assure their accumulated lifetime productivity rather than risking it for short-term gain in the current season. They don't do this because they are clairvoyant, and I doubt they have any consciousness of what is at stake. They do it because in the population of boobies, genes inclining females to play the long game long ago outcompeted genes for short-termism. And this happened because, in a bird that can expect to live 9 years (and up to 25 years!), genes that promote the maximization of *lifetime* reproductive success get into more bodies of the next generation than genes which sacrifice the future for short-term success.

So, yes, female blue-foots do cut back on infidelity when the food limitation of El Niño constrains their activities, but they cut back on reproductive behavior with their partners, too. Infidelity is not a luxury activity they indulge in only in good times, but a regular component of reproductive striving in good and poor conditions. And austere conditions, rather than actually hobbling boobies, incline them to prudently postpone investing in reproduction.

Final Word on Ecological Influences

The search for an effect of El Niño on female infidelity reminds us that all animals are economists who garner resources and deploy them so as to maximize their lifetime production of viable offspring. It's no surprise that a long-lived animal free of predation during adulthood responds to resource shortage by cutting back on reproductive effort, both with its partner and with extra partners. The effect of El Niño on infidelity is, in the end, mundane: austere feeding conditions limit reproductive activity with extra partners just as they limit it with partners.

More intriguing and harder to interpret are the constraints of habitat structure and social environments on the incidence of both sexes' production of extra-pair offspring, an issue closer to my old friend's intuitions about Londoners. The observed correlations, open as usual to alternative interpretations, imply that the infidelity of paired blue-foot females is constrained by nesting in places where habitat structure limits the transit of foraying males or exposes the males to parasites. Female infidelity is also constrained by nesting in neighborhoods where there are too few pre-laying females to attract foraying males or so many sexually active males that foraying males will be deterred. Added to this, there are hints that females may do their own foraying to proactively access extra males and that foraying females, too, may find crowded neighborhoods too well defended to be penetrable.

Might it be that patterns of nest aggregation in the booby colony reflect not only choice (by males) of ideal spots for pairs of partners to incubate eggs and raise chicks but also spots for encountering and engaging with extra-pair partners on the side? Before the sociobiological revolution, no one would have thought it, but a strong case has been made for the hypothesis that female birds oblige males to nest in places where the females will be able to access extra partners.[26] Females could do this by selectively bonding with males on territories that have attractive neighbors and ignoring males with mediocre neighbors. That female choosiness would channel the evolution in males of nesting near the competition! We wanted, more than anything else, to test our inferences about neighborhood effects on infidelity by remotely tracking boobies foraying for extra-pair liaisons in the colony, but it was not to be; we had the technology and a well-trained student lined up but failed to raise the funding.

Why Are Females Unfaithful?

For males it seems simple enough. Although they may be in for some surprises, most behavioral ecologists are confident about the function of male philandering: as anticipated by Trivers,[2] male birds are trying to get away with siring additional offspring they won't even have to feed—freebies, almost. On the other hand, the widespread infidelity of female birds is one of the great enigmas of behavioral biology because it comes with costs, and evidence of a benefit to female reproductive success is thin on the ground. Unlike males, females can only rarely wriggle out of financing the creation and growth of all their offspring. And the likely costs of females' infidelity to their reproductive success could easily exceed the benefits that have so far been mooted. The costs include increased risk of sexually transmitted disease (about which we know little), and risk of the partner either deserting the female or trimming his investment in their joint offspring. Which benefits of female infidelity could be great enough to justify incurring these risks?

Let's consider the five functional hypotheses for the infidelity of female birds that are, on the face of them, candidates for explaining this behavior in blue-foots. First, liaisons with attractive males may permit females to evaluate candidates for partner switching in case things fail to work out with the current partner or to identify candidate partners for future reproductive attempts. That is, liaisons may be interviews that enable mutual assessment and familiarity and create the vibes for new relationships. Second, females might establish relationships with paired males to facilitate dumping eggs in their nests.[27] A male who has copulated with an extra female may be less motivated to expel her when she cheekily enters his territory to lay an egg in his nest for him and his partner to take care of. Third, sperm from an extra male could function as a fertility back-up in case the partner's sperm are inviable, or it could increase genetic variability among the female's offspring. Fourth, by copulating with a neighbor, a female may give him an incentive for helping her out in some way, for example, abstaining from infanticide in the event of the female's chick blundering into his territory a few weeks later—because it might be his own progeny.[28,29] A rule of thumb for tolerating intruding infants if you copulated with their mother may sound far-fetched, but it's not unrealistic; a male booby could rise to the cognitive challenge of tolerating chick trespassers at a particular boundary, and of course, no comprehension of the outcome of tolerance is required. Fifth, the most widely *believed* explanation of promiscuity in female birds—the good genes hypothesis—holds that by copulating with attractive males of higher quality than their partners, females increase the genetic quality of their next offspring. In the colony pairing market, low-quality females very likely end up with low-quality partners, but the superior genes of high-quality males may be available to the females, or rather to their next offspring, via sexual liaisons. Despite its strong intuitive appeal, this hypothesis has stubbornly resisted consistent confirmation.[30]

Switching

There is evidence in favor of the first two hypotheses. As we saw in Chapter 7's bobonovelas, 6 percent of booby partnerships are dissolved by switching, and switching is usually preceded by courting and copulating with an extra partner. Chains of social causation can be complex and responsibility for break-ups can be hard to assign, especially when both partners are unfaithful or one's infidelity seems to provoke the other's desertion. The point though is that in diverse circumstances paired females can find themselves unpaired and in need of a partner. Lest that happen, it's convenient to have an additional relationship on the back burner, so you can save the season by abruptly upgrading it into a bond. Most female blue-foots are not going to switch, but it may be prudent for the average paired female to dedicate some of her time to dallying with a promising extra male or two, just in case.

Egg-Dumping

This parasitical cuckoo-like behavior was occasionally documented by Marce and Diana in the busy neighborhoods where they just sat and watched.[31] Dumping an egg in another pair's nest is not an easy trick because there is necessarily strong selection on both sexes to avoid being parasitized; so a would-be parasite needs to fool two boobies who habitually expel from their territory any animal that dares to enter. Nonetheless, blue-foot females can sometimes pull it off and several times we've caught them in the act. The classic example I'll describe illustrates how unseeing natural selection can fit animals out with unwitting cunning in the form of a complex, routine coordinated by rules of thumb. Like infants controlling and killing their siblings, adult females can be disturbingly effective while knowing not what they do.

The protagonist was a bonded female who regularly courted and copulated with three males, including the male of the pair on the neighboring territory—just 3 m away—and, of course, her own partner. During the 3 days before she made her move, she copulated with both of them. Then, one day after that male neighbor and his partner lost their single egg to a gull, she calmly walked onto their territory at 4:30 pm when they were away hunting, and laid an egg in their newly empty nest scrape. Next, rather than making herself scarce, she crouched over the egg in an incubation posture and waited.

After 75 minutes, the female neighbor returned to her territory only to discover another female on her nest and went ballistic, grunting excitedly, yes-headshaking, and jabbing into the dumper's face. But the dumper was cool. Despite being caught in an act of blatant territorial invasion, compounded by incubating on the resident pair's nest scrape, the dumper did not immediately scamper away. As is typical of a dumper, she first stood her ground, vigorously defending by displaying and jabbing, before gradually backing off. After 5 minutes of this confrontation, the male neighbor returned to the territory, too, and joined in the eviction of the dumper. However, the dumper was also his extra-pair partner, so, typically, his eviction efforts were ambivalent: rather than making a headlong dash and raining jabs on her, he stood in the middle of the territory, jabbing ineffectually into space.

What happened next was critical, and neatly consummated the crime. Under threat of eviction by two irate residents, the dumper walked away from nest and egg and flew off to the ocean. Finally able to access her nest, the resident female settled on the egg and incubated it; I imagine she felt relieved and satisfied that finally, all was as it should be. A few hours later she graciously rose from the egg to allow her partner to take his turn at incubating. Forty-three hours later, during her incubation turn, she laid a second egg—second in the nest and second in her own laying sequence, and 100 hours after that she made it a complete clutch of three. All was normal, with both partners coordinating their hunting and incubating round the clock, except that one of the three eggs in their nest was not hers and may not have been his either.

Our interpretation, after witnessing six females pull off the same trick, is that female boobies inherit a repertoire of dumping tactics that they deploy, when circumstances are favorable, to parasitize the parental care of breeding pairs. A dumper chooses the right time, when the victims are about to lay or have recently lost their clutch, so that they will be hormonally primed to incubate. She chooses the right moment, when both victims are out at sea or busy with their backs turned fighting neighbors, to avoid being caught red-handed. And she contrives adoption of her egg by incubating it until the resident arrives and then resisting eviction until the resident has fought long enough to naturally "resume" incubating, as one does after evicting an intruder. Sometimes the dumping repertoire is richly rewarded. Only 2 percent of hatchlings in the colony are not the progeny of their supposed mothers, but in each case, a dumper got away with deviously outsourcing 40 days of incubation and about 4 months of round-the-clock defense and feeding.

But the prospective dumper's masterstroke is probably compromising the nest defense of the males they will parasitize by conducting an extra-pair relationship with them. All six dumpers copulated with their male victims before dumping, and those males were noticeably half-hearted about repelling them, although two showed unambiguous resolve by breaking the dumped egg at the first opportunity. It has been suggested that male birds willingly conspire in dumping by their extra partners, presumably to obtain superior maternal genes for their offspring.[32] However, now that we've witnessed the interactions of blue-foot dumpers and their victims, I believe that males are motivated to resist dumping and that they resist because it doesn't benefit them. Why else would males sometimes break dumped eggs, and females confine their dumping to moments when males are unaware of it? However, male resistance to the machinations of their extra-pair partners falters because they are simultaneously attracted to them. Selection may well have equipped females to play on that susceptibility by courting and copulating with target males. Rather than the extra-pair relationship bribing the male into collusion in dumping on the basis that he may be the father, the relationship weakens but doesn't quite abolish his resolve to expel the extra-pair female.

Good Genes for Offspring

The potential of female infidelity for facilitating switching and dumping was revealed to us unexpectedly by simple watching. Exploring infidelity's potential for upgrading the genetic quality of offspring—the appealing but hard-to-prove good genes hypothesis—called for a more proactive approach. We rose to the challenge by carrying out two ad hoc studies analyzing whether female preference among extra-pair males depends on either the extra males' quality relative to the male partner, or on the extra males' age relative to the male partner, and wrapped up with a final study testing whether extra-pair offspring are better-quality boobies than within-pair offspring.

For the question of relative quality, Lynna scrutinized the choosiness of females in her sample of 55 paired females who directed some of their courtship at extra males. Contrary to her prediction, in these threesomes (female + partner + extra male) neither the extra males merely courted nor the extra males copulated with were systematically superior to the females' own partners, not at least for any of our measures of quality (body size, body condition, foot color, age, or previous breeding success). Likewise, the probability that a female courting with an extra male would also copulate with him, which half of them did, was unrelated to any measure of her partner's quality or the relative quality of partner and extra male.[8] Results of these painstaking comparisons were disappointing, to say the least; they yielded no hint that females selectively target superior males for promiscuous attention. But further drilling into those data showed they were consistent with a subtle version of the good genes hypothesis that takes into account the uncertainties females face when evaluating males.[33] As predicted by that model, female boobies *who were paired with attractive males* copulated with their extra-pair partners only if these were more attractive—greener feet, that is—but females with ordinary partners did not discriminate in this way. This observation was reassuring and it kept the good genes hypothesis alive for boobies, but it was only one part of the picture and, of course, it fell short of showing that this choosiness confers any benefit on the females' offspring. We were left dangling.

And that, as it turned out, was it for sitting and watching bobonovelas. We would continue studying the function of female infidelity, but mostly by mining the database rather than watching neighborhoods. That observational work had been intermittent across a 22-year span, starting with Marce's 13 pairs of partners in 1990 and ending with Lynna's last observation stint in 2012. I'd had the good fortune to recruit promising graduate students—Diana was another—who were capable of organizing research teams and watching social behavior with sufficient concentration and consistency to record high-quality data. I had done little neighborhood watching myself, and by the time Lynna completed her fieldwork I was 66 years old and grateful to be delegating that role to another generation. Working at the computer in Mexico City on manuscripts and database analyses was more and more appealing, and as we

Fig. 8.4 Team in transition: Hugh is currently stepping back, and Sergio Ancona (Cheko) will coordinate the booby research program going forward, with Cris's continuing support.
Photo by Sergio Ancona.

accumulated ever more information on the boobies' lives, our team, which now included researchers who were ex-students and *their* students (grand-students?), was basing most of its research reports on stored data.

After 2012, I still enjoyed setting up camp every year, supervising and joining in experiments and observations, and launching each season's nest monitoring by training a new crop of volunteers. But it was a relief every season, after getting the projects going, to say adios to the island team and head for the altiplano. Age, not effort, had taken its toll. I still reveled in riding my mountain bike on steep, rocky trails in the cool pine forests of the mountains south of Mexico City, but the tropical heat of Isla Isabel slowed me down and sapped my strength. Despite some troublesome old climbing injuries, in the early morning breeze, I could saunter contentedly along the beach and pick my way happily through the forest searching for nests. It's always a real joy to work your way along the eastern edge of the colony after the sun emerges from the ocean beside the Las Monas spires, when the waves are rinsing and slapping at the wet sand and the boobies' white breasts radiate gold. However, several hours of crouching and crawling beside nests in sultry late-morning heat, rising unsteadily from nest after nest, was leaving me spent and lethargic. Increasingly (and guiltily) I depended on our fieldwork manager, Cris, not only to organize the equipment and supplies but also to do the heavy carrying and the crouching, banding, and measuring while I tagged along noting the data. Every long-term island research program needs a Cris! Unwavering and tireless, she has sustained our program and its complex logistics for 30 years. (Though at midnight once, after a 17-hour day of monitoring and camp chores, she dropped off at the camp table once, out for the count with her forehead resting on a pile of datasheets while the team worked on.)

Parental Age Combination

Our second study took us a step closer to embracing the good genes hypothesis by following a quite different and, as it turned out, meandering route. It uncovered evidence for a new, tailor-made version of the hypothesis that puts parental ages at center stage; and it persuaded us that some females enhance the quality of their offspring by manipulating the *age combinations* of their *biological* parents through infidelity. This all started when Ale turned her attention to testing an old chestnut of a theoretical idea, namely that females should choose old males as extra-pair partners because old males' long survival in a dangerous and challenging world demonstrates the high quality of the males themselves and certifies the quality of their genes.[34–36] Other things being equal, offspring sired by a male older than a female's own partner should on average get better genes and eventually yield more grand-offspring. On the other hand, with increasing age, the fertility of male vertebrates falls off and the genes they contribute to their offspring accumulate mutational flaws, including in blue-foots.[37,38] So a female's dalliance with an old male is likely to be a double-edged sword: her offspring may very well get from him both better genes and damaged genes.

The evidence on female birds' preference for old males and the superiority of old males' genes is equivocal and shaky. Field studies have suggested that females of some species do indeed prefer to grant paternity of their offspring to extra males who are old.[39–41] However, rigorous studies of two very different birds recently showed that the net effect of old males' genes on offspring is negative.[42] On balance, it's looking like old guys inseminate shoddier genetic material than we used to believe.

Our understanding of the importance of partners' ages may have been hampered not just by scarce and contradictory data but also by paying insufficient attention to female age. Despite the obvious desirability of taking all ages into account, ages of all three members of infidelity threesomes in wild populations are seldom known, simply because most wild birds are mobile enough to escape multiyear monitoring. Female age could be relevant though, for at least two reasons. First, a double dose of damaged genes from two old parents could be especially harmful to an offspring even if, as is thought, the genes provided by old females have less accumulated damage than those provided by males.[43,44] Second, the remarkable ability of female germ cells to repair genetic damage in male germ cells after fertilization probably declines as females get older,[45,46] implying that the harmful effects of male age on offspring development could depend on female age.

Ale stepped lightly onto this quagmire of parental age effects on development with a descriptive study that ignored offspring quality and the whole issue of genetic and other mechanisms—that would come later if we found something that needed explaining. Our "population in a jar" of mostly known-age boobies would allow her to ask how the ages of both partners—female and male—affect the probability of the female granting paternity to an extra male. A descriptive study to get baseline data on the basics of age combinations couldn't possibly let us down and would surely raise new, well-founded questions that could take us deeper. With luck, we would find out the ages of some extra-pair sires and be able to compare them with the males they cuckold. For all this, Ale had paternity data on 350 blood-sampled families for which both parents' ages were known, and the mothers' and fathers' ages ranged from 2 to 25 years.

Those families told an intriguing and ground-breaking story of females flexibly adjusting their infidelity to their own ages and the ages of their partners. Specifically, in young females (2–4 years old), the *younger* their partners the more likely they are to grant paternity to a different male; whereas in old females (8 years old or more) the *older* their partners the more likely they'll make that move.[10] Otherwise stated, both young and old female blue-foots are inclined to deny paternity to partners similar in age to themselves. After some head-scratching, we thought this could mean that offspring themselves are worse off, for whatever reason, when their two biological parents are of similar age. This new hypothesis predicts that extra-pair sires should be different in age from the females that choose them and from those females' partners. But how old are those sires, in fact? Only for young mothers was our sample adequate to answer this question, and the answer confirmed the prediction: the young mothers in Ale's sample who denied paternity to their young partners granted it instead to extra-pair sires who were, on average, 6 years older than those partners.

These unique findings spoke to females strategically (and slyly) manipulating the age combinations of their offspring's biological parents, and they begged for follow-up. Most urgent of all when the news on age combinations broke, we just had to find out whether the offspring of two young parents or two old parents are indeed less viable than the offspring of parents of mixed ages. This was an extraordinary question to ask because biologists usually think of the two biological parents' ages having genetic effects on offspring that are independent and additive. There are exceptions though, and for some human birth defects thought to be genetic, the predisposing effect of one parent's age depends on the other parent's age.[47,48] Anyway, luxury of luxuries, in the database we already had the data to answer this question, so Cris threw herself into mining the data and building a series of statistical models. These asked how two bonded partners' *combined ages* affect their fledglings' probability of graduating into the breeding population, a sterling test of viability that six out of ten fledglings fail. Well over 3,300 fledglings from seven generations were available for analysis, the offspring of close to 2,000 individual females and males that were between 1 and 21 years old when they raised them.[49]

The results made for a gratifying confirmation of our predictions. In the Isla Isabel population, the fledglings least likely to ever reappear in the colony as breeders are those with two young parents or two old parents, and those most likely to do so have one young parent and one old parent. It matters little whether the old parent is the mother or the father: fledglings produced by very young mothers are twice as likely to graduate as breeders if their fathers are very old; fledglings produced by very young fathers are 50 percent more likely to graduate if their mothers are very old.[50]

These effects of parental ages on offspring viability are probably strong enough to have driven the evolution by natural selection of opposite-age mate preferences in both sexes. Given some choice in the hurly-burly of the colony mating market, boobies may in their youth favor the elderly for both pairing and extra-pair relationships, and in old age gravitate toward youthful mates. We need to remember though, that at pairing more is at stake than genetic material for offspring; a booby needs a partner who can defend territory, incubate eggs, and feed growing chicks, tasks at which mid-aged boobies likely outperform both the inexperienced young and the declining old.[51] And the mate a booby prefers may seldom be the one it gets; old boobies, for example, are always a small minority of adults, so young boobies who prefer to pair with them will normally be disappointed. Most young boobies will have to make do with mid-aged or young partners simply because there are plenty of them around. Quite likely, the rules of thumb of different-aged boobies for mate choice at pairing and during infidelity, and in different social circumstances, are complex.

To our great frustration, we've been unable to come up with a neat experimental test of which male ages particular females prefer. If our subjects were small song-birds, we might be able to cage females temporarily with candidate male partners of different ages and observe which ones they court with most enthusiasm, but such manipulations of colony-nesting, seagoing, plunge-diving birds as big as geese are unthinkable, especially when you know that cages induce lassitude in them. Of course, we can *wonder* about preferences, and when I watch pairs of boobies pos-turing and parading on the beach or forest floor, earnestly evaluating each other's feet and displays, I often wonder how much additional information is available be-yond the cues in their webs to age, health, and recent sabbatical experience, and how flexibly the cues are being used. In the context of choosing extra-pair sires though, it does seem that females, young and old, probably opt for males whose ages contrast with their own.

Quite *how* parental age combinations influence offspring's prospects for gradu-ating was a wide-open question and of burning interest. We could imagine individ-uals with different-aged parents having superior immune systems, greater prowess at plunge-diving, or superior social skills, endowed by unidentified mechanisms of genetic inheritance or bi-parental care. As a shot in the dark, we chose to test immu-nity because we could readily make indirect comparisons of chicks' susceptibility to ectoparasites by counting the ticks on their bodies. Numerous ticks in a chick's axillas could signify weak immune responsiveness, assuming that bloodsucking elicits in-flammation of local tissue and this immune response slows blood extraction, pro-longing each tick's visit. Alternatively, tick infestation could reflect poor parental care. For example, parents might let chicks down by locating their nests in terrain ridden with ticks or by grooming them carelessly, and mothers might shortchange their off-spring during laying by underdosing eggs with maternal antibodies. Either way, if chicks with different-aged parents harbored fewer argasid tick larvae, it would get us started on tracing causal pathways linking parental age combinations and the success of fledglings at becoming breeders.

Ale was game to add tick counting to her list of marketable research skills. She made daytime counts of larvae in 5 cm^2 circles under the downy white wings of chicks in 35 broods, scooping each of them from under its protesting parents three times: when it was 2, 4, and 6 weeks old. Her data showed that young blue-foot chicks with different-aged parents are burdened by fewer ticks than chicks with two young parents or two old parents.[52] In nine out of ten cases, these parents were the biological parents, so it is their ages that influence infestation, either through some inherited factor or through the parental care they jointly provide.

This finding gives a pointer, albeit open to diverse interpretations, to the mech-anism by which combined parental ages affect chicks, and it edges us closer to under-standing what blue-foot females gain through their infidelity. Interestingly, the fact that some females manipulate the age combination of their offspring's *biological* par-ents by obtaining sperm and genes from extra males, who contribute no parental care at all, implies that the benefit of different-aged parents to offspring is mediated not by care but by inheritance. Other mechanisms of inheritance such as yolk constituents cannot be discounted, but it could be the combination of genes from a young parent with genes from an old parent that makes for the most viable fledglings.

Our good genes quest continued. We plunged deeper with a field experiment designed to test this last suggestion. It asked whether the superiority of fledglings with different-aged parents derives from the genes parents provide to offspring or the care parents bestow on offspring during incubation and brood care.[53] After considerable planning, recruitment of collaborators, and elaboration of theoretical predictions, Cris and I swapped clutches reciprocally between scores of pairs with different age combinations. Early in each of two successive seasons, we estimated the ages of eggs from the angle at which they floated in water, measured and marked them, and ferried them in heated boxes between nests, to which we gained access by pushing their vociferous, lunging parents back with forked sticks. Our camp buzzed with activity as five students advised (mostly remotely) by four UNAM researchers measured egg volumes, incubation temperatures, hatching success, hatch weights, growth, innate and acquired immunity, susceptibility and resistance to ticks, and survival through fledging. In our camp seminars, we were discussing the predicted effects of swapping clutches between pairs with different age combinations, and there was tangible suspense. We had momentum.

But the clutch-swapping experiment was doomed. All the while, under our noses but beyond our ken, unsuspected numbers of milk snakes were prowling through the colony at night and feasting on our experimental subjects, stealthily engulfing and digesting nearly half of the annual crop of blue-foot hatchlings. We aborted clutch-swapping in the second year when we confirmed that snake predation had not abated and our samples of swapped clutches would be inadequate. Then we suspended the experiment more definitively when the COVID-19 pandemic froze most ecological fieldwork in Mexico in early 2020. Worldwide, three-quarters of long-term ecological monitoring studies halted. At the insistence of Cris, Cheko, and our institute's director, we kept our long-term program alive by developing travel and camping protocols for skeletal teams to monitor the colony, with the logistical support of fishermen. But the future is uncertain. Currently, we see field experiments as an unwise luxury, and we're apprehensive about the future of Isla Isabel's marine birds under increasing pressure from snakes.

Year-to-year ecological variability in natural habitats often frustrates ecologists' studies and experiments. It's part of the variety in nature that we want to inventory and explain, but it can leave us high and dry. Over the decades, El Niño events have occasionally obliged us to abruptly abandon planned studies and switch to different research questions, or even to different species. The current surge in snake numbers is different though, unwittingly caused by well-intentioned conservation measures. This time a component of the national park authority's island restoration program backfired, shocking ecologists and putting the future of the booby colony at risk. No one predicted that the elimination of the island's long-standing population of invasive black rats would put the marine birds at risk. Quite the opposite! In 1995 when

Fig. 8.5 A pair of courting blue-foots pause to briefly examine a prowling milk snake.
Photo by Hugh Drummond.

Cris and I inaugurated a program of feral cat eradication on Isla Isabel, a precursor of the national policy of restoring marine islands by removing invasive plants and animals, prospects for Mexico's native island ecosystems looked better than ever. After 3 years of Cris's tenacious efforts, the cats were gone and ecosystem benefits were evident: the gull colony was spared sprees of surplus killing, the spinytail iguana population boomed impressively, Clark's spiny lizards became more conspicuous, and the decline of the sooty tern colony was arrested. Then after the park authority hired professional island restorers to eliminate thickets of pineapples, banana trees, and sugarcane, along with invasive grasses and a stand of lemon trees, native vegetation recolonized areas where the exotics had kept it at bay, and the island *felt* more pristine, more authentic.

The prosecution's case against the glamorous red, black, and yellow serpents? Ever since Ceci and Alicia inaugurated the systematic study of the booby colony in 1981, we've spotted the odd milk snake coiled underneath a brooding blue-foot or brown booby parent at dawn or dusk, methodically ingesting a hatchling while its protector twitched nervously, unprepared to thwart a snake. When chick disappearance became more common, to the point of seriously affecting the colony's annual crop of fledglings and undermining our clutch-swap experiment, we were obliged to study it.

During the boobies' hatching period in 2018, the second year of the experiment, a pair of snake monitors wearing red-light headlamps strolled twice nightly in single file along a 900 m path through the booby colony, capturing an average of seven snakes per night and massaging their stomachs to elicit regurgitation of prey. More than 200 snakes, mostly just over 1 m long, were marked that year to permit the identification of individuals. The day after capture, monitors released the glossy, good-natured snakes (they don't bite even when stretched out for measuring or massaged to obtain stomach contents) at forested spots more than half a kilometer away in two separate drainages, to test their ability to find breeding boobies and our ability to keep them away from the colony.

As we feared, the released snakes promptly returned to the part of the booby colony where we had captured them, indicating an impressive ability either to home to where they come from or to locate newly hatched boobies. Steep rocky ascents and descents were no obstacle. Seven hundred recaptures of those marked individuals were made during the 4-month study, some snakes showing up near the monitoring path just a few hours after release! Undeterred by repeated captures, manhandling, detention in cloth bags, and translocation across watersheds, the two most persistent snakes—one of each sex—spent the greater part of the whole 4 months, not in the drainages where we released them, but in the booby colony close to our monitoring path.

Whenever our monitors came upon a milk snake, it was prowling solitarily among the boobies' nests. The boobies were incubating, brooding, or simply standing beside their partners, apparently ignoring the snakes, whose lateral undulations carried them silently over sand, soil, and leaf litter, tongue-flicking, probing for odors of hatchlings. Captured snakes' stomachs were nearly always empty, implying that Isla Isabel milk snakes seek refuge in crevices after ingesting prey, but 28 individuals regurgitated nearly intact prey, all of them booby hatchlings 1–3 days old.[54] Beyond that age, chicks are too bulky to swallow.

We plundered the booby database to trace the incidence, over the decades, of sudden disappearances of hatchlings from nests in circumstances implying snake predation. Across 31 years of booby monitoring, 16 percent of nearly 27,000 blue-foot hatchlings in our study areas were predated by milk snakes, but the proportion predated depended on which species of invasive mammals inhabited the island at the time.[54] To interpret the proportions, you need to know that feral cats on Pacific islands mostly eat mammals, especially rats, and that black rats on Pacific islands severely depress populations of some snakes, including milk snakes, by gnawing and feeding on them. In the 7 years from 1989 to 1995, when there were cats and rats on Isla Isabel, 5 percent of blue-foot hatchlings were predated by snakes. Then cats were eliminated and during the next 14 years, a mere 2 percent of hatchlings were lost to snakes. In the absence of cats, rats must have multiplied and increased their persecution of milk snakes, curtailing the snakes' predation on hatchlings.

If island restoration efforts had ended then, the boobies might still be rejoicing and our experiment might have worked out, but the restorers followed up with a helicopter-based poisoning program in 2010 that instantly annihilated the island's rat population and led to disaster. The proportion of hatchlings annually predated by snakes climbed during the next decade to about 50 percent, presumably because the snake population burgeoned after their predators were gone.

It's a sorry but fascinating story. Removal of one introduced mammal brought partial relief to the blue-foots; removal of the second introduced mammal prejudiced blue-footed boobies disastrously and may, for all we know, be similarly impacting other ground-nesting prey of the island's milk snakes such as brown boobies, sooty terns, and whiptail lizards. Restoration ecologists know that removal from islands of top predators like feral cats can "release" predators of the next level, such as introduced rats, allowing them to propagate and prejudice native species. To avoid such ecosystem upsets, they counsel removing cats and rats simultaneously. Understandably, no one foresaw that the successive removal of introduced cats and rats *from a species-rich tropical island* would release a lower-level *native* predator to prey on the offspring of a charismatic top predator of the ocean ecosystem: a bizarre but, with hindsight, logical outcome.

Good Genes—Last Try

A more critical test of the good genes hypothesis was the thesis project of Juan Pablo Ramírez, a brilliant undergraduate who, after graduating, contrived to ride out the pandemic by pursuing a master's degree in behavioral ecology at universities in Europe. Juan Pablo sought hard evidence to confirm or reject a prediction that goes to the heart of the hypothesis: that chicks sired by extra partners are superior to chicks sired by females' regular partners. By following up the within-pair and extra-pair chicks uncovered by Ale's paternity study, he dealt a blow to the hypothesis, as it applies to blue-foots. Information in our database revealed that the 55 chicks sired by extra partners did no better on average than more than 700 chicks sired by mothers' regular partners.[55] They grew no faster and were no more likely to fledge, they were no more likely to ever graduate as breeders, and they did not graduate at earlier ages. More damning still, during their first 10 years of life they neither bred more often nor produced more eggs or fledglings. So having your mother's extra partner for a sire makes you no better at growing and fledging, nor at competing in the adult realm for territories and partners, or producing offspring.

A good genes chauvinist could object that offspring of extra partners might yet shine in old age (an analysis we haven't done yet); or that, for all we know, mothers might be masking the genetic superiority of extra-pair offspring by investing less in them, selfishly diverting the benefit brought by extra-pair sires from the offspring to themselves. (Interpreting observations can get pretty complicated!) That would be to clutch at straws though. For the time being, Juan Pablo's study has persuaded me that genetic upgrade of offspring is not a *general*, colony-wide consequence of extra-pair paternity. However, it leaves open the possibility that particular extra-pair pairings, such as those of females with attractive partners and those of young and old adults, could furnish genetic upgrades.

Final Word on Function

Although we struggle to understand its adaptiveness in birds generally,[30,56] female infidelity is too big a phenomenon to write off as a nonadaptive behavior that is inconsequential or even dysfunctional. For pair-bonded female blue-foots in crowded neighborhoods at least, it is self-evidently an important activity. They maintain prolonged relationships with one to three extra partners, interacting frequently and energetically with them. And it's all voluntary. Multiple paternity of avian broods is sometimes attributed to forced copulations by extra males, or to females submitting to extra males' coercion to mitigate the costs of resisting. But female boobies are substantially bigger and stronger than males, and we witnessed their transparently willing participation in extra-pair copulations and relationships. Roughly half of blue-foot females copulate with extra males, with increasing frequency as laying approaches, and 11 percent of them grant paternity to an extra male. Such behaviors are not indulged in lightly. Their seriousness in family life is reflected in the marked (but not total) reluctance of females to copulate with extra males when their partners are around, and by the willingness of wised-up male partners to destroy suspicious eggs in their own nests. If females run this risk, then it's highly likely that on average a female's infidelity somehow increases the number of grand-offspring she produces, most likely by boosting the number or viability of offspring she fledges or by increasing her own longevity.

We made a case for a particular type of infidelity benefitting some females by upgrading the genetic quality of their offspring, but it's not conclusive and there are loose ends. Young females paired with young males seem to use extra-pair liaisons to arrange for their offspring to have biological parents with the optimal age combination—young plus old, and old females paired with old males may be up to a similar trick. The offspring of different-aged partners certainly seem to be of high quality: they bear few ectoparasites, hinting at superior immunity, and they are especially likely to become adult breeders. However, for two reasons the case is not closed: first, because we haven't shown that the extra-pair offspring of similar-aged partners outperform their within-pair half-siblings; and second, because we can't be sure that the superiority of offspring with different-aged parents is due to their genes rather than the parental care they receive, or a combination of genes and parental care. The milk snakes kept us from answering that question.

We also found circumstantial evidence for other plausible functions of female infidelity: facilitating switching and egg-dumping. Paired females sometimes pair anew with their extra partners, which might increase their reproductive success or, in the case of abandoned females, salvage their opportunity to reproduce that season. And dumpers seem to weaken the nest defense of host males by establishing sexual relationships with them. Finally, we shouldn't forget fertility assurance: in the Isla Isabel colony, hatching failure through infertility is common, so it's highly likely that stored sperm from extra males sometimes save the day when the sperm of male partners fail to do their job.[57]

Overall, I suspect infidelity is worthwhile for blue-foot females because, depending on circumstances, it can serve a mixed bag of reproductive functions. Many individuals may pay a net cost for their infidelity, through wasted energy, venereal infections, retaliatory desertion by partners, or egg destruction by partners, but the average female's infidelity surely boosts her lifetime reproductive success somehow or other.

Male boobies seem to do rather well out of their infidelity, managing to parasitize the parental care of their rivals in 11 percent of nests, but the costs incurred by parasitic males at their own nests have not been quantified. Maybe unfaithful males risk desertion, parasite infestation or venereal disease, and neglect of defense responsibilities during their forays might make their territories vulnerable to incursions. Certainly, it would be a mistake to idealize male blue-foots as conquering studs when 11 percent of them are duped into the gross error of raising other males' offspring and others are no doubt driven to mistakenly sacrifice their own offspring by tossing or puncturing the wrong eggs. We expect natural selection to have fine-tuned males' infanticidal behavior so that on average it makes a positive contribution to their lifetime reproductive success, and that may well be the case. Even so, every time a male makes the wrong call, he destroys his own daughter or son, and I can't help suspecting that some males who fail to repel extra-pair partners who dump in their nests are suckers.

One reason the infidelities and countermeasures of blue-footed boobies intrigue and entertain us is that booby behavior resembles human behavior. Likewise, the competition between booby siblings echoes our children's rather different competitiveness. Are these just amusing coincidences? Embarrassed and evasive smirks may be our natural response to this question, but shouldn't we also try to get to the bottom of it? Shouldn't we take seriously the idea that similar behavior could have evolved in both species under the influence of similar selective pressures? Whether we warm to it or not, the hypothesis of similarity due to convergent evolution in boobies or their ancestors and humans or their ancestors is plausible. In the next and final chapter, we'll examine its merits by taking a serious look at the many parallels between boobies and humans modern and ancient.

9

Are Humans Similar?

What Do We Expect of Humans?

After-dinner tales of boobies' sibling strife, partner relationships, and infidelity get people's attention. People want to know how and why chicks fight, how parents deal with sib fighting, what females and males get out of their dalliances, and how they both deal with a partner's infidelity and jealousy. Similarly, after Hamilton's and Trivers's theories in the 1960s and 1970s (Chapters 1, 5, and 8) raised the possibility that sibling conflict,[1–4] parent–offspring conflict, and sexual conflict are evolved behaviors widespread across nature, my colleagues piled in: professors reoriented their research from behavioral mechanisms to sociobiology, graduate students flocked to sociobiology programs, and for decades animal behavior conferences buzzed with talk about selfishness, altruism, and adaptive strategies. Ethologists rebranded themselves as sociobiologists or behavioral ecologists. Dawkins's million-copy book *The Selfish Gene*,[5] extending and popularizing theories about family cooperation and conflict (and much more) was translated into 25 languages and raced through 40 editions.

Fascinating as the theories are, profound as their implications may be, there may have been more to their widespread acceptance than meets the eye. Sure, some readers are simply intrigued and amused by the idea of animals deceiving and slugging it out with kin. But some may also feel that the occurrence of selfishness, imposition, and deception in other species relieves us of some guilt. *Animal behavior might explain human misbehavior.* And behind such thoughts lurks the suspicion of common causality: that similar behavior in humans and other species such as boobies could have evolved through natural selection and for similar reasons.

It's a jarring and demeaning idea for some but deserves to be taken seriously. We should at least consider the possibility that both species—blue-foots and humans—oppress their siblings and cheat on their partners because in the distant, and even recent, pasts of both species, genes linked to those behaviors outcompeted genes for fair and loyal dealing. Since Darwin explained that humans are an evolved species, too, there have always been people who indignantly insist that human social behavior is above such biological influence because humans inhabit a world of reason, morality, and culture. Yet the theoretical proposals for sibling conflict, parent-offspring conflict, and partner conflict discussed in Chapters 1, 5, and 8 are applicable to all animal species that reproduce sexually and care for their offspring, so evolved family conflict in humans is entirely on the cards. Putting aside species chauvinism, there is no reason to suppose that the evolution of human family behavior has ever been exempt

Blue-Footed Boobies. Hugh Drummond, Oxford University Press. © Oxford University Press 2023.
DOI: 10.1093/oso/9780197629840.003.0010

from the pressures of natural selection, though there has been scope for cultural in-novation to modify selective pressures and bring about quantitative changes in our behavioral tendencies, particularly during the last 10,000 years. Some animal species, like most snakes and pelagic fish, may be exempt because their lifestyles provide scant opportunity for interaction between mothers and fathers, between siblings, or be-tween parents and offspring. But in a monogamous, group-living primate with over-lapping offspring that require many years of care, intense family cooperation is the name of the game, and evolved conflict among family members is expected.

What's more, it's evident that humans come into this world prepared to express some naturally selected social inclinations. Human minds, far from being blank slates,[6] host a collection of evolved urges and emotions that develop under environ-mental influences in every individual, along with a unique and impressive rationality (with its own biological biases). For example, natural selection has equipped us with inclinations to hold back from the edges of precipices and flee from predators. Why then wouldn't it have equipped us with a tool kit of urges and emotions for dealing with social challenges, including conflicts of interest between siblings, between par-ents and their offspring, and between the sexes?

However, rather than booby-like rules of thumb that dictate specific actions, we expect human youngsters to have natural inclinations to do what it takes to outcom-pete siblings for food and care. Similarly, we expect human adults to experience urges and emotions that motivate the strategic choice of mates, artful courting, discrete extra-pair liaisons, and measured investment in offspring. Such urges do not depend on language, and they influence thinking and motivate behavior at all ages.[7,8]

In Steven Pinker's conceptualization,[6] they are deep, probably universal, mecha-nisms of mental computation that generate behaviors that vary across cultures and contexts and *have to be learned*. For example, a man's urge to increase his status could provoke him to hunt and kill valuable prey for his band, lead a raid against hostile neighbors, or host a sumptuous banquet. The mechanisms are probably shaped and calibrated by whole suites of genes of small effect,[9] the hidden agents that don't con-trol but fine-tune human strategies for cooperation and conflict.

Aren't humans exceptional though? Yes, undoubtedly. Humans are truly excep-tional animals in many ways. Their outsized brains allow them to store and process reams of information, communicate complex thoughts using the symbols and syntax of language, and, paired with their social inclinations, cooperate with other humans on scales varying from the family and neighborhood to nationwide and even glob-ally. These capacities have allowed them to re-engineer their feeding ecology, their living spaces, and the surface of the planet; accumulate cultural innovation in every-thing they do; and adopt lifestyles unimaginable in any other species. But it doesn't necessarily follow that under the cumulative cultural change of increasingly com-plex civilization humans have evolved to leave their animal nature behind. Even if their intelligence often enables humans to govern and even override their natural inclinations by rational deliberation—a truism, surely—they are not exempt from evolutionary processes that keep them on track to reproduce and pass on their genes. The evidence for this is there in the natural inclinations that influence our social behavior and in our conscious struggles to cope with them.

To encourage reflection on the origins of our behavior, in this chapter I will hold up blue-footed boobies as a mirror to humans. Taking the two major themes of the book, I'll argue that sibling conflict and sexual conflict manifest in human families and blue-foot families for similar reasons. Naturally, the behavior is different in form between two species so unlike in their ecology and their physical and mental capacities, but it serves similar functions. After sketching out the sorts of conflict behavior an evolutionary biologist might expect to see in a species as extraordinary as *Homo sapiens*, we'll peer into the scientific literature to find out what actually goes on in human families. The more human family conflict looks like a different version of booby family conflict, albeit more variable and engineered differently for the characteristics and circumstances of a group-living hominin, the more we may be tempted to infer common biological causation. On this basis, I will argue that humans resemble boobies in ways that suggest *convergent evolution*. That is, in response to similar selective pressures, each of these species or its ancestors evolved similarly conflictive family behavior, just as the need to get airborne resulted in birds, bats, and some insects independently evolving wings and flight.

What should we expect of the descendants of the tool-using bands of humans who gathered plant foods and hunted animals during 200,000 years of the Pleistocene, transitioned to agricultural food production roughly 10,000 years ago, and adopted industrial manufacturing, capitalism, and city dwelling during the last 200 years? Monogamous partnerships along with extended dependence of offspring on care and provisioning by parents and others set humans up for the types of conflict we analyzed in blue-footed boobies: between siblings, between parents and their offspring, and between partners. Monogamy is the human way. It prevailed in historical times and could well have prevailed among our hunter-gatherer ancestors, along with some polygyny (one male/multiple females) and polyandry (one female/multiple males). Although humans are flexible in their mating behavior, monogamy predominates in all contemporary societies, including those that permit polygyny, in most of which fewer than one male in ten is polygynous.[10–14] Human partnerships, which arise spontaneously through courtship or are arranged by parents or other clan members, are of variable duration. Females produce offspring every few years, and overlapping offspring share in the care and resources that caretakers provide.

We expect human family conflicts to be expressed in more variable and nuanced ways than those of boobies because humans live in cooperative social groups and they are psychologically sophisticated and highly flexible in their social behavior. Several lines of evidence, including observations on contemporary hunter-gatherers, suggest that during the Pleistocene humans lived in egalitarian bands with a culture of food sharing and, like silver-backed jackals and Florida scrub-jays, were *cooperative breeders*, meaning that children were fed and cared for not only by parents but also by kin, especially siblings, grandmothers, and aunts.[15–18] In these societal and family contexts, the evolution of cooperation, mutual understanding, and empathy were promoted by three mechanisms of natural selection—mutual benefit, kin-selected altruism, and reciprocal altruism,[19,20] making *Homo sapiens* the planet's most quintessentially cooperative primate. Nonetheless, selection for sibling conflict and sexual conflict would have persisted.

We must bear in mind that, like boobies, humans evolved to survive and breed not in any Garden of Eden but in fluctuating and often stressful ecological circumstances. For example, thousands of generations of our ancestors faced food shortages in the wildly fluctuating climate of Africa in the Pleistocene,[21] and then hundreds more generations lived through the famines ushered in by the invention of agriculture.[22] During such privation in particular, natural selection would have been unrelenting in equipping adults with strategies to produce offspring and ensure their growth and survival and equipping children with strategies for getting a greater share of provisioned food than their siblings. It would be remarkable if the psychological underpinnings—emotions and urges—of our ancestral social behaviors had been lost during the relative eye blink of the last several thousand years. So we expect to see them influencing the social behavior of all extant societies, including traditional societies (hunter-gatherers and peoples who have transitioned to pastoralism and agriculture) and those with industrial economies.

How might children get their own way in competition for food with siblings? Their own way, of course, is not utter selfishness. Under Hamiltonian inclusive fitness theory, it means taking priority over a sib whenever the selfish benefit obtained will be, on average, more than half as great as the average cost to that sib (Chapter 1). Obviously, dependent children won't bring this about by hammering at each other's heads; plausible tactics for them to prevail include imposition, coercion, persuasion, deception, and aggression, depending on the circumstances. Naturally, wheedling extra investment from parents and other caretakers is expected, too, for example, by charm, tantrums, simulating illness, or faking the dependence and vulnerability of a younger child.[4] As to which resources will be contested, I always think first of food because the growth and development of progeny is fueled by food, but warmth, space, protection, social support, and diverse physical resources (e.g., animal skins and weapons) could have been key during the Pleistocene when our ancestors inhabited a variety of ecosystems and habitats. Therefore, children's competitive urges may not be specifically tied to food but more general so that they can usefully attach to whatever counts locally. Among the breathtaking cultural variety of the present, we should probably anticipate that competitive urges will attach to different resources in different societies depending on variation in such factors as weather, clothing, housing, food, transport, education, entertainment, and games.

However, we need to always keep in mind that positive interactions between siblings are expected, too. A child should give generously to a sib whenever the cost incurred is less than half the benefit to the sib, and children have an interest in establishing relationships of mutual benefit and reciprocal exchange with sibs and, to a lesser extent, others. They might well be aggressive, but they are also expected to be sensitive, affectionate, and kind to their closest kin and those who treat them well.

What are human females and males likely to get up to in their mating behavior? If females can profit from extra-pair behavior, for example, in some of the ways that blue-foot females are suspected of profiting (Chapter 8), then they, too, are expected to court and copulate with extra partners, if they can get away with it. We saw that, in theory, the infidelity of a female booby might facilitate partner-switching, provide fertilization backup, secure better genes for her offspring, or induce the extra male to be kind to her offspring if it blundered into his territory. All of those functions are on the table for humans, too, although kindness to offspring needs to be a broader concept including economic assistance: extra males might provision a female or her offspring. And if, like paired males of most monogamous bird species, human males follow a mixed reproductive strategy (Chapter 8), then we expect them to court and copulate not only with their partners but also with extra females, in order to sire offspring at other males' expense. As for both partners' resources for coming out ahead in such conflict by persuading, seducing, deceiving, or coercing each other, well humans are consummate mind-readers and actors and they're skilled in self-deception,[23] so the expected limit is the sky.

To detect similarities between humans and boobies we draw on behavioral ecology, anthropology, psychology, and sociology. There's a technical problem though, because different disciplines tend to ask different questions and use different methods of research, so the information from different fields is only broadly and patchily comparable. For example, anthropologists don't do much experimenting and developmental psychologists who study conflict resolution don't ask whether human children develop dominance relationships or how the conflict between siblings affects their access to food or survival. On the other hand, data riches arise from a diversity of methods. For example, humans are uniquely able to provide data by verbally reporting on their actions and thoughts and on the lives of their families, so interview-based studies of evolutionary psychologists have accumulated a wealth of information on urges, emotions, and private behavior. In comparison, our direct observations and experiments on what boobies actually do in nature are often more precise and trustworthy, but too laborious for probing as widely into behavioral inclinations and tactics. The net result is an uneven patchwork of the descriptive and experimental studies of wild boobies we've already seen plus the more varied social science data on human behavior and psychology that follow. You will make your own judgment on whether that patchwork allows us to conclude that the family conflict of the two species, although different in many ways, is similar and similarly explained by evolutionary theory.

First, we'll look through a behavioral ecologist's lens at the conflict between siblings between birth and the end of adolescence and between reproductive partners in industrial societies. Each of these reviews will end with a mention of how the same conflict manifests in traditional societies. The latter are of special interest for gaining evolutionary perspective because their behavior is arguably the closest approximation we have to how siblings and partners interacted over the eons when the emotions and urges of *Homo sapiens* evolved. Ecologically and culturally, they may have departed less from ancestral conditions and conserved more customs than the societies we are most familiar with. On the other hand, social behavior in industrial societies tells us how those same inclinations develop and play out when communities are composed of vastly more than a few dozen members, when resources are often plentiful, when fertility is controlled by contraceptives, when states provide care and education, and when some social behavior is regulated by law.

Throughout these two reviews, readers should keep a mental tally of whether and how human behavior echoes the booby behavior described in earlier chapters. It's unlikely that similarities in the family behavior of two markedly different species from lineages (mammals and birds) that separated more than 300 million years ago are due to both of them retaining characteristics of their last common ancestor. Similarities in family behavior of humans and boobies more likely arose independently in their two lineages through convergent evolution. And those similarities may have been maintained into the present because both species continue to face similar selective pressures and have evolved to maximize the inclusive fitness of individuals.

Sibling Conflict

In his book *Sibling Aggression: Assessment and Treatment*, American psychologist and family therapist Jonathan Caspi,[24] no biologist himself, describes the dark side of contemporary human sibling relationships thus:

> Both sibling violence and abuse entail a range of physical and verbal acts perpetrated with the intent to do harm, and involve the same behaviors.... "Violence" reflects mutual or bidirectional aggression, in which both siblings' aim is to harm the other, in the context of a perceived egalitarian relationship. "Abuse" is unidirectional hostility where one sibling seeks to overpower the other via a reign of terror and intimidation and reflects an asymmetrical power arrangement.... Siblings overwhelmingly perceive their violent interactions to be mutual ... but what may appear to be a mutually violent encounter may, upon closer inspection, reveal a clear perpetrator with a victim defending himself or herself.... "Relational aggression" are behaviors that intentionally hurt others through relationships (e.g., gossip, peer exclusion, distributing embarrassing pictures), and some evidence suggests that it occurs between siblings more than physical or verbal violence.

Lest you be tempted to dismiss Caspi's professional opinion as an exaggeration, it may be salutary to consider what we know about public attitudes to sibling violence. Citing Caspi's book and eight supporting studies, two British psychologists, Roxanne Khan and Paul Rogers,[25] preface their analysis of public perceptions of sibling violence with these sobering comments:

Sibling violence is widely tolerated and commonly thought to be symptomatic of most, if not all, sibling relationships. This normalization—hence minimization—of sibling violence may result from its pervasiveness and/or the misperception that physical conflict resolution is, for children at least, character building. Consequently, parents are not always motivated to intervene when sibling violence occurs. Such parental inaction is likely to vicariously reinforce its personal, familial, and social acceptability. Similarly, the language often used to describe sibling violence (e.g., as "rivalry" and "horseplay") reflects further minimization of violence into minor altercations with seemingly little impact on victims. As such, siblings who report sibling violence victimization are more likely to be blamed either for provoking their assailant and/or for not defending themselves. To this end, as sibling violence is not only accepted but also expected, it tends to be normalized by family members as well as health professionals.

Like me, readers may be too young to remember when wife-beating was declared in the 1930s to be as American as apple pie, but I remember when "corporal punishment" of school children was normal in England, and when parents there commonly beat their children with slaps, canes, or belts, doing so out of a sense of duty. All so twentieth century, it sounds now, but the beating still goes on in private. Verbally and physically assaulted wives and children are all around us, although violence has declined under the influence of legal prohibition, education, and change in attitudes. Maybe sibling relations are the next domain where a new awareness will dawn, attitudes will change, and the violence we currently trivialize will be noticed and viewed as unacceptable. We can and do rein in our unacceptable tendencies, including those that are rooted in our biology. But how common is sibling aggression?

Incidence of Aggression

Social scientists working to non-evolutionary agendas in industrial societies have incidentally documented tendencies consistent with the broad expectations of family conflict theory. Foremost is a clutch of population-level studies of the prevalence of violence in American families. The first probing census of family violence in the United States—*Behind Closed Doors: Violence in the American Family*, by Strauss, Gelles, and Steinmetz[26]—was based on interviews 40 years ago of a nationally representative sample of over 700 families with two or more children. This shocking and controversial book, still cited today and reprinted in 2017, revealed that violence between siblings was *the most prevalent category of family violence* and that many families readily acknowledged it as a normal aspect of their relationships. Every year only 2 percent of parents punched, kicked, or bit each other, compared to 42 percent of 3- to 17-year-old children doing similar things to their siblings. Violence between siblings peaked at age 3–4 years (a common pattern), but even at 15–17 years two-thirds of children were hitting their sibs at least once a year, and the hitters, mostly boys, were hitting 19 times. Variation among families was great and some families were extreme, one mother commenting, "It's a wonder they're not all bruised and bloody. They must get tired of yelling at each other. They fight all the time. Anything can be a problem, it's just constant, but I understand that this is normal. I talk to other people, and their children are the same way."

Ten years after *Behind Closed Doors,* a follow-up study based on more than 8,000 American families reported that 40 percent of 3- to 17-year-old children had engaged in severe acts of physical violence against a sibling in the last year. Probably of greater significance was that *85 percent of children regularly used psychological aggression on their sibs,* meaning verbal and emotional aggression involving such behaviors as taunting, denigration, and threatening.[26] One wonders whether physical violence is the tip of an iceberg of more widespread and continuous psychological hostility.

Famous though they briefly were, those censuses failed to dent the popular myth of all-loving brothers and sisters, and in 2013, when "sibling aggression remained an un-recognized form of violence by researchers and the general public,"[27] more rigorous interview sampling of children and their guardians across the United States again confirmed that sibling aggression was commonplace in American homes. Among close to 2,000 children aged 1 month through 17 years, 40 percent of boys and 35 per-cent of girls had been victimized by a sibling during the previous year. Most of those had suffered physical assault, but a minority had suffered theft or vandalism of their property, and a smaller minority reported psychological aggression, including name-calling, belittling, spiteful teasing, and threats to injure them, their pets, or their prop-erty. Most affected were younger siblings, children close in age, and preadolescents, and rates of physical injury increased with age, from 4 percent of 2- to 5-year-olds to 23 percent of 14- to 17-year-olds.[27] Bullying—subjection to severe aggression sev-eral times a week—was suffered at this time, less than 10 years ago, by 11 percent of 12-year-olds.

Overall, censuses in industrial countries do not paint a picture of ubiquitous violent aggressiveness but are suggestive of a widespread propensity for siblings to interact aggressively. Aggressiveness could have been underestimated because interviewees are likely to minimize their behavior and censuses often omit the more subtle forms of aggression. *Relational aggression,* the insidious form of aggression highlighted by Jonathan Caspi, escaped those censuses. It occurs when a youngster acts deviously to isolate their sibling socially, for example, by starting rumors, persuading others to ignore the sibling, restricting the sibling's participation in social events, pranking the sibling in public places, or posting humiliating information or images of the sibling.

Competing for Property or Food?

Studies of diverse societies—contemporary Britain, a preindustrial population of the United States, and 27 developing countries of sub-Saharan Africa—hint strongly that food is a *limiting resource* in human families under a variety of ecological circumstances. When there are more children in a family, they tend to grow more slowly and fewer of them make it through to adulthood, later-borns being the most likely to die.[28-30] This tantalizing observation implies competition for food, although sadly it tells us nothing about *how* children compete for it or parents' roles in the allocation of food among their children.

However, developmental psychologists analyzing sibling conflict in contemporary Western societies have pointed the finger at a quite different *casus belli*: competition throughout childhood over property.[31-34] For example, in an observational study of 40 Canadian families with two children aged 2–5 years, a silent female observer ("Pretend the lady isn't there") systematically recorded interactions during 9 hours spread across several days when either the mother or both parents were present in the home and the children were not engaging with video games or television.[35] Both sibs frequently violated family rules of possession, sharing, and ownership, provoking *more than six conflicts per hour* on average, as seen in other studies.[36] Elder siblings were the most frequent violators, specializing in bossing their siblings, verbal and physical aggression, physical imposition, excluding, tattling, and nagging. Younger siblings, for their part, specialized in interfering with their sibling's play and damaging family property. Parents intervened to prohibit aggression and insist on the children's property rights, especially when the victim's reaction was emotional. More often than not, elder siblings got their way, but when parents intervened, outcomes were more even. Significantly, in my opinion, property rights enjoyed the special respect of both generations: parents were particularly assiduous at remedying property violations, and children generally upheld rules for both ownership and right to possession even when parents didn't intervene. On the other hand, for children to share property with their sibs, parental intervention was often needed.

Similar themes emerged from interviews of more than a hundred dyads of American preadolescents.[37] Both siblings claimed that their conflicts centered on physical aggression and sharing personal possessions rather than parental attention. Older sibs usually got their way, and many conflicts were resolved by parental intervention. Contrasting these dyads implicitly with non-sibling dyads, the researchers suggested: "It is possible that quarreling over rights and property is a unique and consistent characteristic of the sibling relationship, at least while both children live at home."

In sum, data on human sibling competition in industrial economies are scarce but loosely consistent with the expectation that competitive urges will attach to different resources in different economic and social contexts. We infer from population studies of growth and mortality that children compete for food, although we don't know how they go about it; and researchers observe in domestic contexts that children compete intensely for property, using verbal and physical aggression, imposition, and other tactics.

Development over Childhood

The existence of an emotion that, on the face of it, is innate and potentially fuels sibling competition across a range of ages and in a variety of social contexts has recently been demonstrated by experimental psychologists. Tests in four industrial countries showed that children in the first year of life, toddlers, and preschoolers protest when their mother selectively directs her attention not to them but to a peer (a lifelike doll) of the same age.[38] At 9–14 months, children slowed or stopped their playing and strove to monopolize interaction with the mother by calling out and gesturing. Not only that, they intruded themselves between mother and doll, clung to the mother, and *swiped at the doll*. Some went as far as a full-blown temper tantrum, including hostile acts against the mother, the doll, or themselves, or they withdrew into themselves, or they froze. "Nascent jealousy"—negative responses to a beloved individual who preferentially directs attention to a rival—can motivate a variety of affiliative, protesting, rejecting, and hostile behaviors.

The tests also revealed 6-month-old babies expressing negativity via sad facial expressions, forward-leaning motor agitation, or mother-directed gazing. And children protested jealously when mothers directed their attention to peers, whether these were younger or older, protesting more vigorously if they had siblings, especially older siblings. Nor were children the only people treated as rivals: protests followed mothers switching attention to their spouses and fathers switching their attention to anyone. From a very early age when their cognitive grasp of competition is limited, children are emotionally primed to compete both peacefully and aggressively with sibs and other rivals. They compete by claiming the attention and acquiescence of caretakers, potentially increasing their access to resources such as breast milk, other foods, and objects at early ages when, historically, infant mortality is high.

Naturalistic observations have shown that the sudden emergence of a baby sibling into the life of an infant is a milestone. Western toddlers commonly respond in problematic and demanding ways. They regress to babbling,[39] a behavior they had grown out of but which may be just the ticket for competing with a baby; they turn aggressive; and they show signs of stress, including disturbance in bodily functions, withdrawal, dependency, and anxiety.[40] These innate tendencies are very likely exacerbated by the reduction in the quality and intensity of maternal attention that follows the birth of a sibling.[41–44] Interestingly, on the birth of a younger sibling, bonobos, our close primate relatives, experience a fivefold, 7-month pulse of cortisol, the mammalian stress hormone.[45] The bonobo pulse is not accompanied by behavioral change and its function is uncertain, but it is thought to be part of an evolutionarily ancient response to the stress that accompanies the birth of a sibling.

But we are neglecting the positive interactions expected between siblings. In sibships of toddlers, along with competitive behavior, there are also rudiments of the positive, altruistic, and reciprocal relationships between siblings predicted by inclusive fitness theory. After 18 months, children sometimes share with their sibs, help them, and even comfort them; and after 2 years, cooperation and conciliation are

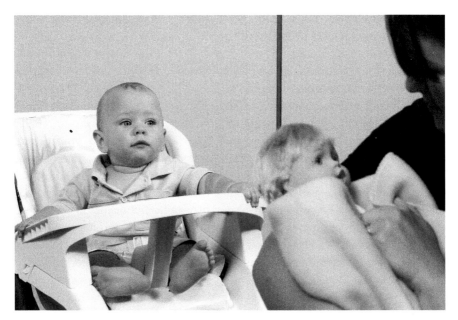

Fig. 9.1 Nascent jealousy. As his mother preferentially directs her attention to a lifelike baby doll, a 6-month-old male responds with a sad expression, mother-directed visual attention, approach posture, and motor agitation (Hart 2022).
Photo by Kenny Braun, courtesy of Sybil Hart, Texas Tech University.

increasingly shown to younger sibs, especially to those who have been cooperative.[46] Left for 4 minutes with their younger sibling, six out of ten preschool girls and four out of ten preschool boys responded to separation from their mothers by caring for their siblings—approaching and hugging them, reassuring them verbally, carrying them, or distracting them with toys.[47] And it's common for first-borns to become surrogate parents who emulate their parents in their care of their younger sibs.[48]

Thereafter, through the age of 5 years, fully one in five interactions between siblings involve intense negative emotions, far higher than in interactions with parents or friends.[40] There's unexplained variation though: some dyads are mostly hostile, irritated, and aggressive, while others are mostly affectionate, cooperative, and supportive, and still others vacillate between those modes. And when parents favor one sibling with more affection and attention than the other, their relationship deteriorates, particularly in situations of stress such as parental separation or illness in the family.

The elder sib's hold over the younger is shaky and partial—younger sibs don't just concede—and slackens with the passing of time. In middle childhood, parents relax their interventions on behalf of younger siblings, who develop the ability to push back increasingly against elder sibs' dominance, obliging them to control their sibs through coercion or rewards. Thereafter, as the difference in power declines, a more egalitarian relationship emerges and sibling influence becomes more bidirectional.[49,50] Conflict peaks at the transition from childhood into adolescence and continues well into adolescence. However, its nature changes as children mature socially and cognitively, and the difference between them in maturity narrows. Conflicts between adolescent siblings are greater when they are similar in age and maturity and increase with poverty and family stress,[51] implying increased competition in leaner circumstances.

And what of dominance per se, the hallmark of blue-foot sibling relationships? In Jonathan Caspi's therapeutic experience with American families, some sibling dyads experience relationships that in my judgment are functionally similar to the dominance-subordination in blue-footed boobies, spotted hyaenas, and common quail. Under what Caspi calls a *reign of terror*, victims end up just as defeated, defanged, and controlled as the typical blue-foot junior. "Psychological aggression reaches the level of abuse where it involves repetitive hurtful themes in order to diminish and overpower siblings. The hurtful words of childhood are often internalized and stay with victims throughout their life.... The threat of property damage is often terrorizing, and involves intent to cause psychological harm and to overpower."[52] The prevalence of these understudied relationships and milder versions of them in humans is unknown, no doubt because we normalize and minimalize them.

Comparing Industrial Societies

Expression of the urge to compete aggressively with siblings for resources undoubtedly varies with the social norms and cultural practices of different societies. It could not be otherwise. But how great is the variation in power imbalance? A sample of well over a hundred children in early and middle childhood and adolescence from each of four variably industrial countries confounded expectations.[53] First, mothers from each country scored power imbalance between their own children on a six-point scale by estimating, for example, how often one child bossed its sibling and ordered it around. Pairs of siblings from Turkey's collectivist culture and Holland's individualist culture showed similar power imbalances. Second, 11-year-olds of different ethnicity resident in the Netherlands scored their own power imbalances with their siblings on a similar scale. Children of all ethnicities showed power imbalance, but collectivistic Moroccan and individualistic Dutch children were almost equal and the Dutch were *less* imbalanced than the collectivist Indians.

There does appear to be cultural variation, but maybe the most significant finding was the existence of a sibling power imbalance in all cultures compared and at all ages tested.

Traditional Societies

In traditional societies, sibling conflict manifests after a mother, on becoming pregnant, suspends breastfeeding of her toddler (3 years old, on average), effectively consigning it to the company of other children in the local playgroup.[54] It's a land-mark event. Toddlers don't take readily to forced weaning, and during weeks or even months, some show signs of depression and malnutrition,[55] responses that, like the stress responses of toddlers in industrial societies (see earlier), could be interpreted either as imposed deficits or as ways of competing for nurture with the upstart rival.

Babies themselves come to this fray fully prepared, not only with insistent, loud, and highly aversive vocalizations (crying) that are most effectively quelled by ad lib-itum suckling but also with personal presentation cards that advertise their worthi-ness. To a mother agonizing over whether to neglect or sacrifice her newborn, the better to nourish her toddler, a baby's *chubbiness* proves that it's a viable and worthy recipient of investment, and may incline her to raise both rather than reducing her "brood" as blue-foots so often do. It's a persuasive pitch because extra body fat bodes well for a baby's ability to stay warm and for the healthy development of its brain;[16] rightly do mothers pay attention. Both crying and chubbiness are competitive tactics if, as seems likely, they evolved partly by inducing mothers to reassign resources from toddlers to babies, to the cost of the former and the benefit of the latter.

However, competition is by no means the whole story because children also treat siblings generously, and care of babies during their first year is often undertaken by 6- to 10-year-old sisters.[56,57] Among the Hadza people of Northern Tanzania, who hunt and gather in savanna woodland, young foragers, besides provisioning them-selves, give to other youngsters in amounts consistent with theoretical expectations. At 7–17 years, they share their spoils with other children, preferring to share with kin and providing to siblings, half-siblings, and cousins in proportion to their related-ness. What's more, there's reciprocity: they are more generous to the kin and non-kin who have previously given them more.[58] It's not surprising then that among the Ifaluk fishing and gardening people of Micronesia and the Mandinka subsistence horticul-turalists in the Gambia, children with elder sisters are the ones most likely to survive through age 5, and in other societies, children with elder sibs become more successful adults.[15,59]

Ambivalence of children toward their sibs is illustrated by the personal story of one of the !Kung San hunter-foragers who inhabit the western edge of the Kalahari desert. Nisa is highly regarded among anthropologists for her independent spirit and col-orful love life, and for her frank and articulate accounts of her own family conflicts. When her mother became pregnant with her little brother Kumsa, Nisa cried almost continually to oblige her mother to carry and suckle her. After his birth, she would sometimes detach him from her sleeping mother's breast in order to suckle, for which she was scolded. Yet on the day of Kumsa's birth, when their mother resolved to bury him because he'd arrived too soon after Nisa for mother to manage both, it was Nisa who declined to bring a digging stick and insisted he be spared.[60,61]

Interestingly, in the child playgroup to which a dethroned toddler is traditionally exiled, and where it integrates into a mixed-sex, mixed-age group of fellows including sibs, cousins, and non-kin, conflict with siblings seems to be light. The typical unsu-pervised playgroup develops a largely harmonious age-based dominance hierarchy that's respected by all its members. Parents grant older children authority to control and discipline younger ones, and in this circumstance, sibling pairs get along remarkably well, with the elder caring for and protecting the younger and the two of them mostly behaving more like allies than rivals.[54] It's been suggested that mothers' absence from playgroups helps take the steam out of rivalry between siblings because they stop competing for her attention.[62] Nonetheless, we expect the fundamental conflict of interests between siblings to manifest in competitive behavior, at least when food is scarce. Might it be that competition is sometimes mediated not so much by interactions between siblings, as by interactions between each of them and their caretakers, the winner being the sibling better able to charm treats from parents and other adults? Something like that goes on in the broods of most birds.

In agrarian and pastoral social groups worldwide, sibling relationships across the adult lifespan have been described as supportive as well as conflictive and *not* qualitatively unlike those of industrial cultures.[63,64] Elder sibs function as caretakers, guides, and educators, and they commonly retain some authority over their sibs throughout adulthood. Sibling conflicts and fights occur, and the relationship can deteriorate into tyranny, usually of the elder over the younger but sometimes in the opposite direction when elder sibs appease their wards to fend off parental disapproval. Throughout adult life, close sibling relationships are a mainstay of families in traditional societies, even though jealousy, competition, and conflict between sibs are common, and despite younger sibs chafing at the demands and conventional authority of elder sibs.

Sibling Conflict in a Nutshell

Although only a small fraction of the observations on human sibling relations in this chapter were collected to address evolutionary questions, the broad patterns that emerge align approximately with expectations of both competition and altruism between children under Hamilton's inclusive fitness theory. We focused mainly on aggressive interactions because those are what we studied in boobies, and we found them in child dyads of all ages. Children evidently enter this world equipped with emotions and urges such as nascent jealousy for competing with siblings and others, and this competitiveness manifests as early as the first year of life. Our best bet is that these inclinations were favored by natural selection because they help children survive and prosper in the company of competitors during their long dependence on caretakers, and quite possibly during adulthood. However, in this extraordinarily flexible primate, the competitive and conflictive behavior and the resources at issue appear to vary greatly with the society in which individuals develop.

In industrial societies, children's evolved emotions and urges incline them to compete by a variety of means. For example, during early and middle childhood they compete for caretakers' attention, they are reluctant to share, and they readily resort to imposition and aggression, including physical and emotional aggression. Elder siblings tend to dominate and get their own way despite sibling resistance and parental efforts to promote and enforce equity. Conflict in early and middle childhood, particularly psychological aggression, seems greater than in traditional societies, possibly exacerbated by the relative isolation of children in their homes, the existence of treasured toys, and two parental practices that are common in industrial societies but absent from traditional societies—insisting that different-aged sibs are equal and giving preferential treatment to the younger one.[54,62] It's tempting to surmise that the balance between selfish and generous interactions with siblings was more harmonious in antiquity than it is in contemporary Western societies.

In traditional societies, girls increase the survival of their infant siblings by holding and carrying them, and insistent socialization in sharing dampens the expression of children's competitive emotions and urges, as does integration in early and middle childhood into a playgroup with an age-based hierarchy in which siblings become allies. The hierarchy seems to stifle or obviate competition, but children might compete with sibs and others at that stage by ingratiating themselves with parents and other caretakers rather than entering into conflict with rivals.

What do siblings compete for? In both types of society, some baby-toddler dyads compete intensely for maternal attention and nurture. Beyond that stage, it remains to be seen how food scarcity stimulates sibling competition in industrial and traditional societies, how competition for food is expressed, and whether dominant siblings win out. However, as far as we can tell, siblings in industrial countries compete in early and middle childhood not for food and parental attention as much as for property and space, and that competition increases when families are under economic or social stress.

Sibling competition and conflict are similar between humans and blue-footed boobies in being common, emerging spontaneously in the very young, initially centered around access to food, sometimes violent, and involving behavioral dominance. However, in blue-foots dominance relationships are highly asymmetrical, they're only expressed through a small repertoire of stereotyped aggressive and submissive behaviors, and they're always about getting food from parents. Blue-foot and human infants both come equipped with arguably innate but developmentally plastic urges and emotions for competing with siblings. But whereas every blue-foot chick sticks to the same repertoire of alternative tactics throughout 3 months of cohabitation, maturing children probably do whatever works in their particular social context to implement their competitive inclinations.

Maybe the most striking difference between sibling conflict in humans and blue-foots concerns the psychological mechanisms involved. Consider food-sharing. Hadza teenagers make conscious, purposeful decisions about offering the food they gather to their young companions on the basis of inclinations influenced by several factors. These presumably include affection, an urge to share, the expectation of the recipient's gratitude, awareness of kinship, familiarity with local customs, the memory of food previously received from the recipient, and maybe the anticipation of future payback. Now dominant booby chicks also share food with their sibs, but they do it by pecking them less frequently when well fed. That's what allows the subordinate chick to beg more and receive more parental regurgitations. By adjusting their rate of pecking to the amount of food they recently ingested, dominants determine what proportion of regurgitations will go to each chick. Functionally they're doing something similar to what Hadza teenagers do—conceding food to rivals—but probably in the absence of empathy or even awareness that they're affecting another individual. I doubt they're even aware of other chicks' interests or other chicks' minds; they're just following their rule book.

Infidelity

Careful quantitative surveys and literature reviews by authoritative academics have shown that infidelity is common in human females and males and that coercive control by males in the context of female infidelity is commonplace:

> Numerous historical, literary, anthropological, and other sources suggest that among humans the temptation to become involved in a sexual relationship outside one's marriage is, and always has been, a widespread phenomenon. Infidelity can take many forms, including one-night stands, passionate love affairs, sexual fantasies about someone else, mate exchange, extradyadic romantic attachments, flirting, and sex with prostitutes (Buunk and Dijkstra, 2006).[65]
>
> … killing is just the tip of the iceberg: For every murdered wife, hundreds are beaten, coerced, and intimidated. Although homicide probably does not often serve the interests of the perpetrator, it is far from clear that the same can be said of sublethal violence. Men … strive to control women, albeit with variable success; women struggle to resist coercion and to maintain their choices. There is brinkmanship and risk of disaster in any such contest, and homicide by spouses of either sex may be considered the slips in this dangerous game (Daly and Wilson, 2017).[66]

But let's start at the beginning, with this hominin's pair-bonding.

Pair-Bonding and Mate Choice

Pair-bonding is just as conspicuous and important in humans as it is in blue-footed boobies, and it involves an evolved psychological mechanism that has been documented in all human cultures where it has been sought. During the first few years at least, many partnerships are fueled and sustained by *romantic love*, a powerful package of emotions that attract partners to each other, keep them together, and motivate courtship and copulation.[11,12] Notorious for its power to transform a person's normal day-to-day functioning and experience, romantic love involves empathy, emotional dependency and reciprocity, and psychophysical states such as exhilaration, euphoria, and the sensation of fusion with the other person; not to mention desire for exclusive sexual union and a feeling that the passion is involuntary and uncontrollable. An exalted state, it surely evolved in both sexes to sustain and facilitate cooperation in the production and care of offspring.[67]

As for the choice of a mate, both sexes have preferences among candidate partners that may well be biologically based but are also socially influenced. Humans don't resort to stereotypical motor coordination like sky-pointing or exhibiting patches of colored skin, but females and males find ways to enhance and display their physical and behavioral attributes, status, and wealth to prospective partners. *From what they tell researchers*, the two sexes in numerous cultures take into account a number of physical and personality traits, including intelligence, kindness, reliability, age, and sense of humor, but differ in the weights they give to some attributes of biological significance. For example, across a variety of cultures, males stand out for their inclination to choose partners whose youth, body shape, beauty, and health testify to their fecundity, while females stand out for their interest in older and taller partners whose financial prospects promise a supply of resources for raising children. These preferences are thought to align broadly with the widespread preeminence of females in engendering and nurturing offspring, and of males in defending and making economic provision for the family.[68,69] However, it must be emphasized that cultural variation in domestic roles of the two sexes is extensive, and female contributions to the family economy often equal or exceed male contributions.

Mixed Strategies

We can't observe and document human infidelity because it goes on mostly in private, but historical accounts, interview-based studies, and paternity analyses have given us a rough idea of its incidence in females and males, and of how this varies among cultures and other contexts. This motley body of research makes a case for biologically based and culturally shaped sexual conflict in human partnerships over mating and paternity.

In recent times in the United States, between 20 and 25 percent of married females and slightly more married males admit to having an extramarital affair during their lifetime, and 2–4 percent of females and males report having extramarital sex in the past year. Infidelity was commonplace in ancient and traditional societies such as the classical Greeks and Romans, the preindustrial Europeans, the historical Japanese, Chinese, and Hindus, the Inuit of the Arctic, Kuikuru of the Amazon Basin, Kofyar of Nigeria, Turu of Tanzania, and numerous other tribal societies; and its incidence varies, as one would expect, with culture, educational level, income, and occupation.[70] To my knowledge, the absence of infidelity has not been reported in any society. What's more, in a band of hunter-gatherers, females' copulations with both their husbands and their extra males peaked around the time of ovulation,[61] an echo of female blue-foots' copulations with partners and extra males peaking just before ovulation.

Extra-Pair Paternity

All blue-foot females court with extra males, and roughly half of them copulate with extra males, yet only 11 percent of booby families include an extra-pair chick. What proportion of babies are sired by extra males in humans, who in many societies frustrate fertilization by using contraceptives? This is a delicate and also ethically challenging matter. Not surprisingly, it has proven difficult to obtain reliable estimates based on representative samples, but a team of Belgian biologists recently cracked it using "genealogical pair analysis." In one especially revealing study,[71] they identified over 500 pairs of contemporary males in the Low Countries who both derive from the same paternal ancestor and who ought, therefore, to have in their respective genomes exact copies of that ancestor's Y chromosome, *provided none of the intervening male ancestors was an extra-pair sire.* The biologists didn't need to look at the ancestors' chromosomes, only those of each pair of contemporaneous males. Discrepancies between the two males' Y-chromosomes, which pass intact from father to son, allowed the researchers to estimate average rates of extra-pair paternity among close to 7,000 male ancestors of the sample.

It turned out that fewer than 2 percent of offspring in each generation were sired by an extra-pair male (with no difference in rate between Protestant Netherlands and Catholic Flanders), a much lower incidence than observed in blue-foots (11 percent) and most species of birds. In addition, the social and economic context counted: the proportion of extra-pair offspring was four times higher in lower socioeconomic classes than in farmers, merchants, and craftsmen, and it increased with population density, being four times higher in cities than in villages. After relative stability across the fifteenth through eighteenth centuries, the rate increased steeply in the late nineteenth century as a booming proletariat crowded into cities over the course of the Industrial Revolution. The researchers' interpretation was that adverse ecological circumstances may increase a female's incentive for seeking the material or social payoffs of extra-pair mating.[71]

Functions of Infidelity

Privacy and research ethics have largely drawn a veil across the reproductive benefits obtained through infidelity, but the aforementioned observations on paternity imply that extra-pair males sometimes *gain additional offspring* at low cost, and it's intuitively likely that females with infertile partners sometimes *achieve fertilization* through infidelity. Whether females' infidelity functions to secure *better or more variable genes* for their offspring remains a fascinating, unanswered question.[72] However, there are grounds for suspecting that infidelity can lead to *mate-switching*, although reproductive consequences are unknown. In a study of close to 200 American college students, 53 percent of women and 60 percent of men confessed to having attempted to lure someone else's mate into a committed relationship, and nearly half of them claimed to have been successful! That sounds like much more switching than goes on in boobies. How the "poached" mates transitioned between their first and second partners is unclear, but one suspects that some were unfaithful before making the definitive switch. A follow-up study of 17,000 students from 50 nations revealed that mate-poaching is prevalent worldwide.[73,74]

And then there's *paternity confusion*, a factor of suspected but unmeasured importance in the lives of humans and speculative importance in blue-foots. The hypothesis proposes that human females copulate with extra males while concealing the paternity of their next offspring in order to elicit contributions from those males to its care and survival—either by providing food for mother or child or by desisting in the infanticide that would free up the female to copulate with them and raise their progeny.

Human females keep paternity secret by hiding their ovulation (no chimpanzee-like rear-end swellings or pheromones in human females) and by being continuously receptive across their whole reproductive cycle.[75,76] In boobies, remember, we speculate that paternity confusion might inhibit a female's extra-pair male neighbor from attacking her trespassing offspring several weeks after copulating with her; in humans paternity confusion could in theory discourage neglect or infanticide and also motivate an extra male to provide food and other support, as we shall see later.

Defenses against Infidelity

Sexual jealousy, a nexus of emotional responses that almost certainly function to protect females and males from desertion by their partners and males from the risk of investing in other males' offspring,[77] is an evolved motivator we can all relate to. Jealousy aroused by suspicion of sexual infidelity is an awesome passion highlighted by playwrights and novelists for its remarkable power to channel angry indignation and provoke protest, imposition, manipulation, and revenge. However, the specific behavioral acts that result from jealousy are variable in the extreme and highly dependent on social learning. A jealous human's native culture and social experience will, together with their circumstances and idiosyncrasies, largely determine whether they sulk, throw a tantrum, turn to aggression or revenge, or abandon their partner, as well as how they go about it.

Both sexes' first line of defense is jealousy-fueled *mate-guarding*, which serves to frustrate partners' liaisons and bonding with rivals and head off their desertion and switching. There is evidence that the jealousy of the two sexes has been differently customized by natural selection to reflect the higher value to males of mates who are fertile and the higher value to females of mates who control resources. For example, according to self-reports, males are especially motivated to guard when their partners are young and physically attractive and when interested rivals are wealthy, whereas females guard more assiduously when their partners are high in wealth or status-striving and when interested rivals are young and attractive.[78,79]

Both sexes admit that their guarding tactics include vigilance, monopolizing their partners' time, threatening, signaling possession of the partner, derogating or attacking rivals, and sexual inducement. In addition, males—the larger and usually more dominant sex—display their wealth, conceal their partners, and threaten or assault rivals, and females make themselves more attractive and flirt with other males.[79] Worldwide, jealous suspicion excites males' sense of proprietary rights over their partners and fuels violent inclinations. For example, in nearly all of 100 Canadian lawsuits over marital beating, males said they acted because their partner's adultery was out of control; in interviews with over 100 battered Scottish wives, half of them explained that their partner's motive for attacking was possessiveness and sexual jealousy; and about half of 31 battered American women in hostels and hospitals said that in a pattern of repeated assaults jealousy was consistently the main incitement. When other battered American wives were asked if they'd ever had sex with another man while cohabiting with their husband, roughly half of the ones who'd been battered and subjected to forced copulation by their partners admitted to adultery.[66] Lamentably, in recent times and despite societal condemnation and legal prohibition, a significant minority of males abuse their partners to control their sexual behavior.

Finally, the history of female *confinement* by males illustrates how male jealousy, initially calibrated by natural selection in the context of egalitarian hunter-gatherer societies can, in the context of agriculture-driven abundance and accumulation of wealth by high-ranking individuals, motivate grotesque abuse. History is crowded with despots who have secreted dozens of young wives and concubines, sometimes guarded by eunuchs, under lock and key in their palaces, and occasionally killed and replaced whole harems after security failures. To this tactic for suppressing female infidelity, males have added a mostly deplorable panoply of customs and devices, including veils, chaperones, purdah, hobbling, chastity belts, and genital mutilation.[72]

Other physical and behavioral characteristics of human males look like *adaptations for winning sperm competition*, an extensively documented function in numerous other vertebrates and invertebrates. First, the gonads of human males are intermediate in size between the relatively large gonads of chimpanzees, whose females mate promiscuously, and the relatively small gonads of gorillas, whose females generally mate with a single male, consistent with the intermediate level of sperm competition human males face.[80] Second, experiments with prosthetic male and female genitalia have tentatively supported the suggestion that the shape of human penises evolved not only for inseminating females but also for the collateral function of displacing semen inserted into them by rivals.[81,82] Third, there is preliminary evidence that males whose sperm are likely to compete inside their partner with other males' sperm increase their rates of both forced and consensual copulations with their partners. Not seen in boobies with unfaithful partners, such retaliatory copulations presumably improve a cuckold's chance of being the one to fertilize.[83,84]

Human males' fourth line of defense against partner infidelity is to *invest less* in offspring sired by rivals by discriminating against suspect offspring. Tellingly, in societies where the certainty of paternity is low, males invest less in offspring,[61] and three questionnaire-based studies in the United States, England, and China showed that fathers invest less time, attention, and resources in offspring they perceive to resemble them less closely.[85–87]

Traditional Societies

Studies of traditional societies have revealed higher rates of extra-pair paternity than were calculated for the Low Countries, implying that infidelity may have been more common or had greater fitness consequences during the Pleistocene, when natural selection was shaping the human mating tendencies we see today. In the sixties, a 9 percent rate was found among females of the *Yanomami*, forest-dwelling hunter-gatherers of the Amazon basin who paired monogamously and polygynously, did not use contraceptives, and commonly had extra-marital affairs.[88]

The semi-nomadic Himba people of Namibia rely mainly on livestock for subsistence and both sexes participate in their care, resulting in lengthy separations of partners with one inhabiting the homestead while the other resides at a cattle post. There's no taboo on talking about infidelity, and mothers confidently identify extra-pair babies, called "omoka," by counting back from the date when their pregnancies become visible. They do not use contraceptives. Based on those identifications, no fewer than 17 percent of births within marriage are attributed by mothers to extra-male sires, and almost one-third of Himba women give birth to at least one omoka in their lifetime.[89]

In this context of uncertain paternity, the mixed reproductive strategy of married Himba males could well maximize their inclusive fitness: they economize on mate-guarding and care of their own offspring, which might not be theirs at all, while applying themselves to siring omokas by dallying with extra females. Effects of the reproductive strategies of Himba females and males on their respective fitnesses have not yet been analyzed, but, superficially at least, they make adaptive sense in the ecological and social context where they are expressed. Certainly, those strategies, incorporating social monogamy by both sexes along with genetic polyandry by some females and genetic polygyny by some males, testify to flexible mating patterns and parental care in preindustrial humans. One final detail of Himba infidelity is suggestive of females in a society of arranged marriages nonetheless managing to choose the sires of their offspring: whereas in arranged marriages nearly a quarter of babies are omekas, in "love matches"—marriages arising from romantic love—none are.

Customs of the cacao-cultivating Bari people in the tropical lowland forests of the Maracaibo Basin give a clue to what could have been a widespread function of female infidelity during the Pleistocene: securing the support of their infants by additional males through *paternity confusion*. When married Bari women give birth, they commonly name any extra males they copulated with during the pregnancy. Under the paternity confusion arising in this system of "partible paternity"—the common belief among traditional peoples of South America that two or more males can be joint sires of a baby—extra-pair partners provide supplements of fish and game to the offspring over the course of its childhood, and supplemented offspring are more likely to survive.[90] Such care by extra-pair partners is probably critical to the survival of offspring of divorced, widowed, and married females in societies where there is no safety net (no societal support of children), especially where hunting (and consequently provision of protein) is confined to males.[91,92]

Defense of paternity by *jealous mate-guarding* seems to be common in traditional societies, but Sarah Hrdy,[61] an evolutionary anthropologist who has thought deeply about human mating systems, suggests that guarding against female infidelity is more relaxed there than in patriarchal societies because extra males contribute to infants' survival when they suspect they are the father. In environments where high male mortality makes divorce and widowhood likely, males may grudgingly put up with their wives' infidelity with extra males because their own offspring can benefit from those males' care. In support of this suggestion, men of the Canela hunter-gatherers of northeastern Brazil deny feeling sexual jealousy and satisfied an anthropologist with years of experience among them that they were sincere in this. However, the Aché hunter-horticulturalist men of eastern Paraguay, who also deny being jealous, beat their wives for infidelity (Hrdy 1999),[61] and Yanomami men resent their partners' infidelity just as Westerners do.[88]

The best evidence of males *investing less* in offspring sired by rivals was obtained in a polygynous population of subsistence agriculturalists of the Sereer, Wolof, and Peuhl people in rural Senegal. Careful measures of fathers' direct and indirect investment in their offspring were positively correlated with independent ratings of the resemblance between fathers and offspring in both body odor and facial features, implying that extra-pair offspring get less paternal investment.[93] Taken with the questionnaire-based studies in the United States, England, and China, this study came close to showing that human males control the risk arising from their partners' infidelity by detecting children they didn't sire and reducing investment in them. Little wonder that people admiring babies comment so often and so naturally that they look more like their official father than their mother, and that the mother and her family are especially active in this regard.[94,95]

Finally, the extreme expression of discrimination by cuckolded males against extra-pair offspring, *infanticide*, although legally suppressed or carried out secretly in most societies, has been documented by data in the ethnographic records of the Human Relations Area Files. In a quarter of 60 societies sampled in Africa, Asia, Europe, North America, South America, and Oceania, adulterous conception was recognized as a motive for infanticide. For example, among the Ojibwa hunter-agriculturalists of Canada, mothers sometimes abort or otherwise do away with extra-pair babies even in the absence of any coercion,[96] possibly aware that such children have little chance of surviving.

Infidelity in a Nutshell

These observations of human pairing, mating behavior, and family interactions ring loudly true to a behavioral ecologist familiar with the most widely accepted theories of the evolution of animal mating systems, parental investment, and sexual conflict. In contemporary, historical, and traditional societies, pairs of partners bond selectively, either under the potent influence of romantic love or in obedience to parents and others. The two partners cohabit and collaborate for years to engender, nourish, and defend highly dependent offspring while they grow and integrate into the wider social group. This demanding collaboration benefits from the close cooperation and fidelity of both partners, but conflicts of interest arise as partners get involved with outsiders and each partner strives to maximize its own inclusive fitness.

A minority of bonded females and males have secret sexual relationships with extra partners, and some of them copulate with extra partners, but the proportion of females that bear extra-pair offspring is low and varies greatly with the culture and with economic circumstances. The powerful sexual jealousy of both sexes and the highly variable mate-guarding it motivates speak clearly to biological adaptation for sexual conflict. In addition, males have physical and behavioral adaptations for sperm competition and for limiting their investment in offspring sired by rivals. By mating with extra females, males are thought to increase the number of offspring they sire and sometimes prepare the ground for mate-switching. By mating with extra males, females are suspected of assuring fertilization, obtaining better genes for their offspring, facilitating mate-switching, and, in traditional societies at least, inducing extra males to provision their offspring or spare them from ill treatment and infanticide. This last function of female infidelity is facilitated by women's distinctive concealed ovulation, which leaves extra males suspecting they are fathers.

The bonding, infidelity, and countermeasures to infidelity of female and male boobies are understood to serve similar functions. However, in boobies, infidelity is more common than in humans and identified countermeasures are different in being less diverse and more stereotyped. For example, cuckolds are not often aggressive to their rivals and don't respond to their partners' infidelity with increased or retaliatory copulations; instead, they dampen their partners' infidelity and reduce their own investment in extra-pair offspring by modifying their timetables of attendance on territory and destroying eggs housing suspicious offspring, as well as switching from partners who are unfaithful. Countermeasures of females have not been identified, but they presumably exist because males indulge in less infidelity when their partners are around, implying that there's a risk of retaliation or abandonment.

Designed for Family Conflict

Ever since an explicitly evolutionary approach to understanding social behavior was adopted in the 1960s and 1970s, the mathematical models of theoreticians have led the way and plodding field workers like me have been closely scrutinizing hundreds of species to see whether animals do what the models predicted, and to learn *how* they go about it. Overall, results have made for a ringing endorsement of theories. Observations and experiments on competition in broods and litters of several avian and mammalian species have increased confidence in inclusive fitness theory and O'Connor's derivative theory of sibling competition and brood reduction. Furthermore, a slew of observations and experiments on female and male roles in the partnerships of scores of birds, mammals, and fishes has built confidence in sexual selection and sexual conflict theory.[97–101] Parent–offspring conflict theory is less secure because researchers have struggled to pull off convincing tests, but, because the theory and some observations are compelling, few behavioral ecologists doubt its validity. In none of these cases can we be sure that the fundamental theory and its derivatives will not eventually need revision and reformulation, but for now they look reassuringly sturdy and they have markedly improved our understanding.

It's no surprise to a behavioral ecologist that blue-footed boobies and humans are similar in their conflictive family behavior because both are long-lived, monogamous, sexually reproducing species that care intensively for highly dependent offspring. But all of us are likely to be intrigued by the different ways conflict is expressed by such obviously different animals: a specialized, colony-living bird that follows rules of thumb and a highly flexible, group-living primate driven by emotions and urges. First, let's recall the conspicuous similarities and differences a behavioral ecologist would see between them in their sibling relations.

All *booby chicks* are congenitally inclined to huddle with their sibs and parents and compete with their sibs for parental food deliveries by begging and by attacking their sibs. Their repertoires are governed by a relatively simple rulebook; during the first 3–4 weeks, they attack and threaten each other until one is a trained dominant and the other a trained subordinate. Dominants limit subordinates' access to their parents' regurgitations by controlling their begging. If a dominant's food consumption and body weight decline too much, it automatically redoubles its attacks to the point where its sibling starves in the nest or flees and probes riskily for adoption into another family. When they're several weeks old, two siblings may briefly pass a twig back and forth now and then, but that's as close as they'll ever get to playing, and their relationship probably ends completely when they both fledge. They will experience no more family interactions until 2–4 years later when, if they managed to survive (most don't!), they bond with partners and create their own families.

Human siblings start out more immature than booby hatchlings and they develop more slowly, but as early as 6 months, babies' nascent jealousy inclines them to compete with siblings for parental nurture using an impressive variety of behaviors—they can signal anguish, freeze, call out, gesture, cling to their parent, intrude between parent and sibling, and even strike rivals. Subsequently, their repertoires expand: their emotions and urges incline them to compete using physical and verbal aggression for property and space, and they come to use relational aggression. Sibling dyads' levels of aggression vary from minimal and occasional to bullying, dangerous violence, and "reigns of terror" in which victims are as cowed and overpowered as blue-foot subordinates. Food clearly is at issue in some austere contexts, but its role in sibling competition and conflict has not been studied. Importantly, human sibling relationships also include dimensions that are largely or wholly absent from booby sibling relationships. Even before they can speak, children perceive other humans, including their siblings, as thinking, feeling beings with intentions, and they establish relationships with them. And besides competing, human siblings play together extensively, they are generous with each other, and they show reciprocity. Siblicide is rare in humans at any age.

Now let's recall the conspicuous similarities and differences a behavioral ecologist would see between boobies and humans in their partnership relations.

Blue-foot adults bond and commit to monogamous reproduction with a partner through a process based on displaying and mutual evaluation of characteristics such as foot color that reveal age, reproductive experience, health, and nutritional condition. However, before their clutches are laid, all females and males court with extra partners and roughly half of them copulate with their extra partners, although only 11 percent of families include an extra-pair chick. Both sexes, especially females, are reluctant to copulate with extra mates in the presence of their own partners, and males deal with infidelity by adjusting their timetables of attendance on territory, occasionally disrupting their partners' copulations with extra males, and destroying eggs they may not have sired.

Human adults bond and commit to monogamous reproduction with a partner through a process powered by romantic love and involving displaying and mutual evaluation of characteristics such as beauty, age, reliability, and resources. However, whereas boobies bond by following a rulebook of stereotyped behaviors, human courtship and bonding are variable in the extreme, although shaped by local custom—humans send love letters and wear perfumes. Approximately a quarter of human females and males admitted to having an extramarital affair (usually clandestine), and the proportion of extra-pair offspring produced by females varies from roughly 2 percent in one industrial society to 17 percent in a traditional society with natural fertility (no use of contraceptives). In humans, jealousy motivates a suite of culturally influenced countermeasures to the infidelity of males and, more especially, females. In marked contrast with female boobies' relative passivity and male boobies' limited repertoire of stereotyped mate-guarding and egg-destroying, human females and males invigilate and threaten their partners, monopolize their partners' time and induce sexual activity with them, as well as signaling possession, and derogating and attacking rivals. In addition, human females make themselves more attractive and flirt with other males, and human males display wealth, and also conceal, confine, or hobble their partners in numerous ways and reduce their investment in offspring they may not have sired.

The biological functions of human and booby infidelity are poorly understood. Unfaithful booby and human males gain additional offspring at low cost (though we don't know whether their infidelity diminishes their reproductive success with their partners). Unfaithful booby and human females may gain fertilization insurance, easier mate-switching, or better genes for offspring; and their infidelity may secure benefits for their offspring by inducing extra males to behave as if they are fathers.

It is a reasonable hypothesis that the sibling relations and partnership relations described in this book are similar and similarly turbulent in boobies and humans because in both cases family members' behavior is a product of natural selection acting not to improve or sustain the species itself, but to maximize the inclusive fitness of individual behavers. Fitness effects are often difficult to analyze with quantitative precision, but in sibling interactions and partnership interactions, the observed mix of cooperation and conflict aligns qualitatively with the predictions of the fitness theories of Hamilton, Trivers, and O'Connor (Introduction) and doesn't align nearly as well with any alternative theory. It makes sense that children threaten, assault, and dominate their younger siblings, as well as showing some restraint, generosity, and reciprocity because the hypothetical selfish genes that underlie those behaviors get themselves into more bodies of the next generation than hypothetical genes for sharing fairly or utter selfishness. Ditto for the genes behind bonding, infidelity, and mate-guarding. That's why the behavioral inclinations are there—it's because through the ages the hypothetical genes that underlie them have engineered their own propagation and persistence in the population.

It may seem paradoxical that a species whose behavior is conscious, thoughtful, and goal-directed, and which has peerless intellectual, scientific, and artistic achievements, is nonetheless inclined to experience family struggles similar to those of a species that follows rules of thumb unthinkingly. But that's the lesson. Similar behavior in humans and blue-foots probably did evolve through similar processes of natural selection, and animal behavior does indeed explain some human misbehavior. That's what you get when intelligence cohabits in the mind with a core of evolved behavioral predispositions and inclinations tied to the overriding function of maximizing reproduction. It explains why, although sometimes we rationally veto our urges, we so often find ourselves accommodating our thoughts to our inclinations and confecting justifications for what our instincts incline us to do, often doubting whether rationality is fully in control.[8,102] In the movie *The African Queen*, there was truth in Katherine Hepburn's riposte to a self-indulgent and self-justifying Humphrey Bogart: "Nature, Mr. Allnut, is what we were put in this world to rise above." Most of us know this, but we don't necessarily realize that some of the inclinations we need to rise above were installed in our minds by nonintuitive processes of natural selection.

Despite sometimes being unconscious,[23] our genetically supported inclinations can often be thoughtfully managed, and for that purpose, we are surely better off if we acknowledge their existence. We can better handle our regrettable urges and emotions if, rather than denying their existence, we look out for them and reflect on them. Surely acknowledgment of questionable inclinations gives our rational and moral selves a leg up in the struggle against them. I believe that acknowledging, since my twenties, the envy and the schadenfreude I occasionally feel for others has been empowering and liberating. From the time I attached an *unworthy biological inclinations* label to them and decided I'm not responsible for my inclinations, only for what I do, I've felt less culpable about my envy and schadenfreude, more guarded about indulging them, and more in control. The same goes for prejudice against outsiders and vengeance after perceived injustice. We evolved to privilege fairly and unfairly our own inclusive fitness interests, and indulge and justify self-serving behavior, but we can aspire to make our moral principles prevail.

And let's not be overly pessimistic about human inclinations! Our evolved nature—human nature—is certainly a questionable, even dangerous asset when our inclinations set us against each other rather than tending toward harmony. However, human nature is not as uniformly nasty and pernicious as some fear, and we do well to occasionally celebrate its valuable and even noble elements. For example, we have evolved empathy and moral emotions like shame, guilt, and indignation. To these, you can add our capacity for respect, admiration, and awe; our reciprocal solidarity with friends; our romantic, parental, filial, and grandparental love (what a package that is!); our vocation for teaching (virtually unknown in other animals); our loyalty to community and empathy with the needy; and our unique capacity to cooperate creatively and constructively with groups of our fellows. *Homo sapiens* is no ordinary beast!

Notes

Introduction

1. Stanford (1999).
2. de Waal (2016).
3. de Waal (2019).
4. The naturalistic fallacy is the idea that what is natural is good or cannot be wrong. It wrongly derives a prescription from a statement of fact.
5. Goodall (1971).
6. van Lawick and Goodall (1979).
7. Schaller (1964).
8. Ardrey (1970).
9. Ardrey (1961).
10. Ardrey (1966).
11. Lorenz (1961).
12. Hamilton (1964a).
13. Hamilton (1964b).
14. Williams (1966).
15. Trivers (1974).
16. Parker (1979).
17. Maynard Smith (1974).
18. Wilson (1975).
19. Dawkins (1976).
20. For a thorough and even-handed review of the sociobiology controversy, see Segerstrale (2000).
21. Segerstrale (2000).
22. Alcock (2001).
23. Grafen (2004).
24. Levin and Grafen (2019).
25. Barlow and Silverberg (1980).
26. Schaller (1964).
27. Rowe (1947).
28. Virgina Morell's book reviews research documenting remarkable cognitive achievements of some corvids and parrots.
29. Morell (2013).
30. The international authority on boobies and gannets, Bryan Nelson, also suggested (2005) that boobies have only modest intelligence compared with other species.
31. Nelson (2005).
32. Pérez-Staples et al. (2013).
33. Nagel (1974).

34. I follow Koch 2019 in using "experience" rather than the common but redundant expression "subjective experience."
35. Koch (2019).
36. Pinker (1997).
37. Damasio (2018).
38. Darwin (1965).
39. Panksepp (2004).
40. For arguments for feelings in other animals, see Safina (2015), Bekoff (2008), and Hauser (2000).
41. Bekoff (2008).
42. Safina (2015).
43. Hauser (2000).
44. Hrdy (1999).
45. Murphy et al. (2014).
46. Cousteau's file, *The Sea Birds of Isabella*, is available at https://www.youtube.com/watch?v= KEE2HpPoUyI.
47. Nelson (1978).
48. Drummond et al. (2010).
49. Lorenz (1973).
50. Oro et al. (2010).
51. Pinson and Drummond (1993).
52. Hughes et al. (2017).
53. Ancona et al. (2010).
54. Wingfield et al. (1999).
55. Ancona et al. (2011).
56. Ancona et al. (2017).

Chapter 1

1. Mock and Parker (1997).
2. Drummond (2006).
3. Forbes (2005).
4. Kölliker et al. (2013).
5. Simmons (1988).
6. Drummond et al. (1986).
7. Drummond et al. (1991).
8. Guerra and Drummond (1995).
9. Grafen (2004).
10. Hamilton (1964a).
11. Hamilton (1964b).
12. Levin and Grafen (2019).
13. Under Hamilton's rule, kin selection causes genes for altruistically helping a relative to increase in frequency when the benefit to the recipient multiplied by their genetic relatedness to the actor is greater than the reproductive cost to the actor. Genetic relatedness is 0.5 between parent and offspring, 0.5 between full siblings, 0.25 between half-siblings, and 0.125 between cousins (Hamilton 1964a).

14. Trivers (1971).
15. Clutton-Brock (2009).
16. Bednekoff (1997).
17. Gould (1982).
18. Anderson (1991).
19. Expulsion pushing has been inferred in Galapagos Blue-foots by Anderson (1995) and Lougheed and Anderson (1999), but eyewitness accounts are lacking.
20. Anderson (1995).
21. Lougheed and Anderson (1999).
22. Osorno and Drummond (2003).
23. Drummond (1987).
24. Tershy et al. (2000).
25. Cohen Fernandez (1988).
26. Lack (1947).
27. Lack (1954).
28. Lack (1968).
29. Mock and Forbes (1995).
30. Trivers (1974).
31. O'Connor (1978).
32. Mock (2004).
33. Caspi and Barrios (2016).
34. Dantchev and Wolke (2018).

Chapter 2

1. Drummond (2006).
2. Kaufmann (1983).
3. Poisbleau et al. (2009).
4. Scheiber et al. (2011).
5. Sharp et al. (2005).
6. Rossi et al. (2017).
7. Leedale and Hatchwell (2020).
8. Holekamp and Strauss (2016).
9. Hock and Huber (2006).
10. Drummond et al. (1991).
11. Drummond and Osorno (1992).
12. The ability of animals to assess relative size is not universally accepted. For contrasting views, see Elwood and Arnot (2012) and Chapin et al. (2019).
13. Elwood and Arnott (2012).
14. Chapin et al. (2019).
15. Dorward (1962).
16. Tershy et al. (2000).
17. Grafen (1987).
18. Smith (1979).
19. Drummond et al. (2003).
20. Hsu et al. (2006).

21. Nuñez-De La Mora et al. (1996).
22. Rutte et al. (2006) provide theoretical and empirical support for the idea that trained losing is an adaptation rather than the physiological imposition of disability.
23. Rutte et al. (2006).
24. The idea of evolvability of trained losing in boobies gains credibility from Oldham et al.'s (2020) finding that aggressiveness and winner-loser effects are largely independent in pigs, and Okada et al.'s (2019) demonstration that the duration of trained loser effects can evolve, with heritability of 17 percent in one species of beetle.
25. Oldham et al. (2020).
26. Okada et al. (2019).
27. Wittig and Boesch (2003).
28. Maestripieri (2003).
29. Anderson et al. (2015).
30. Gregg et al. (2018).
31. Valderrábano-Ibarra et al. (2007).
32. Benavides and Drummond (2007).

Chapter 3

1. Ploger and Mock (1986).
2. Mock and Lamey (2016).
3. Fujioka (1985).
4. Drummond (2006).
5. Mock and Ploger (1987).
6. Gonzalez-Voyer and Drummond (2007).
7. Mock (1987).
8. Pinson and Drummond (1993).
9. Safriel (1981).
10. Miller (1973).
11. Layne (1982).
12. Millar and Young (2003).
13. Spellerberg. (1971).
14. Procter (1975).
15. Li et al. (2004).
16. Yu et al. (2015).
17. Yu et al. (2006).
18. Roulin (2002).
19. Roulin and Dreiss (2012).
20. Scheiber et al (2011).
21. Hernández-Reyes et al. (2017).
22. Hudson and Trillmich (2008).
23. Frank et al. (1991).
24. Smale et al. (1995).
25. Hofer and East (1997).
26. Golla et al. (1999).
27. Hartsock and Graves (1976).

28. Fraser (1975).
29. Fraser and Thompson (1991).
30. de Passillé et al. (1988).
31. Hudson et al. (2016).
32. Bekoff (1978).
33. Bekoff et al. (1981).
34. Fox and Clark (1971).
35. Drummond et al. (2000).
36. Gonzalez-Voyer et al. (2007).
37. Redondo et al. (2019).

Chapter 4

1. Mock et al. (1987).
2. Guerra and Drummond (1995).
3. Drummond (2001a).
4. Drummond (2001b).
5. Drummond and Garcia Chavelas (1989).
6. Machmer and Ydenberg (1998).
7. Cook et al. (2000).
8. Fujioka (1985).
9. Creighton and Schnell (1996).
10. Rodriguez-Girones et al. (1996).
11. Osorno and Drummond (2003).
12. Mock and Lamey (1991).
13. Drummond and Rodríguez (2009).
14. Mock (1984).
15. Mock (1985).
16. Gonzalez-Voyer et al. (2007).
17. Mock and Parker (1998).
18. Pinson and Drummond (1993).
19. Gonzalez-Voyer and Drummond (2007).

Chapter 5

1. Trillmich and Wolf (2008).
2. Trivers (1974).
3. Dawkins and Carlisle (1976).
4. Alexander (1974).
5. Kilner and Hinde (2012).
6. Trivers (2002).
7. Trivers and Hare (1976).
8. Emlen et al. (1995).
9. Drummond et al. (1986).
10. Anderson and Ricklefs (1995).
11. Mock (1987).

12. Drummond (1987).
13. Schaller (1964).
14. Wolf and Brodie (1998).
15. Without luck, the conflict is not resolved over evolutionary time and the two parties—parents and offspring—may be landed with continual and escalated behavioral conflict.
16. Stamps et al. (1978).
17. Kölliker et al. (2012).
18. Hinde et al. (2010).
19. Roulin and Dreiss (2012).
20. Drummond et al. (2008).
21. Clifford and Anderson (2001).
22. Osorno and Drummond (1995).
23. Rodriguez-Girones et al. (1996).
24. Drummond et al. (1991).
25. Hahn (1981).
26. Mock and Ploger (1987).
27. Viñuela (1999).
28. Merkling et al. (2014).
29. Gilby et al. (2011).
30. Torres and Drummond (1999).
31. Briskie et al. (1994).
32. Boncoraglio et al. (2009).
33. Kilner and Drummond (2007).
34. Drummond and Burghardt (1983).

Chapter 6

1. Drummond and Ancona (2015).
2. Metcalfe and Monaghan (2001).
3. Blount et al. (2006).
4. Criscuolo et al. (2011).
5. Nuñez-de La Mora et al. (1996).
6. Drummond et al. (2011).
7. Drummond et al. (2003).
8. Muehlenbein and Watts (2010).
9. Naguib and Gil (2005).
10. Naguib et al. (2006).
11. Carmona-Isunza et al. (2013).
12. Drummond and Rodríguez (2013).
13. Drummond and Osorno (1992).
14. Sánchez-Macouzet and Drummond (2011).
15. Simmons (1988).
16. Wolke et al. (2015).
17. Hawley (1999).
18. Buhrmester and Furman (1990).
19. Tucker and Updegraff (2010).

20. Dantchev et al. (2019).
21. Caspi and Barrios (2016).
22. Ernst and Angst (1983).
23. Schooler (1972).
24. Bleske-Rechek and Kelley (2014).
25. Harris (1998).
26. Rohrer et al. (2015).
27. Damian and Roberts (2015).
28. Rohrer et al. (2017).
29. Damian and Roberts (2015).
30. Ancona and Drummond (2013).
31. Ancona et al. (2017).

Chapter 7

1. Trivers (1974).
2. Osorno (1999).
3. Fromonteil et al. (2023).
4. Nelson (1978).
5. Lorenz (1966).
6. Dawkins and Krebs (1978).
7. Servedio et al. (2013).
8. Velando et al. (2014).
9. Velando et al. (2006).
10. Velando et al. (2010).
11. Segami et al. (2021).
12. Ploming (2018).
13. Velando et al. (2005).
14. We can't be sure whether the benefit to chicks from handsome foster fathers arose from them providing high-quality care themselves or from their partners doing so on perceiving that handsomeness. Female birds sometimes invest more in offspring at the egg formation stage or during brood care if the father is attractive. For example, zebra finch mothers feed their chicks more when their partners are made to look more handsome and, for all we know, blue-foot mothers may also invest generously in offspring whose fathers are of visibly high quality.
15. Burley (1977).
16. Torres and Velando (2003).
17. Velando et al. (2006).
18. Torres and Velando (2005).
19. Pérez-Staples et al. (2013).
20. Choudhury (1995).
21. Black (1996).
22. Dhondt (2002).
23. Sánchez-Macouzet et al. (2014).
24. Drummond et al. (2016).
25. Pruitt and Carnevale (1993).

26. Pruitt and Rubin (1986).
27. Castillo Alvarez and Chavez Peon Hoffmann-Pinther (1983).
28. Gonzalez del Castillo and Osorno (1987).
29. Ortega et al. (2017).
30. Guerra and Drummond (1995).
31. Ancona et al. (2012).
32. Velando and Alonso-Alvarez (2003).
33. Blue-foot mothers compensated only partially for the provisioning shortfall, in approximate accord with sexual conflict theory.
34. Lessells (2012).
35. Harrison et al. (2009).
36. Houston and Davies (1985).
37. Jehl and Murray (1986).
38. Andersson (1994).
39. Blomqvist et al. (1997).
40. Székely et al. (2000).
41. Serrano-Meneses and Székely (2006).
42. Forty-three percent of fledglings are females.
43. Torres and Drummond (1999).
44. Torres and Drummond (2009).

Chapter 8

1. Marler (1956).
2. Trivers (1972).
3. Parker (1979).
4. Chapman et al. (2003).
5. Birkhead and Moller (1992).
6. Black (1996).
7. Osorio-Beristain and Drummond (1998).
8. Kiere and Drummond (2014).
9. Smith (1984).
10. Ramos et al. (2014).
11. Pérez-Staples and Drummond (2005).
12. Wright and Cotton (1994).
13. Westneat and Sargent (1996).
14. Kempenaers et al. (1998).
15. Osorio-Beristain and Drummond (2001).
16. Ramos et al. (2014).
17. Charmantier and Blondel (2003).
18. Rubenstein (2007).
19. Ramos and Drummond (2017).
20. Pizzari et al. (2004).
21. Thuman and Griffith (2005).
22. Kim et al. (2007).
23. Foerster et al. (2003).

24. Westneat and Stewart (2003).

25. Kiere and Drummond (2016).

26. Wagner (1998).

27. Davies (2000).

28. Assistance by males to neighboring extra-pair partners was observed in pied flycatchers. Copulating with female neighbors made males more likely to help them repel predators, to the probable benefit of their extra-pair young.

29. Krams et al. (2022).

30. Forstmeier et al. (2014).

31. Osorio-Beristain et al. (2006).

32. Griffith et al. (2004).

33. Hasson and Stone (2011).

34. Manning (1985).

35. Kokko (1998).

36. Brooks and Kemp (2001).

37. Johnson and Gemmell (2012).

38. Velando et al. (2011).

39. Cleasby and Nakagawa (2012).

40. Akçay and Roughgarden (2007).

41. Moller and Ninni (1998).

42. Segami et al. (2021).

43. Ellegren (2007).

44. Wilson Sayres and Makova (2011).

45. Velando et al. (2008).

46. Hamatani et al. (2004).

47. Fish et al. (2003).

48. Bille et al. (2005).

49. Of course, roughly 11 percent of those fledglings were sired by unidentified extra-pair males of unknown ages, but we felt entitled to ignore that influence, on the assumption that the influence of the other 89 percent of families would be great enough to swamp their effects in the analyses.

50. Drummond and Rodríguez (2015).

51. Ortega et al. (2017).

52. Ramos and Drummond (2018).

53. Effects of genes here would necessarily be confounded with effects of egg constituents, such as nutrients and hormones.

54. Ortega et al. (2021).

55. Ramírez Loza (2019).

56. Lifjeld, et al. (2019).

57. Reding (2015).

Chapter 9

1. Hamilton (1964a).

2. Hamilton (1964b).

3. Trivers (1972).

4. Trivers (1974).
5. Dawkins (1976).
6. Pinker (2002).
7. Zajonc (1980).
8. Haidt (2012).
9. Plomin (2018).
10. Daly and Wilson (1983).
11. Fisher (1994).
12. Fisher (1998).
13. Walker et al. (2011).
14. Schacht and Kramer (2019).
15. Hrdy (2009).
16. Hrdy (2016a).
17. Hrdy (2016b).
18. Hrdy and Burkart (2020).
19. There are three main mechanisms of natural selection behind the evolution of cooperative helpfulness in humans. In *mutual benefit*, when two or more individuals cooperate in some activity—for example, fishing by dragging a net through the water—each of them gains more than they would by fishing separately, resulting in the propagation of genes for cooperation. As we saw in Chapter 1, in *kin-selected altruism*, an individual propagates its own helping gene by helping a relative when the benefit to the relative devalued by their relatedness is greater than the cost to itself. A gene for *reciprocal altruism* is propagated if an individual bearing the gene helps another when flush in return for recompense when needy.
20. Clutton-Brock (2009).
21. Potts and Sloan (2010).
22. Mithen (2003).
23. Trivers (2011).
24. Caspi (2011).
25. Khan and Rogers (2015).
26. Straus et al. (1980).
27. Tucker et al. (2013).
28. Lawson and Mace (2008).
29. Lawson et al. (2012).
30. Penn and Smith (2007).
31. Steinmetz (1977).
32. Dunn and Munn (1987).
33. Salmon and Hehman (2014).
34. Myers and Bjorklund (2018).
35. Ross et al. (1994).
36. Perlman and Ross (1997).
37. McGuire et al. (2000).
38. Hart (2022).
39. Hrdy and Burkart (2022).
40. Dunn (2003).
41. Dunn et al. (1981).

42. Dunn and Munn (1985).
43. Stewart et al. (1987).
44. Dunn and Kendrick (1980).
45. Behringer et al. (2022).
46. Dunn et al. (1986).
47. Stewart and Marvin (1984).
48. Sulloway (2008).
49. Buhrmester and Furman (1990).
50. Campione-Barr (2017).
51. Tippett and Wolke (2015).
52. Caspi and Barrios (2016).
53. Buist et al. (2017).
54. Lancy (2022).
55. Konner (2017).
56. Henry et al. (2017).
57. Marlowe (2017).
58. Crittenden and Zes (2015).
59. Hrdy (2017).
60. Shostak (1991).
61. Hrdy (1999).
62. Harris (1998).
63. Cicirelli (1991).
64. Cicirelli (1995).
65. Buunk and Dijkstra (2006).
66. Daly and Wilson (2017).
67. Bode and Kushnick (2021).
68. Buss (2018).
69. Walter et al. (2020).
70. Tsapelas et al. (2010).
71. Larmuseau et al. (2019).
72. Wilson and Daly (1992).
73. Schmitt et al. (2004).
74. Schmitt and Buss (2001).
75. Hrdy (1979).
76. Opie (2019).
77. Archer (2013).
78. Buss (2000).
79. Buss (2002).
80. Harcourt et al. (1995).
81. Gallup and Burch (2004).
82. Gallup and Burch (2006).
83. Shackelford et al. (2006).
84. Goetz and Shackelford (2006).
85. Burch and Gallup (2000).
86. Apicella and Marlowe (2007).
87. Yu et al. (2019).

88. Neel and Weiss (1975).
89. Scelza (2011).
90. Beckerman et al. (1998).
91. Hrdy (1981).
92. Hrdy (2000).
93. Alvergne et al. (2009).
94. Daly and Wilson (1982).
95. Regalski and Gaulin (1993).
96. Daly and Wilson (1984).
97. Birkhead and Moller (1992).
98. Mock and Parker (1997).
99. Alcock (2001).
100. Forbes (2005).
101. Royle et al. (2012).
102. Haidt (2007).

Glossary

Altruism In the jargon of behavioral ecology, an animal behaves altruistically to another when the help it provides confers a benefit to the other and a cost to itself. Ordinarily, natural selection does not favor altruism because by definition it reduces the altruist's fitness, but there are many circumstances where altruism is favored, for example, under kin selection or when there is reciprocity.

Benefit The benefit of a behavior (or other contribution to reproduction) is the positive impact it has on the behaver's fitness. For example, when a booby lays an egg or raises a brood, that effort increases the number of viable offspring it produces over its lifetime.

Bobonovela The name used in our lab for the amorous and gossip-worthy goings-on in booby colony neighborhoods when breeders are engaged in the courtship, pair formation, infidelity, and partner-switching that precedes laying.

Brood reduction An avian brood is reduced when one or more of its members dies, usually the youngest. *Adaptive brood reduction* involves the death of one or more nestlings to benefit the survivors, who may get more food, or the parents, who may work less hard to feed a smaller brood. Brood reduction of a species is *obligate* when the youngest nestlings in nearly all broods are automatically killed and *facultative* when the youngest nestling's survival depends on the adequacy of parental food provision.

Cloaca The aperture at the posterior end of a male or female bird's digestive tract where fecal material is discharged and ejaculates pass from male to female.

Coevolution Coevolution occurs when two actors (e.g., two species, the two sexes, or parent and offspring) interact consequentially with each other in such a way that evolutionary change in the behavior of one provokes an evolutionary response in the other. It is a process of correlated adaptive responses. When the interaction is conflictive, the process is called antagonistic coevolution.

Cost The cost of a behavior (or other contribution to reproduction) is the negative impact it has on the behaver's fitness. For example, when a booby lays a clutch or raises a brood, that effort is likely to reduce its fitness by prejudicing its subsequent survival or reproductive performance. Of course, the benefit of laying a clutch or raising a brood, an increase in fitness, is normally greater than the cost.

Cuckold A monogamously paired male whose partner has a sexual relationship with an extra male.

Dominance An animal dominates another when it aggressively controls aspects of the other's behavior, obliging it to display submission or flee, and restricting its access to resources such as space and food. Control may be exercised by physical violence or threats. The psychological mechanism behind dominance can be a *personal dominance-subordination relationship*, *trained winning and trained losing*, or mere *assessment* of the differential aggressive potential of the contenders.

Egg-dumping Female blue-foots dump when they secretly lay an egg in the empty, unguarded nest scrape of a pair of blue-foots on the verge of laying. With luck, the pair treat it as their own, incubate it, and raise the parasitic nestling.

Extra-pair When an individual in a monogamous reproductive partnership engages sexually with an extra partner, that partner is an *extra-pair partner*, the sexual interaction between them is *extra-pair behavior*, and any resulting offspring are *extra-pair offspring*.

Fitness *Lifetime fitness* usually refers to the number of viable offspring an individual produces over its lifetime. *Fitness* sometimes means the same but is also used to refer to components of lifetime fitness, such as the number of eggs laid or fledglings produced in a particular year.

Foray The excursion of a paired booby into the colony in search of extra-pair mates.

Genome The whole set of genes in each cell of an animal. The individual animal's complete complement of hereditary information in the form of DNA.

Germ cells The gene-bearing cells that give rise to gametes, that is, to the oocytes (unfertilized eggs) of females and the sperm of males.

Habitat structure The physical characteristics of an animal's habitat, including vegetation and geology.

Immune system The entire network of organic processes that protect the individual from infection by pathogens of all kinds and other agents.

Inclusive fitness The sum of (a) the number of viable offspring produced by an individual and (b) increments in its kin's production of viable offspring due to the individual's altruism, multiplied by their coefficient of relatedness; minus any components of (a) due to the altruism of kin, multiplied by their coefficient of relatedness. To simplify, in Chapter 1 and elsewhere, no mention is made of the need to deduct the altruistic contributions of kin when calculating an individual's inclusive fitness.

Industrial society A human society in which mass production enabled by technological innovation and machinery supports a large population. Industrial societies are characterized by urbanization, administrative/political hierarchy, and division of labor.

Kin selection The process of natural selection by which altruistic helping of kin evolves. When kin selection results in animals helping kin such as offspring, siblings, or cousins, they do so because the cost to their own fitness is less than the benefit to their kin multiplied by their coefficient of relatedness.

Mating system The main mating systems are, in highly schematic terms, monogamy, polygyny, polyandry, and promiscuity. However, there is variability among species and within some species; and within a species, one sex can be monogamous and the other polygamous (polygynous or polyandrous). More broadly, mating system refers to the number of mates obtained, how they are obtained, and the relative participation of the partners in parental care.

Nascent jealousy A human infant's negative responses to a beloved individual who preferentially directs attention to a rival. The responses include protesting, rejecting, and hostile behaviors including hostility toward the rival, as well as affiliative gestures to the beloved individual.

Nest attendance A bird attends its brood, nest, or territory by spending time on the territory. It may defend the territory, court and copulate with its partner or extra partners, incubate eggs or care for nestlings, or just loaf there.

Parading A blue-foot courtship display involving slow, exaggerated foot lifting, either on the spot or as part of forward walking.

Parental care Any behavior of a parent that benefits its offspring, including incubation, food provision, brooding, nest repair, and defense of nestlings.

Parental investment Any material or behavioral contribution to an offspring by a parent that improves its probability of surviving and reproducing. Parental investment connotes a cost to the parent—a reduction in its capacity to invest in other offspring.

Partible paternity The belief in many societies of lowland South America where females often mate polyandrously that more than one male sexual partner can contribute to the formation of a fetus; that is, a child can have more than one biological father.

Paternity confusion A male's uncertainty over whether he is the father of particular offspring, arising from polyandrous mating of his bonded partner or extra partner, or incomplete information about her mating with other males and the timing of her fertility.

Philopatry An animal's tendency to return to the locality of its birth for reproduction.

Polyandry The mating of a female with more than one male.

Polygyny The mating of a male with more than one female.

Reciprocity One animal altruistically helps a needy other in return for a reciprocal favor when the former is in need. One of the few mechanisms of natural selection by which altruism between strangers or kin may evolve. Cheating may stand in the way of its evolution, and *Homo sapiens* may be the only animal with the requisite cognitive abilities to make it work.

Relatedness The *coefficient of relatedness* between two relatives is the probability that they share any particular gene due to its inheritance within the family. The value is 0.5 for full siblings, 0.5 for parent and offspring, 0.25 for half-siblings, and 0.125 for cousins.

Relational aggression Aggression that intentionally hurts others by socially isolating them, for example, by spreading rumors, excluding them from activities, and pranking them in public places.

Reproductive success A measure of the success of a breeding attempt, usually in terms of the number of fledglings or next-generation breeders produced.

Sexual conflict In monogamous reproductive partnerships, although male and female interests overlap, there can be scope for sexual conflict of interests over particular factors, such as extra-pair mating by either partner or relative contributions of the partners to parental care. This can give rise to behavioral conflict between the sexes or some sort of evolved solution.

Sperm competition The competition inside a multiply inseminated female between the ejaculates of two or more males for fertilizing the female's eggs. Success may depend on the timing and sequence of ejaculates and the relative numbers of sperm in them, as well as possible manipulation of them by the female's internal organs and physiology.

Strategy A pattern of behavior that tends to produce a predictable outcome favorable to the fitness of the behaver. The outcome is the *function* of the strategy. Individual behavioral components of the pattern are termed *tactics*. Neither term, strategy or tactic, implies intentionality or comprehension.

Subordination   An animal is subordinate to another when the other's attacks or threats oblige it to signal submission or retreat and limit its access to resources such as space and food. The psychological mechanism behind subordination can be a personal dominance-subordination relationship, trained winning and trained losing, or mere assessment of the difference in aggressive potential between the two individuals.

Switching   During the period preceding laying, a male or female booby switches when it abandons its partner to pair with another booby.

Tactic   A component of a strategy, usually a particular action.

Territory   The patch of land that a booby or pair of boobies defend against the intrusion of boobies and other animals; it is where pairs lay and incubate their eggs and raise their nestlings.

Traditional society   A human society oriented toward the past; one in which custom and habit rule behavior, and roles and status are assigned by category, for example, sexual identity and age. Traditional societies have no political leadership and few social institutions.

Yes-headshaking   A blue-foot display involving frenzied nodding and whistling (males) or grunting (females) while leaning toward the target of the display. The display is used for courtship, greeting partners, and repelling intruders.

References

Akçay, E., and J. Roughgarden. 2007. "Extra-Pair Paternity in Birds: Review of the Genetic Benefits." *Evolutionary Ecology Research* 9 (5): 855–68.

Alcock, J. 2001. *The Triumph of Sociobiology*. New York: Oxford University Press.

Alexander, R. D. 1974. "The Evolution of Social Behavior." *Annual Review of Ecology and Systematics* 5: 325–83.

Alvergne, A., Ch. Faurie, and M. Raymond. 2009. "Father-Offspring Resemblance Predicts Paternal Investment in Humans." *Animal Behaviour* 78 (1): 61–69.

Ancona, S., I. Calixto-Albarrán, and H. Drummond. 2012. "Effect of El Niño on the Diet of a Specialist Seabird, *Sula nebouxii*, in the Warm Eastern Tropical Pacific." *Marine Ecology Progress Series* 462: 261–71.

Ancona, S., and H. Drummond. 2013. "Life History Plasticity of a Tropical Seabird in Response to El Niño Anomalies during Early Life." *PLoS ONE* 8 (9). https://doi.org/10.1371/journal.pone.0072665.

Ancona, S., H. Drummond, C. Rodríguez, and J. Zúñiga-Vega. 2017. "Long-Term Population Dynamics Reveal That Survival and Recruitment of Tropical Boobies Improve after a Hurricane." *Journal of Avian Biology* 48 (2): 320–32.

Ancona, S., H. Drummond, and J. Zaldívar-Rae. 2010. "Male Whiptail Lizards Adjust Energetically Costly Mate Guarding to Male-Male Competition and Female Reproductive Value." *Animal Behaviour* 79 (1): 75–82.

Ancona, S., S. Sánchez-Colón, C. Rodríguez, and H. Drummond. 2011. "El Niño in the Warm Tropics: Local Sea Temperature Predicts Breeding Parameters and Growth of Blue-Footed Boobies." *Journal of Animal Ecology* 80 (4): 799–808. https://doi.org/10.1111/j.1365-2656.2011.01821.x.

Anderson, D. J. 1991. "Parent Blue-Footed Boobies Are Not Infanticidal." *Ornis Scandinavica* 22 (2): 169–70.

Anderson, D. J. 1995. "The Role of Parents in Sibilicidal Brood Reduction of Two Booby Species." *Auk* 112 (4): 860–69. https://doi.org/10.2307/4089018.

Anderson, C., J. A. D. Hildreth, and L. Howland. 2015. "Is the Desire for Status a Fundamental Human Motive? A Review of the Empirical Literature." *Psychological Bulletin* 141 (3): 574–601. https://doi.org/10.1037/a0038781.

Anderson, D. J., and R. E. Ricklefs. 1995. "Evidence of Kin-Selected Tolerance by Nestlings in a Sibilicidal Bird." *Behavioral Ecology and Sociobiology* 37 (3): 163–68. https://doi.org/10.1007/BF00176713.

Andersson, M. 1994. *Sexual Selection*. Vol. 72. Princeton, NJ: Princeton University Press. https://doi.org/10.1515/9780691207278/html.

Apicella, C. L., and F. W. Marlowe. 2007. "Men's Reproductive Investment Decisions." *Human Nature* 18 (1): 22–34. https://doi.org/10.1007/BF02820844.

Archer, J. 2013. "Can Evolutionary Principles Explain Patterns of Family Violence?" *Psychological Bulletin* 139 (2): 403–40. https://doi.org/10.1037/a0029114.

Ardrey, R. 1961. *African Genesis: A Personal Investigation into the Animal Origins and Nature of Man*. New York: Atheneum.

Ardrey, R. 1966. *The Territorial Imperative: A Personal Inquiry into the Animal Origins of Property and Nations*. New York: Atheneum. https://doi.org/10.1525/aa.1969.71.5.02.

Ardrey R. 1970. *The Social Contract. Anthropological XIV*. New York: Atheneum.

Barlow, W. G., and J. Silverberg. 1980. *Sociobiology: Beyond Nature/Nature?* 1st ed. New York: Routledge.

Beckerman, S. R., R. Lizarralde, C. Ballew, S. Schroeder, C. Fingleton, A. Garrison, and H. Smith. 1998. "The Bari Partible Paternity Project: Preliminary Results." *Current Anthropology* 39 (1): 164–67. https://doi.org/10.1086/204702.

Bednekoff, P. A. 1997. "Mutualism among Safe, Selfish Sentinels: A Dynamic Game." *The American Naturalist* 150 (3): 373–92.

Behringer, V., A. Berghänel, S. M. Lee, B. Fruth, and G. Hohmann. 2022. "Transition to Siblinghood Causes a Substantial and Long-lasting Increase in Urinary Cortisol Levels in Wild Bonobos." *Elife* 11: e77227. https://doi: 10.7554/eLife.77227.

Bekoff, M. 1978. "Behavioral Development in Coyotes and Eastern Coyotes." In *Biology, Behavior, and Management*, edited by Marc Bekoff, 97–126. New York: Academic Press.

Bekoff, M. 2008. *The Emotional Lives of Animals*. Novato, CA: New World Library.

Bekoff, M., M. Tyrrell, V. E Lipetz, and R. Jamieson. 1981. "Fighting Patterns in Young Coyotes: Initiation, Escalation, and Assessment." *Aggressive Behavior* 7: 225–44.

Benavides, T., and H. Drummond. 2007. "The Role of Trained Winning in a Broodmate Dominance Hierarchy." *Behaviour* 144 (9): 1133–47.

Bille, C., A. Skytthe, W. Vach, L. B. Knudsen, A-M. N. Andersen, J. C. Murray, and K. Christensen. 2005. "Parent's Age and the Risk of Oral Clefts." *Epidemiology* 16 (3): 311–16. https://doi.org/10.1097/01.ede.0000158745.84019.c2.

Birkhead, T. R, and A. P. Moller. 1992. *Sperm Competition in Birds: Evolutionary Causes and Consequences*. London: Academic Press.

Black, J. M. 1996. *Partnerships in Birds: The Study of Monogamy*. Oxford: Oxford University Press.

Bleske-Rechek, A., and J. A. Kelley. 2014. "Birth Order and Personality: A within-Family Test Using Independent Self-Reports from Both Firstborn and Laterborn Siblings." *Personality and Individual Differences* 56 (1): 15–18. https://doi.org/10.1016/j.paid.2013.08.011.

Blomqvist, D., O. C. Johansson, U. Unger, M. Larsson, and L. Åke Flodin. 1997. "Male Aerial Display and Reversed Sexual Size Dimorphism in the Dunlin." *Animal Behaviour* 54 (5): 1291–99.

Blount, J. D., N. B. Metcalfe, K. E. Arnold, P. F. Surai, and P. Monaghan. 2006. "Effects of Neonatal Nutrition on Adult Reproduction in a Passerine Bird." *Ibis* 148 (3): 509–14. https://doi.org/10.1111/j.1474-919X.2006.00554.x.

Bode, A., and G. Kushnick. 2021. "Proximate and Ultimate Perspectives on Romantic Love." *Frontiers in Psychology* 12 (April): 1–29.

Boncoraglio, G., M. Caprioli, and N. Saino. 2009. "Fine-Tuned Modulation of Competitive Behaviour According to Kinship in Barn Swallow Nestlings." *Proceedings of the Royal Society B: Biological Sciences* 276 (1664): 2117–23. https://doi.org/10.1098/rspb.2009.0085.

Briskie, J. V., C. T. Naugler, and S. M. Leech. 1994. "Begging Intensity of Nestling Birds Varies with Sibling Relatedness." *Proceedings of the Royal Society B: Biological Sciences* 258 (1351): 73–78. https://doi.org/10.1098/rspb.1994.0144.

Brooks, R., and D. J. Kemp. 2001. "Can Older Males Deliver the Good Genes?" *Trends in Ecology and Evolution* 16 (6): 308–13. https://doi.org/10.1016/S0169-5347(01)02147-4.

Buhrmester, D., and W. Furman. 1990. "Perceptions of Sibling Relationships during Middle Childhood and Adolescence." *Child Development* 61 (5): 1387–98.

Buist, K. L., A. Metindogan, S. Coban, S. Watve, A. Paranjpe, H. M. Koot, Pol van Lier, S. J. T. Branje, and W. H. J. Meeus. 2017. "Cross-Cultural Differences in Sibling Power Balance and Its Concomitants across Three Age Periods." *New Directions for Child and Adolescent Development* 2017 (156): 87–104. https://doi.org/10.1002/cad.20199.

Burch, R. L., and G. G. Gallup. 2000. "Perceptions of Paternal Resemblance Predict Family Violence." *Evolution and Human Behavior* 21 (6): 429–35. https://doi.org/10.1016/S1090-5138(00)00056-8.

Burley, N. 1977. "Parental Investment, Mate Choice, and Mate Quality (Sexual Selection/Sociobiology/Columba Livia)." *Proceedings of the National Academy of the Sciences of the United States of America* 74 (8): 3476–79.

Buss, D. M. 2000. *The Dangerous Passion Why Jealousy Is as Necessary as Love and Sex.* New York: Free Press.

Buss, D. M. 2002. "Human Mate Guarding." *Neuro Endocrinology Letters* 23 (Suppl 4): 23–29.

Buss, D. M. 2018. "The Evolution of Love in Humans." In *The New Psychology of Love*, edited by R. J. Sternberg and K. Sternberg, 42–63. 2nd ed. New York: Cambridge University Press. https://doi.org/10.1017/9781108658225.004.

Buunk, A. P., and P Dijkstra. 2006. "Temptations and Threat: Extradyadic Relations and Jealousy." In *The Cambridge Handbook of Personal Relationships*, edited by L. A. Vangelisti and D. Perlman, 533–56. New York: Cambridge University Press. https://doi.org/10.1017/cbo9780511606632.

Campione-Barr, Nicole. 2017. "The Changing Nature of Power, Control, and Influence in Sibling Relationships." *New Directions for Child and Adolescent Development* 156: 7–14. https://doi.org/10.1002/cad.

Carmona-Isunza, M. C., A. Núñez-de la Mora, and H. Drummond. 2013. "Chronic Stress in Infancy Fails to Affect Body Size and Immune Response of Adult Female Blue-Footed Boobies or Their Offspring." *Journal of Avian Biology* 44 (4): 390–98. https://doi.org/10.1111/j.1600-048X.2013.00057.x.

Caspi, J. 2011. *Sibling Aggression: Assessment and Treatment.* New York: Springer.

Caspi, J., and V. R. Barrios. 2016. "Destructive Sibling Aggression." In *The Wiley Handbook on the Psychology of Violence*, edited by A. Carlos Cuevas and Callie Marie Rennison, 297– 323. Chichester, UK: Wiley-Blackwell.

Castillo Álvarez, A., and M. C. Chávez Peon Hoffmann-Pinther. 1983. *Ecología Reproductiva e Influencia Del Comportamiento En El Control Del Número de Crías En El Bobo de Patas Azules Sula nebouxii En La Isla Isabel, Nayarit.* Undergraduate Thesis, Mexico D.F: Universidad Nacional Autónoma de México.

Chapin, K. J., Paulo Enrique Cardoso Peixoto, and Mark Briffa. 2019. "Further Mismeasures of Animal Contests: A New Framework for Assessment Strategies." *Behavioral Ecology* 30 (5): 1177–85. https://doi.org/10.1093/beheco/arz081.

Chapman, T., G. Arnqvist, J. Bangham, and L. Rowe. 2003. "Sexual Conflict." *Trends in Ecology and Evolution* 18 (1): 41–47.

Charmantier, A., and J. Blondel. 2003. "A Contrast in Extra-Pair Paternity Levels on Mainland and Island Populations of Mediterranean Blue Tits." *Ethology* 109: 351–63. www.blackwell.de/synergy.

Choudhury, S. 1995. "Divorce in Birds: A Review of the Hypotheses." *Animal Behaviour* 50 (2): 413–29. https://doi.org/10.1006/anbe.1995.0256.

Cicirelli, V. G. 1991. "Sibling Relationships in Cross-Cultural Perspective." *Marriage and Family Review* 16 (3–4): 291–310. https://doi.org/10.1300/J002v16n03_05.

Cicirelli, V. G. 1995. *Sibling Relationships across the Life Span.* New York: Springer. https://doi.org/10.1007/978-1-4757-6509-0.

Cleasby, I. R., and S. Nakagawa. 2012. "The Influence of Male Age on Within-Pair and Extra-Pair Paternity in Passerines." *Ibis* 154 (2): 318–24. https://doi.org/10.1111/j.1474-919X.2011.01209.x.

Clifford, L. D., and D. J. Anderson. 2001. "Experimental Demonstration of the Insurance Value of Extra Eggs in an Obligately Siblicidal Seabird." *Behavioral Ecology* 12 (3): 340–47. https://doi.org/10.1093/beheco/12.3.340.

Clutton-Brock, T. 2009. "Cooperation between Non-Kin in Animal Societies." *Nature* 462 (7269): 51–57. https://doi.org/10.1038/nature08366.

Cohen Fernández, E. J. 1988. *La Reducción de La Nidada En El Bobo Café (Sula leucogaser nesiotes, Heller and Snodgras 1901).* Undergraduate Thesis, Mexico D.F: Universidad Nacional Autónoma de México.

Cook, M. I., P. Monaghan, and M. D. Burns. 2000. "Effects of Short-Term Hunger and Competitive Asymmetry on Facultative Aggression in Nestling Black Guillemots *Cepphus grylle.*" *Behavioral Ecology* 11 (3): 282–87. https://doi.org/10.1093/beheco/11.3.282.

Creighton, J. C., and G. D. Schnell. 1996. "Proximate Control of Siblicide in Cattle Egrets: A Test of the Food-Amount Hypothesis." *Behavioral Ecology and Sociobiology* 38 (6): 371–77. https://doi.org/10.1007/s002650050254.

Criscuolo, F., P. Monaghan, A. Proust, J. Škorpilová, J. Laurie, and N. B. Metcalfe. 2011. "Costs of Compensation: Effect of Early Life Conditions and Reproduction on Flight Performance in Zebra Finches." *Oecologia* 167 (2): 315–23. https://doi.org/10.1007/s00442-011-1986-0.

Crittenden, A. N., and D. A. Zes. 2015. "Food Sharing among Hadza Hunter-Gatherer Children." *PloS One* 10 (7): e0131996. https://doi.org/10.1371/journal.pone.0131996.

Daly, M., and M. Wilson. 1982. "Whom Are Newborn Babies Said to Resemble?" *Ethology and Sociobiology* 3 (2): 69–78. https://doi.org/10.1016/0162-3095(82)90002-4.

Daly, M., and M. Wilson. 1983. *Sex, Evolution and Behavior.* 2nd ed.. Boston: Willard Grant Press.

Daly, M., and M. Wilson. 1984. "A Sociobiological Analysis of Human Infanticide." In *Infanticide: Comparative and Evolutionary Perspectives*, edited by G. Hausfater and S. B. Hrdy, 487–502. New York: Aldine de Gruyter.

Daly, M., and M. Wilson. 2017. *Homicide.* New York: Routledge.

Damasio, A. 2018. *The Strange Order of Things: Life, Feeling, and the Making of Cultures.* New York: Pantheon.

Damian, R. I., and B. W. Roberts. 2015a. "Settling the Debate on Birth Order and Personality." *Proceedings of the National Academy of Sciences of the United States of America* 112 (46): 14119–20. https://doi.org/10.1073/pnas.1519064112.

Damian, R. I., and B. W. Roberts. 2015b. "The Associations of Birth Order with Personality and Intelligence in a Representative Sample of U.S. High School Students." *Journal of Research in Personality* 58: 96–105. https://doi.org/10.1016/j.jrp.2015.05.005.

Dantchev, S., M. Hickman, J. Heron, S. Zammit, and D. Wolke. 2019. "The Independent and Cumulative Effects of Sibling and Peer Bullying in Childhood on Depression, Anxiety, Suicidal Ideation, and Self-Harm in Adulthood." *Frontiers in Psychiatry* 10: 1–12. https://doi.org/10.3389/fpsyt.2019.00651.

Dantchev, S., and D. Wolke. 2018. "Trouble in the Nest: Antecedents of Sibling Bullying Victimization and Perpetration." *Developmental Psychology* 55 (5): 1059–71. https://doi.org/10.1037/dev0000700.

Darwin, A. C. 1965. *The Expression of the Emotions in Man and Animals.* Chicago: University of Chicago Press.

Davies, N. B. 2000. *Cuckoos, Cowbirds and Other Cheats. Cuckoos, Cowbirds and Other Cheats.* London: Poyser. https://doi.org/10.5040/9781472597472.

Dawkins, R. 1976. *The Selfish Gene.* Oxford: Oxford University Press.

Dawkins, R., and T. R. Carlisle. 1976. "Parental Investment, Mate Desertion and a Fallacy." *Nature* 262: 131–32.

Dawkins, R., and J. R. Krebs. 1978. "Animal Signals: Information or Manipulation?" In *Behavioural Ecology and Evolutionary Approach*, edited by J. R. Krebs and N. B. Davis, 282–309. Oxford: Blackwell.

Dhondt, A. A. 2002. "Changing Mates." *Trends in Ecology and Evolution* 17 (2): 55–56. https://doi.org/10.1016/S0169-5347(01)02407-7.

Dorward, D. F. 1962. "Comparative Biology of the White Booby and the Brown Booby *Sula* spp. at Ascension." *Ibis* 103 B (2): 174–220. https://doi.org/10.1111/j.1474-919X.1962.tb07244.x.

Drummond, H. 1987. "Review of POC and Brood Reduction in the Pelecaniformes." *Colonial Waterbirds* 10 (1): 1–15.

Drummond, H. 2001a. "A Revaluation of the Role of Food in Broodmate Aggression." *Animal Behaviour* 61 (3) : 517–26. https://doi.org/10.1006/anbe.2000.1641.

Drummond, H. 2001b. "The Control and Function of Agonism in Avian Broodmates." *Advances in the Study of Behavior* 30: 261–301.

Drummond, H. 2006. "Dominance in Vertebrate Broods and Litters." *Quarterly Review of Biology* 81 (1): 3–32.

Drummond, H., and S. Ancona. 2015. "Observational Field Studies Reveal Wild Birds Responding to Early-Life Stresses with Resilience, Plasticity, and Intergenerational Effects." *Auk* 132 (3): 563–76. https://doi.org/10.1642/AUK-14-244.1.

Drummond, H., and G. M. Burghardt. 1983. "Geographic Variation in the Foraging Behavior of the Garter Snake, *Thamnophis elegans*." *Behavioral Ecology and Sociobiology* 12 (1): 43–48. https://doi.org/10.1007/BF00296931.

Drummond, H., and C. García Chavelas. 1989. "Food Shortage Influences Sibling Aggression in the Blue-Footed Booby." *Animal Behaviour* 37: 806–19. https://doi.org/10.1016/0003-3472(89)90065-1.

Drummond, H., E. González, and J. L. Osorno. 1986. "Parent-Offspring Cooperation in the Blue-Footed Boody (Sula nebouxii): Social Roles in Infanticidal Brood Reduction." *Behavioral Ecology and Sociobiology* 19 (5): 365–72. https://doi.org/10.1007/BF00295710.

Drummond, H., and J. L. Osorno. 1992. "Training Siblings to Be Submissive Losers: Dominance between Booby Nestlings." *Animal Behaviour* 44: 881–93.

Drummond, H., J. L. Osorno, R. Torres, C. García Chavelas, and H. Merchant. 1991. "Sexual Size Dimorphism and Sibling Competition: Implications for Avian Sex Ratios." *The American Naturalist* 138 (3): 623–41.

Drummond, H., A. G. Ramos, O. Sánchez-Macouzet, and C. Rodríguez. 2016. "An Unsuspected Cost of Mate Familiarity: Increased Loss of Paternity." *Animal Behaviour* 111: 213–16. https://doi.org/10.1016/j.anbehav.2015.10.019.

Drummond, H., and C. Rodríguez. 2009. "No Reduction in Aggression after Loss of a Broodmate: A Test of the Brood Size Hypothesis." *Behavioral Ecology and Sociobiology* 63 (3): 321–27. https://doi.org/10.1007/s00265-008-0664-7.

Drummond, H., and C. Rodríguez. 2013. "Costs of Growing up as a Subordinate Sibling Are Passed to the Next Generation in Blue-Footed Boobies." *Journal of Evolutionary Biology* 26 (3): 625–34. https://doi.org/10.1111/jeb.12087.

Drummond, H., and C. Rodríguez. 2015. "Viability of Booby Offspring Is Maximized by Having One Young Parent and One Old Parent." *PLoS ONE* 10 (7): 1–14. https://doi.org/10.1371/journal.pone.0133213.

Drummond, H., C. Rodríguez, and D. Oro. 2011. "Natural 'Poor Start' Does Not Increase Mortality over the Lifetime." *Proceedings of the Royal Society B: Biological Sciences* 278 (1723): 3421–27. https://doi.org/10.1098/rspb.2010.2569.

Drummond, H., C. Rodríguez, and H. Schwabl. 2008. "Do Mothers Regulate Facultative and Obligate Siblicide by Differentially Provisioning Eggs with Hormones?" *Journal of Avian Biology* 39 (2): 139–43. https://doi.org/10.1111/j.0908-8857.2008.04365.x.

Drummond, H., C. Rodríguez, A. Vallarino, C. Valderrábano, G. Rogel, and E. Tobón. 2003. "Desperado Siblings: Uncontrollably Aggressive Junior Chicks." *Behavioral Ecology and Sociobiology* 53: 287–296. https://doi.org/10.1007/s00265-002-0571-2.

Drummond, H., R. Torres, and V. V. Krishnan. 2003. "Buffered Development: Resilience after Aggressive Subordination in Infancy." *The American Naturalist* 161 (5): 794–807. https://doi.org/10.1086/375170.

Drummond, H., R. Torres, C. Rodríguez, and S. Y. Kim. 2010. "Is Kin Cooperation Going on Undetected in Marine Bird Colonies?" *Behavioral Ecology and Sociobiology* 64 (4): 647–55. https://doi.org/10.1007/s00265-009-0882-7.

Drummond, H., E. Vázquez, S. Sánchez-Colón, M. Martínez-Gómez, and R. Hudson. 2000. "Competition for Milk in the Domestic Rabbit: Survivors Benefit from Littermate Deaths." *Ethology* 106 (6): 511–26. https://doi.org/10.1046/j.1439-0310.2000.00554.x.

Dunn, J. 2003. "Emotional Development in Early Childhood: A Social Relationship Perspective." In *Handbook of Affective Sciences*, edited by R. Davidson, J. K. R. Scherer, and H. H. Goldsmith, 40: 332–46. Oxford: Oxford University Press. https://doi.org/10.5860/choice.40-3702.

Dunn, J., and C. Kendrick. 1980. "The Arrival of a Sibling: Changes in Patterns of Interaction between Mother and First Born Child." *Journal of Child Psychology and Psychiatry* 2: 119–32.

Dunn, J., C. Kendrick, and R. MacNamee. 1981. "The Reaction of First-Born Children to the Birth of a Sibling: Mothers' Reports." *Child Psychology & Psychiatry & Allied Disciplines* 22 (1): 1–18.

Dunn, J., and P. Munn. 1985. "Becoming a Family Member: Family Conflict and the Development of Social Understanding in the Second Year." *Child Development* 56 (2): 480–92.

Dunn, J., and P. Munn. 1986. "Siblings and the Development of Prosocial Behavior." *International Journal of Behavioral Development* 9 (3): 265–84. https://eric.ed.gov/?id=EJ344348.

Dunn, J., and P. Munn. 1987. "Development of Justification in Disputes with Mother and Sibling." *Developmental Psychology* 23 (6): 791–98. https://doi.org/10.1037/0012-1649.23.6.791.

Ellegren, H. 2007. "Characteristics, Causes and Evolutionary Consequences of Male-Biased Mutation." *Proceedings of the Royal Society B: Biological Sciences* 274 (1606): 1–10. https://doi.org/10.1098/rspb.2006.3720.

Elwood, R. W., and G. Arnott. 2012. "Understanding How Animals Fight with Lloyd Morgan's Canon." *Animal Behaviour* 84 (5): 1095–102. https://doi.org/10.1016/j.anbehav.2012.08.035.

Emlen, S. T., P. H. Wrege, and N. J. Demong. 1995. "Making Decisions in the Family: An Evolutionary Perspective." *American Scientist* 83 (2): 148–57.

Ernst, C., and J. Angst. 1983. *Birth Order. Its Influence on Personality. Psychological Medicine.* Berlin: Springer-Verlag.

Fish, H., G. Hyun, R. Golden, T W. Hensle, C. A. Olsson, and G. L. Liberson. 2003. "The Influence of Paternal Age on Down Syndrome." *The Journal of Urology* 169 (6): 2275–78. https://doi.org/10.1097/01.ju.0000067958.36077.d8.

Fisher, H. E. 1994. "The Nature of Romantic Love." *Journal of NIH Research* 5: 59–64.

Fisher, H. E. 1998. "Lust, Attraction, and Attachment." *Human Nature* 9 (1): 23–52.

Foerster, K., K. Delhey, A. Johnsen, J. T. Lifjeld, and B. Kempenaers. 2003. "Females Increase Offspring Heterozygosity and Fitness through Extra-Pair Matings." *Nature* 425 (6959): 714–17. https://doi.org/10.1038/nature01969.

Forbes, S. 2005. *A Natural History of Families.* Princeton, NJ: Princeton University Press.

Forstmeier, W., S. Nakagawa, S. C. Griffith, and B. Kempenaers. 2014. "Female Extra-Pair Mating: Adaptation or Genetic Constraint?" *Trends in Ecology and Evolution* 29 (8): 456–64. https://doi.org/10.1016/j.tree.2014.05.005.

Fox, M. W., and A. L. Clark. 1971. "The Development and Temporal Sequencing of Agonistic Behavior in the Coyote (*Canis intrans*)." *Zeitschrift Für Tierpsychologie* 28 (3): 262–78.

Frank, G. L., E. S. Glickman, and P. Licht. 1991. "Fatal Sibling Aggression, Precocial Development, and Androgens in Neonatal Spotted Hyaenas." *Science* 252: 702–4.

Fraser, D. 1975. "The Effect of Straw on the Behaviour of Sows in Tether Stalls." *Animal Production* 21 (1): 59–68. https://doi.org/10.1017/S0003356100030415.

Fraser, D., and B. K. Thompson. 1991. "Behavioral Ecology and Sociobiology Armed Sibling Rivalry among Suckling Piglets." *Behavioral Ecology and Sociobiology* 29: 9–15.

Fromonteil, S., L. Winkler, L. Marie-Orleach, and T. Janicke. 2023. "Sexual Selection in Females across the Animal Tree of Life." *PLoS Biology* 21 (1): e3001916. https://doi.org/10.1371/journal.pbio.3001916

Fujioka, M. 1985. "Food Delivery and Sibling Competition in Experimentally Even-Aged Broods of the Cattle Egret." *Behavioral Ecology and Sociobiology* 17 (1): 67–74. https://doi.org/10.1007/BF00299431.

Gallup, G. G., and R. L. Burch. 2004. "Semen Displacement as a Sperm Competition Strategy in Humans." *Evolutionary Psychology* 2: 12–23. https://doi.org/10.1007/978-0-387-28039-4_14.

Gallup, G. G., and R. L. Burch. 2006. *The Semen-Displacement Hypothesis: Semen Hydraulics and the Intra-Pair Copulation Proclivity Model of Female Infidelity. Female Infidelity and Paternal Uncertainty: Evolutionary Perspectives on Male Anti-Cuckoldry Tactics.* Cambridge: Cambridge University Press. https://doi.org/10.1017/CBO9780511617812.007.

Gilby, A. J., M. C. Mainwaring, and S. C. Griffith. 2011. "The Adaptive Benefit of Hatching Asynchrony in Wild Zebra Finches." *Animal Behaviour* 82 (3): 479–84. https://doi.org/10.1016/j.anbehav.2011.05.022.

Goetz, A. T., and T. K. Shackelford. 2006. "Sexual Coercion and Forced In-Pair Copulation as Anti-Cuckoldry Tactics in Humans." *Human Nature* 17 (3): 265–82. https://doi.org/10.1017/CBO9780511617812.005.

Golla, W., H. Hofer, and M. L East. 1999. "Within-Litter Sibling Aggression in Spotted Hyaenas: Effect of Maternal Nursing, Sex and Age." *Animal Behaviour* 58: 715–26.

González del Castillo, E. C, and J. L. Osorno. 1987. *Dinámica de La Territorialidad En Una Colonia de Bobo de Patas Azules, Sula nebouxii En Isla Isabel, Nayarit, México.* Undergraduate Thesis, Mexico, D. F.: Universidad Nacional Autónoma de México.

Gonzalez-Voyer, A., and H. Drummond. 2007. "Is Broodmate Aggression Really Associated with Direct Feeding?" *Behaviour* 144 (4): 373–92. https://doi.org/10.1163/156853907780756049.

Gonzalez-Voyer, A., T. Székely, and H. Drummond. 2007. "Why Do Some Siblings Attack Each Other? Comparative Analysis of Aggression in Avian Broods." *Evolution* 61 (8): 1946–55. https://doi.org/10.1111/j.1558-5646.2007.00152.x.

Goodall, J. 1971. *In the Shadow of Man.* Boston: Houghton Mifflin.

Gould, S. J. 1982. "Darwinism and the Expansion of Evolutionary Theory." *Science* 216 (4544): 380–87. https://doi.org/10.1126/science.7041256.

Grafen, A. 1987. "The Logic of Divisively Asymmetric Contests: Respect for Ownership and the Desperado Effect." *Animal Behaviour* 35: 462–67.

Grafen, A. 2004. "William Donald Hamilton." *Biographical Memoirs of Fellows of the Royal Society* 50: 109–32.

Gregg, A., C. Sedikides, and A. Pegler. 2018. "Self-Esteem and Social Status: Dominance Theory and Hierometer Theory?" In *Encyclopedia of Evolutionary Psychological Science*, edited by T. K. Shackelford and V. A. Weekes-Shackelford, 1–6. New York: Springer. https://doi.org/10.1007/978-3-319-16999-6_1450-1.

Griffith, S. C., B. E. Lyon, and R. Montgomerie. 2004. "Quasi-Parasitism in Birds." *Behavioral Ecology and Sociobiology* 56 (3): 191–200. https://doi.org/10.1007/s00265-004-0766-9.

Guerra, M., and H. Drummond. 1995. "Reversed Sexual Size Dimorphism and Parental Care: Minimal Division of Labor in the Blue-Footed Booby." *Behaviour* 132 (7–8): 479–96.

Hahn, D. C. 1981. "Asynchronous Hatching in the Laughing Gull: Cutting Losses and Reducing Rivalry." *Animal Behaviour* 29: 421–27.

Haidt, J. 2007. "The New Synthesis in Moral Psychology." *Science* 316 (5827): 998–1002. https://doi.org/10.1126/science.1137651.

Haidt, J. 2012. *The Righteous Mind: Why Good People Are Divided by Politics and Religion.* New York: Vintage.

Hamatani, T., G. Falco, M. G. Carter, H. Akutsu, C. A. Stagg, A. A. Sharov, D. B. Dudekula, V. VanBuren, and M. S. H. Ko. 2004. "Age-Associated Alteration of Gene Expression Patterns in Mouse Oocytes." *Human Molecular Genetics* 13 (19): 2263–78. https://doi.org/10.1093/hmg/ddh241.

Hamilton, W. D. 1964a. "The Genetical Evolution of Social Behaviour. I." *Journal Theoretical Biology* 7: 1–16.

Hamilton, W. D. 1964b. "The Genetical Evolution of Social Behaviour. II." *Journal of Theoretical Biology* 7: 17–52.

Harcourt, A. H., A. Purvist, and L. Liles. 1995. "Sperm Competition: Mating System, Not Breeding Season, Affects Testes Size of Primates." *Functional Ecology* 9 (3): 468–76.

Harris, J. R. 1998. *The Nurture Assumption: Why Children Turn Out the Way They Do. Politics and the Life Sciences.* New York: Free Press. https://doi.org/10.1017/s0730938400008984.

Harrison, F., Z. Barta, I. Cuthill, and T. Székely. 2009. "How Is Sexual Conflict over Parental Care Resolved? A Meta-Analysis." *Journal of Evolutionary Biology* 22 (9): 1800–1812. https://doi.org/10.1111/j.1420-9101.2009.01792.x.

Hart, S. L. 2022. "Jealousy and the Terrible Twos." In *Evolutionary Perspectives on Infancy*, edited by S. L. Hart and D. F. Bjorklund, 325–47. Cham, Switzerland: Springer. https://doi.org/10.1007/978-3-030-76000-7_15.

Hartsock, T. G., and H. B. Graves. 1976. "Neonatal Behavior and Nutrition-Related Mortality in Domestic Swine." *Journal of Animal Science* 42 (1): 235–41. https://doi.org/10.2527/jas1976.421235x.

Hasson, O., and L. Stone. 2011. "Why Do Females Have So Few Extra-Pair Offspring?" *Behavioral Ecology and Sociobiology* 65 (3): 513–23. https://doi.org/10.1007/s00265-010-1104-z.

Hauser, D. M. 2000. *Wild Minds: What Animals Really Think.* London: Penguin Books.

Hawley, P. H. 1999. "The Ontogenesis of Social Dominance: A Strategy-Based Evolutionary Perspective." *Developmental Review* 19 (1): 97–132. https://doi.org/10.1006/drev.1998.0470.

Henry, I. P., A. G. Morelli, and Z. E. Tronick. 2017. "Child Caretakers Among Efe Foragers of the Ituri Forest." In *Hunter-Gatherer Childhoods: Evolutionary, Developmental and Cultural Perspectives*, edited by S. B. Hewlett and E. M. Lamb, 191–213. New York: Routledge.

Hernández-Reyes, R., C. Rodríguez, and H. Drummond. 2017. "Aggressive Defense of Food by Precocial Chicks Varies with Its Concentration in Space." *Behaviour* 154 (2): 163–70. https://doi.org/10.1163/1568539X-00003416.

Hinde, C. A., R. A. Johnstone, and R. M. Kilner. 2010. "Parent-Offspring Conflict and Coadaptation." *Science* 327 (5971): 1373–76. https://doi.org/10.1126/science.1186056.

Hock, K., and R. Huber. 2006. "Modeling the Acquisition of Social Rank in Crayfish: Winner and Loser Effects and Self-Structuring." *Behaviour* 143 (3): 325–46. https://doi.org/10.1299/kikaic.77.1491.

Hofer, H., and M. L. East. 1997. "Skewed Offspring Sex Ratios and Sex Composition of Twin Litters in Serengeti Spotted Hyaenas (*Crocuta Crocuta*) Are a Consequence of Siblicide." *Applied Animal Behaviour Science* 51 (3–4): 307–16. https://doi.org/10.1016/S0168-1591(96)01113-6.

Holekamp, K. E., and E. D. Strauss. 2016. "Aggression and Dominance: An Interdisciplinary Overview." *Current Opinion in Behavioral Sciences* 12: 44–51. https://doi.org/10.1016/j.cobeha.2016.08.005.

Houston, A. I., and N. B. Davies. 1985. "The Evolution of Cooperation and Life History in the Dunnock, Prunella Modularis." In *Behavioural Ecology: Ecological Consequences of Adaptive Behaviour*, edited by R. M. Sibly and R. H. Smith, 471–87. Oxford: Blackwell.

Hrdy, S. B. 1979. "Infanticide among Animals: A Review, Classification, and Examination of the Implications for the Reproductive Strategies of Females." *Ethology and Sociobiology* 1 (1): 13–40. https://doi.org/10.1016/0162-3095(79)90004-9.

Hrdy, S. B. 1981. *Woman That Never Evolved*. Cambridge, MA: Harvard University Press.

Hrdy, S. B. 1999. *Mother Nature: Maternal Instincts and How They Shape the Human Species.* New York: Ballantine.

Hrdy, S. B. 2000. "The Optimal Number of Fathers: Evolution, Demography, and History in the Shaping of Female Mate Preferences." *Annals of the New York Academy of Sciences* 907: 75–96. https://doi.org/10.1111/j.1749-6632.2000.tb06617.x.

Hrdy, S. B. 2009. *Mother and Others: The Evolutionary Origins of Mutual Understanding.* Cambridge, MA: Belknap Press.

Hrdy, S. B. 2016a. "Development plus Social Selection in the Emergence of 'Emotionally Modern' Humans." In *Childhood: Origins, Evolution, and Implications*, edited by C. L. Meehan and N. A. Crittenden, 11–44. Santa Fe: School for Advanced Research Press & University of New Mexico Press.

Hrdy, S. B. 2016b. "Mothers and Others." *Natural History Magazine*, 46–61. https://www.naturalhistorymag.com/picks-from-the-past/11440/mothers-and-others.

Hrdy, S. B. 2017. "Comes the Child before Man: Cooperative Breeding and Prolonged Postweaning Dependence Shaped Human Potential." In *Hunter-Gatherer Childhoods*, edited by S. B. Hewlett and E. M. Lamb, 65–91. New York: Routledge.

Hrdy, S. B., and J. M. Burkart. 2020. "The Emergence of Emotionally Modern Humans: Implications for Language and Learning." *Philosophical Transactions of the Royal Society B: Biological Sciences* 375 (1803): 15–18. https://doi.org/10.1098/rstb.2019.0499.

Hrdy, S. B, and J. M Burkart. 2022. "How Reliance on Allomaternal Care Shapes Primate Development with Special Reference to the Genus Homo." In *Evolutionary Perspectives on Infancy. Evolutionary Psychology*, edited by S. L. Hart and D. F. Bjorklund, 161–88. Cham, Switzerland: Springer. http://www.springer.com/series/10583.

Hsu, Y., R. L. Earley, and L. L. Wolf. 2006. "Modulation of Aggressive Behaviour by Fighting Experience: Mechanisms and Contest Outcomes." *Biological Reviews of the Cambridge Philosophical Society* 81 (1): 33–74. https://doi.org/10.1017/S146479310500686X.

Hudson, R., H. G. Rödel, M. Trejo-Elizalde, L. Arteaga, G. A. Kennedy, and B. P. Smith. 2016. "Pattern of Nipple Use by Puppies: A Comparison of the Dingo (*Canis dingo*) and the Domestic Dog (*Canis familiaris*)." *Journal of Comparative Psychology* 130 (3): 269–77. https://doi.org/10.1037/com0000023.

Hudson, R., and F. Trillmich. 2008. "Sibling Competition and Cooperation in Mammals: Challenges, Developments and Prospects." *Behavioral Ecology and Sociobiology* 62 (3): 299–307. https://doi.org/10.1007/s00265-007-0417-z.

Hughes, B. J., G. R. Martin, A. D. Giles, and S. J. Reynolds. 2017. "Long-Term Population Trends of Sooty Terns *Onychoprion fuscatus*: Implications for Conservation Status." *Population Ecology* 59 (3): 213–24. https://doi.org/10.1007/s10144-017-0588-z.

Jehl, J. R., and B. G. Murray. 1986. "The Evolution of Normal and Reverse Sexual Size Dimorphism in Shorebirds and Other Birds." *Current Ornithology* 3: 1–86. https://doi.org/10.1007/978-1-4615-6784-4_1.

Johnson, S. L., and N. J. Gemmell. 2012. "Are Old Males Still Good Males and Can Females Tell the Difference? Do Hidden Advantages of Mating with Old Males Off-Set Costs Related to Fertility, or Are We Missing Something Else?" *Bioessays* 34 (7): 609–19. https://doi.org/10.1002/bies.201100157.

Kaufmann, J. H. 1983. "On the Definitions and Functions of Dominance and Territoriality." *Biological Review* 58: 1–20.

Kempenaers, B., Richard B. Lanctot, and R. J. Robertson. 1998. "Certainty of Paternity and Paternal Investment in Eastern Bluebirds and Tree Swallows." *Animal Behaviour* 55 (4): 845–60. https://doi.org/10.1006/anbe.1997.0667.

Khan, R., and P. Rogers. 2015. "The Normalization of Sibling Violence: Does Gender and Personal Experience of Violence Influence Perceptions of Physical Assault against Siblings?"

Journal of Interpersonal Violence 30 (3): 437–58. https://doi.org/10.1177/088626051 4535095.

Kiere, L. M., and H. Drummond. 2014. "Extrapair Behaviour Reveals Flexible Female Choosiness and Mixed Support for Classic Good Genes in Blue-Footed Boobies." *Animal Behaviour* 95: 145–53. https://doi.org/10.1016/j.anbehav.2014.07.007.

Kiere, L. M., and H. Drummond. 2016. "Female Infidelity Is Constrained by El Niño Conditions in a Long-Lived Bird." *Journal of Animal Ecology* 85 (4): 960–72. https://doi.org/10.1111/1365-2656.12537.

Kilner, R. M., and H. Drummond. 2007. "Parent-Offspring Conflict in Avian Families." *Journal of Ornithology* 148 (2): 241–46. https://doi.org/10.1007/s10336-007-0224-3.

Kilner, R. M, and C. A. Hinde. 2012. "Parent-Offspring Conflict." In *The Evolution of Parental Care*, edited by Nick J. Royle, T. Per Smiseth, and M. Kölliker, 119–32. Oxford: Oxford University Press.

Kim, S. Y., R. Torres, C. A. Domínguez, and H. Drummond. 2007. "Lifetime Philopatry in the Blue-Footed Booby: A Longitudinal Study." *Behavioral Ecology* 18 (6): 1132–38. https://doi.org/10.1093/beheco/arm091.

Koch, C. 2019. *The Feeling of Life Itself: Why Consciousness Is Widespread but Can't Be Computed*. Cambridge, MA: The MIT Press.

Kokko, H. 1998. "Good Genes, Old Age and Life-History Trade-Offs." *Evolutionary Ecology* 12: 739–50.

Kölliker, M., A. Roulin, and A. N. Dreiss. 2013. "Sibling Competition and Cooperation over Parental Care." In *The Evolution of Parental Care*, edited by J. N. Royle, T. P. Smiseth, and M. Kölliker, 133–49. Oxford: Oxford: University Press. https://doi.org/10.1093/acprof:oso/9780199692576.003.0008.

Kölliker, M., N. Royle, and T. P Smiseth. 2012. "Parent-Offspring Co-Adaptation." In *The Evolution of Parental Care*, edited by J. N. Royle, T. P. Smiseth, and M. Köliker, 285–303. Oxford: Oxford University Press. https://www.researchgate.net/publication/237077774.

Konner, M. 2017. "Hunter-Gatherer Infancy and Childhood: The !Kung and Others." In *Hunter-Gatherer Childhoods: Evolutionary, Developmental and Cultural Perspectives*, edited by B. S Hewlett and M. E. Lamb, 19–64. New York: Routledge.

Krams, I. A., A. Mennerat, T. Krama, R. Krams, D. Elferts, S. Luoto, J. M. Rantala, and S. Eliassen. 2022. "Extra-Pair Paternity Explains Cooperation in a Bird Species." *Proceedings of the National Academy of Sciences of the United States of America* 119: 1–5. https://doi.org/10.1073/pnas.2112004119/-/DCSupplemental.Published.

Lack, D. 1947. "The Significance of Clutch-Size." *Ibis* 89: 302–52.

Lack, D. 1954. *The Natural Regulation of Animal Numbers*. Oxford: Clarendon Press.

Lack, D. 1968. *Ecological Adaptations for Breeding in Birds*. London: Methuen.

Lancy, D. F. 2022. *The Anthropology of Childhood: Cherubs, Chattels, Changelings. The Anthropology of Childhood*. Cambridge: Cambridge University Press. https://doi.org/10.1017/9781108943000.

Larmuseau, M. H. D., P. van den Berg, S. Claerhout, F. Calafell, A. Boattini, L. Gruyters, M. Vandenbosch, K. Nivelle, R. Decorte, and T. Wenseleers. 2019. "A Historical-Genetic Reconstruction of Human Extra-Pair Paternity." *Current Biology* 29 (23): 4102–7. https://doi.org/10.1016/j.cub.2019.09.075.

Lawson, D. W., A. Alvergne, and M. A. Gibson. 2012. "The Life-History Trade-off between Fertility and Child Survival." *Proceedings of the Royal Society B: Biological Sciences* 279 (1748): 4755–64. https://doi.org/10.1098/rspb.2012.1635.

Lawson, D. W., and R. Mace. 2008. "Sibling Configuration and Childhood Growth in Contemporary British Families." *International Journal of Epidemiology* 37 (6): 1408–21. https://doi.org/10.1093/ije/dyn116.

Layne, N. J. 1982. "Status of Sibling Aggression in Florida Sandhill Cranes." *Journal of Field Ornithology* 53: 272–74.

Leedale, A. E., J. Li, and B. J. Hatchwell. 2020. "Kith or Kin? Familiarity as a Cue to Kinship in Social Birds." *Frontiers in Ecology and Evolution* 8: 77. https://doi.org/10.3389/fevo.2020.00077.

Lessells, C. M. 2012. "Sexual Conflict." In *The Evolution of Parental Care*, edited by J. Nick, P. Royle, and M. Kölliker, 150–70. Oxford: Oxford University Press.

Levin, S. R., and A. Grafen. 2019. "Inclusive Fitness Is an Indispensable Approximation for Understanding Organismal Design." *Evolution* 73 (6): 1066–76. https://doi.org/10.1111/evo.13739.

Li, X., D. Li, Z. Ma, T. Zhai, and H. Drummond. 2004. "Ritualized Aggression and Unstable Dominance in Broods of Crested Ibis (*Nipponia nippon*)." *Wilson Bulletin* 116 (2): 172–76. https://doi.org/10.1676/02-079.

Lifjeld, J. T., J. Gohli, T. Albrecht, E. Garcia-del-Rey, L. E. Johannessen, O. Kleven, P. Z. Marki, T. C. Omotoriogun, M. Rowe, and A. Johnsen. 2019. "Evolution of Female Promiscuity in Passerides Songbirds." *BMC Evolutionary Biology* 19: 169. https://doi.org/10.1186/s12 862-019-1493-1.

Lorenz, K. 1961. *King Solomon's Ring*. London: Methuen.

Lorenz, K. 1966. *On Aggression*. New York: Harcourt, Brace and World.

Lorenz, K. 1973. "The Fashionable Fallacy of Dispensing with Description." *Naturwissenschaften* 60: 1–19.

Lougheed, L. W., and D. J. Anderson. 1999. "Parent Blue-Footed Boobies Suppress Siblicidal Behavior of Offspring." *Behavioral Ecology and Sociobiology* 45 (1): 11–18. https://doi.org/10.1007/s002650050535.

Machmer, M. M., and R. C. Ydenberg. 1998. "The Relative Roles of Hunger and Size Asymmetry in Sibling Aggression between Nestling Ospreys, Pandion Haliaetus." *Canadian Journal of Zoology* 76 (1): 181–86. https://doi.org/10.1139/z97-183.

Maestripieri, D. 2003. *Primate Psychology*. Cambridge, MA: Harvard University Press.

Manning, J. T. 1985. "Choosy Females and Correlates of Male Age." *Journal of Theoretical Biology* 16: 349–54.

Marler, P. 1956. "Behaviour of the Chaffinch Fringilla Coelebs." *Behaviour*. Supplement 5: III–V, VII–VIII, 1–184.

Marlowe, W. F. 2017. "Who Tends Hadza Children." In *Hunter-Gatherer Childhoods: Evolutionary, Developmental and Cultural Perspectives*, edited by S. B. Hewlett and E. M. Lamb, 177–90. New York: Routledge.

McGuire, S., B. Manke, A. Eftekhari, and J. Dunn. 2000. "Children's Perceptions of Sibling Conflict during Middle Childhood: Issues and Sibling (Dis) Similarity." *Social Development* 9 (2): 173–90.

Merkling, T., L. Agdere, E. Albert, R. Durieux, S. A. Hatch, E. Danchin, and P. Blanchard. 2014. "Is Natural Hatching Asynchrony Optimal? An Experimental Investigation of Sibling Competition Patterns in a Facultatively Siblicidal Seabird." *Behavioral Ecology and Sociobiology* 68 (2): 309–19. https://doi.org/10.1007/s00265-013-1646-y.

Metcalfe, N. B., and P. Monaghan. 2001. "Compensation for a Bad Start: Grow Now, Pay Later?" *Trends in Ecology and Evolution* 16 (5): 254–60. https://doi.org/10.1016/S0169-5347(01)02124-3.

Millar, C. D., and E. C. Young. 2003. "Siblicidal Brood Reduction in South Polar Skuas." *New Zealand Journal of Zoology* 30 (2): 79–93. https://doi.org/10.1080/03014223.2003.9518327.

Miller, R. S. 1973. "The Brood Size of Cranes." *The Wilson Bulletin* 85 (4): 436–41. https://www.jstor.org/stable/4160389.

Mithen, S. 2003. *After the Ice: A Global Human History, 20000–5000 BC*. Cambridge, MA: Harvard University Press.

Mock, D. W. 1984. "Siblicidal Aggression and Resource Monopolization in Birds." *Science* 225 (4663): 731–33.

Mock, D. W. 1985. "Siblicidal Brood Reduction: The Prey-Size Hypothesis." *American Naturalist* 125 (3): 327–43.

Mock, D. W. 1987. "Siblicide, Parent-Offspring Conflict, and Unequal Parental Investment by Egrets and Herons." *Behavioral Ecology and Sociobiology* 20 (4): 247–56. https://doi.org/10.1007/BF00292177.

Mock, W. D. 2004. *More Than Kin and Less Than Kind: The Evolution of Family Conflict.* Cambridge, MA: Harvard University Press.

Mock, D. W., and L. S. Forbes. 1995. "The Evolution of Parental Optimism." *Trends in Ecology and Evolution* 10 (3): 130–34. https://doi.org/10.1016/S0169-5347(00)89014-X.

Mock, D. W., and T. C. Lamey. 1991. "The Role of Brood Size in Regulating Egret Sibling Aggression." *The American Naturalist* 138 (4): 1015–26.

Mock, D. W., T. C. Lamey, Ch. F. Williams, and A. Pelletier. 1987. "Flexibility in the Development of Heron Sibling Aggression: An Intraspecific Test of the Prey-Size Hypothesis." *Animal Behaviour* 35 (5): 1386–93. https://doi.org/10.1016/S0003-3472(87)80011-8.

Mock, D. W., and G. A. Parker. 1997. *The Evolution of Sibling Rivalry.* Oxford: Oxford University Press.

Mock, D. W, and B. J. Ploger. 1987. "Parental Manipulation of Optimal Hatch Asynchrony in Cattle Egrets: An Experimental Study." *Animal Behaviour* 35: 150–60.

Moller, P. A., and P. Ninni. 1998. "Sperm Competition and Sexual Selection: A Meta-Analysis of Paternity Studies of Birds." *Behavioral Ecology and Sociobiology* 43: 345–58.

Morell, V. 2013. *Animal Wise: How We Know Animals Think and Feel.* New York: Broadway Books.

Muehlenbein, M. P., and D. P. Watts. 2010. "The Costs of Dominance: Testosterone, Cortisol and Intestinal Parasites in Wild Male Chimpanzees." *BioPsychoSocial Medicine* 4 (1): 1–12. http://www.bpsmedicine.com/content/4/1/21.

Murphy, E., R. E. Nordquist, and F. J. van der Staay. 2014. "A Review of Behavioural Methods to Study Emotion and Mood in Pigs, *Sus scrofa.*" *Applied Animal Behaviour Science* 159: 9–28. https://doi.org/10.1016/j.applanim.2014.08.002.

Myers, A. J., and D. F Bjorklund. 2018. "An Evolutionary Perspective of Rivalry in the Family." In *The Psychology of Rivalry*, edited by L. Hart and N. A. Jones, 1–33. New York: Nova Science.

Nagel, T. 1974. "What Is It Like to Be a Bat?" *Source: The Philosophical Review* 83 (4): 435–50.

Naguib, M., and D. Gil. 2005. "Transgenerational Effects on Body Size Caused by Early Developmental Stress in Zebra Finches." *Biology Letters* 1 (1): 95–97. https://doi.org/10.1098/rsbl.2004.0277.

Naguib, M., A. Nemitz, and D. Gil. 2006. "Maternal Developmental Stress Reduces Reproductive Success of Female Offspring in Zebra Finches." *Proceedings of the Royal Society B: Biological Sciences* 273 (1596): 1901–5. https://doi.org/10.1098/rspb.2006.3526.

Neel, J. V., and K. M. Weiss. 1975. "The Genetic Structure of a Tribal Population, the Yanomama Indians." *XII Biodemographic Studies. Biological Anthropology* 42: 25–51.

Nelson, B. 1978. *The Sulidae: Gannets and Boobies.* Oxford: Oxford University Press.

Nelson, B. 2005. *Pelicans, Cormorants and Their Relatives: Pelecanidae, Sulidae, Phalacrocoracidae, Anhingidae, Fregatidae, Phaethontidae.* Oxford: Oxford University Press.

Nuñez-De La Mora, A., H. Drummond, and J. C. Wingfield. 1996. "Hormonal Correlates of Dominance and Starvation-Induced Aggression in Chicks of the Blue-Footed Booby." *Ethology* 102 (9): 748–61. https://doi.org/10.1111/j.1439-0310.1996.tb01164.x.

O'Connor, R. J. 1978. "Brood Reduction in Birds: Selection for Fratricide, Infanticide and Suicide?" *Animal Behaviour* 26 (1): 79–96. https://doi.org/10.1016/0003-3472(78)90008-8.

Okada, K., Y. Okada, S. R. X. Dall, and D. J. Hosken. 2019. "Loser-Effect Duration Evolves Independently of Fighting Ability." *Proceedings of the Royal Society B: Biological Sciences* 286: 20190582. https://doi.org/10.1098/rspb.2019.0582.

Oldham, L., I. Camerlink, G. Arnott, A. Doeschl-Wilson, M. Farish, and S. P. Turner. 2020. "Winner-Loser Effects Overrule Aggressiveness during the Early Stages of Contests between Pigs." *Scientific Reports* 10 (1): 1–13. https://doi.org/10.1038/s41598-020-69664-x.

Opie, C. 2019. "Monogamy and Infanticide in Complex Societies." *Royal Anthropological Institute* 45: 59–74.

Oro, D., R. Torres, C. Rodríguez, and H. Drummond. 2010. "Climatic Influence on Demographic Parameters of a Tropical Seabird Varies with Age and Sex." *Ecology* 91 (4): 1205–14.

Ortega, S., C. Rodríguez, B. Mendoza-Hernández, and H. Drummond. 2021. "How Removal of Cats and Rats from an Island Allowed a Native Predator to Threaten a Native Bird." *Biological Invasions* 23 (9): 2749–61. https://doi.org/10.1007/s10530-021-02533-4.

Ortega, S., O. Sánchez-Macouzet, A. Urrutia, C. Rodríguez, and H. Drummond. 2017. "Age-Related Parental Care in a Long-Lived Bird: Implications for Offspring Development." *Behavioral Ecology and Sociobiology* 71 (9): 132. https://doi.org/10.1007/s00265-017-2364-7.

Osorio-Beristain, M., and H. Drummond. 1998. "Non-Aggressive Mate Guarding by the Blue-Footed Booby: A Balance of Female and Male Control." *Behavioral Ecology and Sociobiology* 43 (4–5): 307–15. https://doi.org/10.1007/s002650050496.

Osorio-Beristain, M., and H. Drummond. 2001. "Male Boobies Expel Eggs When Paternity Is in Doubt." *Behavioral Ecology* 12 (1): 16–21. https://doi.org/10.1093/oxfordjournals.beheco.a000373.

Osorio-Beristain, M., D. Pérez-Staples, and H. Drummond. 2006. "Does Booby Egg Dumping Amount to Quasi-Parasitism?" *Ethology* 112 (7): 625–30. https://doi.org/10.1111/j.1439-0310.2006.01201.x.

Osorno, J. L. 1999. "Offspring Desertion in the Magnificent Frigatebird: Are Males Facing a Trade-Off between Current and Future Reproduction?" *Journal of Avian Biology* 30 (4): 335–41. https://www.jstor.org/stable/3677005.

Osorno, J. L., and H. Drummond. 1995. "The Function of Hatching Asynchrony in the Blue-Footed Booby." *Behavioral Ecology and Sociobiology* 37 (4): 265–73. https://doi.org/10.1007/BF00177406.

Osorno, J. L., and H. Drummond. 2003. "Is Obligate Siblicidal Aggression Food Sensitive?" *Behavioral Ecology and Sociobiology* 54 (6): 547–54. https://doi.org/10.1007/s00265-003-0667-3.

Panksepp, J. 2004. *Affective Neuroscience: The Foundations of Human and Animal Emotions*. Oxford: Oxford University Press.

Parker, G. A. 1979. "Sexual Selection and Sexual Conflict." In *Sexual Selection and Reproductive Competition in Insects*, edited by M. S. Blum and N. A. Blum, 123–66. New York: Academic Press.

Parker, G. A., and M. R. Macnairi. 1978. "Models of Parent-Offspring Conflict. I. Monogamy." *Animal Behaviour* 26: 97–110.

Passillé, A. M., J. Rushen, and T. G. Hartsock. 1988. "Ontogeny of Teat Fidelity in Pigs and Its Relation to Competition at Suckling." *Canadian Journal of Animal Science* 67 (2): 325–38.

Penn, D. J., and K. R. Smith. 2007. "Differential Fitness Costs of Reproduction between the Sexes." *Proceedings of the National Academy of Sciences of the United States of America* 104 (2): 553–58.

Pérez-Staples, D., and H. Drummond. 2005. "Tactics, Effectiveness and Avoidance of Mate Guarding in the Blue-Footed Booby (*Sula nebouxii*)." *Behavioral Ecology and Sociobiology* 59 (1): 115–23. https://doi.org/10.1007/s00265-005-0016-9.

Pérez-Staples, D., M. Osorio-Beristain, C. Rodríguez, and H. Drummond. 2013a. "Behavioural Roles in Booby Mate Switching." *Behaviour* 150: 337–57. https://doi.org/10.1163/15685 39X-00003055.

Pérez-Staples, D., M. Osorio-Beristain, C. Rodríguez, and H. Drummond. 2013b. "Behavioural Roles in Booby Mate Switching." *Behaviour* 150: 1–21. https://doi.org/10.1163/1568539X-00003055.

Perlman, M., and H. S. Ross. 1997. "Who's the Boss? Parents' Failed Attempts to Influence the Outcomes of Conflicts between Their Children." *Journal of Social and Personal Relationships* 14 (4): 463–80. https://doi.org/10.1177/0265407597144003.

Pinker, S. 1997. *How the Mind Works*. New York: W. W. Norton & Company.

Pinker, S. 2002. *The Blank Slate: The Modern Denial of Human Nature*. New York: Viking Press.

Pinson, D., and H. Drummond. 1993. "Brown Pelican Siblicide and the Prey-Size Hypothesis." *Behavioral Ecology and Sociobiology* 32 (2): 111–18. https://doi.org/10.1007/BF00164043.

Pizzari, T., H. Lovlie, and C. K. Cornwallis. 2004. "Sex-Specific, Counteracting Responses to Inbreeding in a Bird." *Proceedings of the Royal Society B: Biological Sciences* 271 (1553): 2115–21. https://doi.org/10.1098/rspb.2004.2843.

Ploger, B. J., and D. W. Mock. 1986. "Role of Sibling Aggression in Food Distribution to Nestling Cattle Egrets (*Bubulcus ibis*)." *The Auk* 103 (4): 768–76.

Ploming, R. 2018. *Blueprint: How DNA Makes Us Who We Are*. Cambridge, MA: MIT Press.

Poisbleau, M., M. Guillemain, L. Demongin, D. Carslake, and J. David. 2009. "Within-Brood Social Status and Consequences for Winter Hierarchies amongst Mallard *Anas platyrhynchos* Ducklings." *Journal of Ornithology* 150 (1): 195–204. https://doi.org/10.1007/s10 336-008-0334-6.

Potts, R., and C. Sloan. 2010. *What Does It Mean to Be Human?* Washington, DC: National Geographic.

Procter, D. L C. 1975. "The Problem of Chick Loss in the South Polar Skua Catharacta *mccormack*." *Ibis* 117 (4): 452–59.

Pruitt, G. D., and J. P. Carnevale. 1993. *Negotiation in Social Conflict*. Pacific Grove, CA: Brooks/ Cole.

Pruitt, G. D., and Z. J. Rubin. 1986. *Social Conflict: Escalation, Stalemate and Settlement*. New York: Random House.

Ramírez Loza, J. P. 2019. *Son de Mayor Calidad Las Crías Extra-Pareja En el Bobo de Patas Azules (Sula nebouxii)*. Undergraduate Thesis, Ciudad de México: Universidad Nacional Autónoma de México.

Ramos, G. A., and H. Drummond. 2017. "Tick Infestation of Chicks in a Seabird Colony Varies with Local Breeding Synchrony, Local Nest Density and Habitat Structure." *Journal of Avian Biology* 48 (4): 472–78. https://doi.org/10.1111/jav.01107.

Ramos, G. A., and H. Drummond. 2018. "Ectoparasite Burden of Blue-Footed Booby Chicks Depends on Combined Parental Ages." *Ibis* 160 (4): 914–18. https://doi.org/10.1111/ ibi.12624.

Ramos, G. A., S. O. Nunziata, S. L. Lance, C. Rodríguez, B. C. Faircloth, P. A. Gowaty, and H. Drummond. 2014a. "Habitat Structure and Colony Structure Constrain Extrapair Paternity Ina Colonial Bird." *Animal Behaviour* 95: 121–27. https://doi.org/10.1016/j.anbe hav.2014.07.003.

Ramos, G. A., S. O. Nunziata, S. L. Lance, C. Rodríguez, B. C. Faircloth, P. A. Gowaty, and H. Drummond. 2014b. "Interactive Effects of Male and Female Age on Extra-Pair Paternity in a Socially Monogamous Seabird." *Behavioral Ecology and Sociobiology* 68 (10): 1603–9. https://doi.org/10.1007/s00265-014-1769-9.

Reding, L. 2015. "Increased Hatching Success as a Direct Benefit of Polyandry in Birds." *Evolution* 69 (1): 264–70. https://doi.org/10.1111/evo.12553.

Redondo, T., J. M. Romero, R. Díaz-Delgado, and J. Nagy. 2019. "Broodmate Aggression and Life History Variation in Accipitrid Birds of Prey." *Ecology and Evolution* 9 (16): 9185–206. https://doi.org/10.1002/ece3.5466.

Regalski, J. M, and S. J. C. Gaulin. 1993. "Whom Are Mexican Infants Said to Resemble? Monitoring and Fostering Paternal Confidence in the Yucatan." *Ethology and Sociobiology* 14 (2): 97–113.

Rodríguez-Girones, M. A., H. Drummond, and A. Kacelnik. 1996. "Effect of Food Deprivation on Dominance Status in Blue-Footed Booby (*Sula nebouxii*) Broods." *Behavioral Ecology* 7 (1): 82–88.

Rohrer, J. M., B. Egloff, and S. C. Schmukle. 2015. "Examining the Effects of Birth Order on Personality." *Proceedings of the National Academy of Sciences of the United States of America* 112 (46): 14224–29. https://doi.org/10.1073/pnas.1506451112.

Rohrer, J. M., B. Egloff, and S. C. Schmukle. 2017. "Probing Birth-Order Effects on Narrow Traits Using Specification-Curve Analysis." *Psychological Science* 28 (12): 1821–32. https://doi.org/10.1177/0956797617723726.

Ross, H. S., R. E. Filyer, S. P. Lollis, M. Perlman, and J. L. Martin. 1994. "Administering Justice in the Family." *Journal of Family Psychology* 8 (3): 254–73. https://doi.org/10.1037/0893-3200.8.3.254.

Rossi, M., R. Marfull, S. Golüke, J. Komdeur, P. Korsten, and B. A. Caspers. 2017. "Begging Blue Tit Nestlings Discriminate between the Odour of Familiar and Unfamiliar Conspecifics." *Functional Ecology* 31 (9): 1761–69. https://doi.org/10.1111/1365-2435.12886.

Roulin, A. 2002. "Short and Long-Term Fitness Correlates of Rearing Conditions in Barn Owls Tyto Alba." *Ardea* 90 (2): 259–67.

Roulin, A., and N. A. Dreiss. 2012. "Sibling Competition and Cooperation over Parental Care." In *The Evolution of Parental Care*, edited by J. N. Royle, T. P. Smiseth, and M. Kölliker, 133–45. Oxford: Oxford University Press.

Rowe, E. G. 1947. "The Breeding Biology of *Aquila verreauxi*." *Ibis* 89 (4): 576–606.

Royle, N. J., P. T. Smiseth, and M. Kölliker. 2012. *The Evolution of Parental Care*. Oxford: Oxford University Press.

Rubenstein, D. R. 2007. "Territory Quality Drives Intraspecific Patterns of Extrapair Paternity." *Behavioral Ecology* 18 (6): 1058–64. https://doi.org/10.1093/beheco/arm077.

Rutte, C., M. Taborsky, and M. W.G. Brinkhof. 2006. "What Sets the Odds of Winning and Losing?" *Trends in Ecology and Evolution* 21 (1): 16–21. https://doi.org/10.1016/j.tree.2005.10.014.

Safina, C. 2015. *Beyond Words: What Animals Think and Feel*. New York: Henry Holt and Company.

Safriel, U. N. 1981. "Social Hierarchy among Siblings in Broods of the Oystercatcher *Haematopus ostralegus*." *Behavioral Ecology and Sociobiology* 9 (1): 59–63. https://doi.org/10.1007/BF00299854.

Salmon, C. A, and J. A. Hehman. 2014. "The Evolutionary Psychology of Sibling Conflict and Siblicide." In *The Evolution of Violence: Evolutionary Psychology,* edited by T. Shackelford, R. Hansen, 137–57. Springer: New York. https://doi.org/10.1007/978-1-4614-9314-3_8.

Sánchez-Macouzet, O., and H. Drummond. 2011. "Sibling Bullying during Infancy Does Not Make Wimpy Adults." *Biology Letters* 7 (6): 869–71. https://doi.org/10.1098/rsbl.2011.0461.

Sánchez-Macouzet, O., C. Rodríguez, and H. Drummond. 2014. "Better Stay Together: Pair Bond Duration Increases Individual Fitness Independent of Age-Related Variation." *Proceedings of the Royal Society B: Biological Sciences* 281: 20132843. https://doi.org/10.1098/rspb.2013.2843.

Scelza, B. A. 2011. "Female Choice and Extrapair Paternity in a Traditional Human Population." *Biology Letters* 7 (6): 889–91. https://doi.org/10.1098/rsbl.2011.0478.

Schacht, R., and K. L. Kramer. 2019. "Are We Monogamous? A Review of the Evolution of Pair-Bonding in Humans and Its Contemporary Variation Cross-Culturally." *Frontiers in Ecology and Evolution* 7: 1–10. https://doi.org/10.3389/fevo.2019.00230.

Schaller, G. B. 1964. "Breeding Behavior of the White Pelican at Yellowstone Lake, Wyoming." *Condor* 66 (1): 3–23.

Schaller, G. B. 1972. *The Natural History of Lions: The Serengeti Lion. A Study of Predator-Prey Relations. Science.* Chicago: University of Chicago Press. https://doi.org/10.1126/science.179.4072.466.

Scheiber, I. B. R., A. Hohnstein, K. Kotrschal, and B. M. Weiß. 2011. "Juvenile Greylag Geese (*Anser anser*) Discriminate between Individual Siblings." *PLoS ONE* 6 (8): e22853. https://doi.org/10.1371/journal.pone.0022853.

Schmitt, D. P., L. Alcalay, J. Allik, A. Angleitner, L. Ault, I. Austers, K. L. Bennett, et al. 2004. "Patterns and Universals of Mate Poaching across 53 Nations: The Effects of Sex, Culture, and Personality on Romantically Attracting Another Person's Partner." *Journal of Personality and Social Psychology* 86 (4): 560–84. https://doi.org/10.1037/0022-3514.86.4.560.

Schmitt, D. P., and D. M. Buss. 2001. "Human Mate Poaching: Tactics and Temptations for Infiltrating Existing Mateships." *Journal of Personality and Social Psychology* 80 (6): 894–917.

Schooler, C. 1972. "Birth Order Effects: Not Here, Not Now." *Psychological Bulletin* 78 (3): 161–75. https://doi.org/10.1037/h0033026.

Segami, J. C., M. I. Lind, and A. Qvarnström. 2021. "Should Females Prefer Old Males?" *Evolution Letters* 24 (5): 507–20. https://doi.org/10.1002/evl3.250.

Segerstrale, U. 2000. *Defenders of the Truth: The Battle for Science in the Sociobiology Debate and Beyond.* New York: Oxford University Press.

Serrano-Meneses, M. A., and T. Székely. 2006. "Sexual Size Dimorphism in Seabirds: Sexual Selection, Fecundity Selection and Differential Niche-Utilization." *Oikos* 113 (3): 385–94. https://doi.org/10.1111/j.0030-1299.2006.14246.x.

Servedio, M. R., T. D. Price, and R. Lande. 2013. "Evolution of Displays within the Pair Bond." *Proceedings of the Royal Society B: Biological Sciences* 280: 20123020. https://doi.org/10.1098/rspb.2012.3020.

Shackelford, T. K., A. T. Goetz, F. E. Guta, and D. P. Schmitt. 2006. "Mate Guarding and Frequent In-Pair Copulation in Humans: Concurrent or Compensatory Anti-Cuckoldry Tactics?" *Human Nature* 17 (3): 239–52. https://doi.org/10.1007/s12110-006-1007-x.

Sharp, S. P., A. McGowan, M. J. Wood, and B. J. Hatchwell. 2005. "Learned Kin Recognition Cues in a Social Bird." *Nature* 434 (7037): 1127–30. https://doi.org/10.1038/nature03522.

Shostak, M. 1991. *Nisa, the Life and Words of a!Kung Woman.* Cambridge, MA: Harvard University Press.

Simmons, R. E. 1988. "Offspring Quality and Evolution of Cainism." *Ibis* 130: 339–57.

Smale, L., K. E. Holekamp, M. Weldele, G. L. Frank, and S. E. Glickmanj. 1995. "Competition and Cooperation between Litter-Mates in the Spotted Hyaena, *Crocuta crocuta.*" *Animal Behaviour* 50: 671–82.

Smith, J. M. 1974. "The Theory of Games and the Evolution of Animal Conflicts." *Journal of Theorical Biology* 47: 209–21.

Smith, J. M. 1979. "Game Theory and the Evolution of Behaviour." *Proceedings of the Royal Society of London B* 205: 475–88. https://royalsocietypublishing.org/.

Smith, L. R. 1984. *Sperm Competition and the Evolution of Animal Mating Systems.* Orlando, FL: Academic Press.

Spellerberg, I. F. 1971. "Breeding Behaviour of the McCormick Skua *Catharacta maccormicki* in Antarctica." *Ardea* 59 (3–4): 189–230.

Stamps, A. J., R. A. Metcalf, and V. Krishnan. 1978. "A Genetic Analysis of Parent-Offspring Conflict." *Behavioral Ecology and Sociobiology* 3: 369–92.

Stanford, C. B. 1999. *The Hunting Apes: Meat Eating and the Origins of Human Behavior.* Princeton, NJ: Princeton University Press.

Steinmetz, S. K. 1977. *The Cycle of Violence: Assertive, Aggressive, and Abusive Family Interaction.* Santa Barbara, CA: Praeger.

Stewart, R. B., and R. S. Marvin. 1984. "Sibling Relations: The Role of Conceptual Perspective-Taking in the Ontogeny of Sibling Caregiving." *Child Development* 55 (4): 1322–32.

Stewart, R. B., L. A. Mobley, S. van Tuyl, and M. A. Salvador. 1987. "The Firstborn's Adjustment to the Birth of a Sibling: A Longitudinal Assessment." *Child Development* 58 (2): 341–55.

Straus, M. A., R. J. Gelles, and S. K. Steinmetz. 1980. *Behind Closed Doors: Violence in the American Family.* New York: Anchor Press.

Sulloway, F. J. 2008. "Birth Order." In *Family Relationships: An Evolutionary Perspective*, edited by C. A. Salmon and T. K. Shackelford, 185–204. New York: Oxford.

Székely, T., J. D. Reynolds, and J. Figuerola. 2000. "Sexual Size Dimorphism in Shorebirds, Gulls, and Alcids: The Influence of Sexual and Natural Selection." *Evolution* 54 (4): 1404–13.

Tershy, R. B., D. Breese, and A. D. Croll. 2000. "Insurance Eggs versus Additional Eggs: Do Brown Boobies Practice Obligate Siblicide?" *The Auk* 117 (3): 817–20.

Thuman, K. A., and S. C. Griffith. 2005. "Genetic Similarity and the Nonrandom Distribution of Paternity in a Genetically Highly Polyandrous Shorebird." *Animal Behaviour* 69 (4): 765–70. https://doi.org/10.1016/j.anbehav.2004.10.003.

Tippett, N., and D. Wolke. 2015. "Aggression between Siblings: Associations with the Home Environment and Peer Bullying." *Aggressive Behavior* 41 (1): 14–24. https://doi.org/10.1002/ab.21557.

Torres, R., and H. Drummond. 1999. "Variably Male-Biased Sex Ratio in a Marine Bird with Females Larger than Males." *Oecologia* 161: 447–48. https://doi.org/10.1007/s00442-009-1378-x.

Torres, R., and H. Drummond. 2009. "Variably Male-Biased Sex Ratio in a Marine Bird with Females Larger Than Males (Erratum)." *Oecologia* 118 (1). https://doi.org/10.1007/s00442 0050698.

Torres, R., and A. Velando. 2003. "A Dynamic Trait Affects Continuous Pair Assessment in the Blue-Footed Booby, *Sula nebouxii.*" *Behavioral Ecology and Sociobiology* 55 (1): 65–72. https://doi.org/10.1007/s00265-003-0669-1.

Torres, R., and A. Velando. 2005. "Male Preference for Female Foot Colour in the Socially Monogamous Blue-Footed Booby, *Sula nebouxii.*" *Animal Behaviour* 69 (1): 59–65. https://doi.org/10.1016/j.anbehav.2004.03.008.

Trillmich, F., and J. B. W. Wolf. 2008. "Parent-Offspring and Sibling Conflict in Galápagos Fur Seals and Sea Lions." *Behavioral Ecology and Sociobiology* 62 (3): 363–75. https://doi.org/10.1007/s00265-007-0423-1.

Trivers, R. L. 1971. "The Evolution of Reciprocal Altruism." *The Quarterly Review of Biology* 46 (1): 35–57. https://doi.org/10.1086/406755.

Trivers, R. L. 1972. "Parental Investment and Sexual Selection." In *Sexual Selection and the Descent of Man*, edited by B. Campbell, 136–79. Chicago: Aldine.

Trivers, R. L. 1974. "Parent-Offspring Conflict." *American Zoologist* 14: 249–64.

Trivers, R. L. 2002. *Natural Selection and Social Theory: Selected Papers of Robert Trivers.* Oxford: Oxford University Press.

Trivers, R. L. 2011. *Deceit and Self-Deception: Fooling Yourself the Better to Fool Others.* London: Penguin Books.

Trivers, R. L., and H. Hare. 1976. "Haplodiploidy and the Evolution of the Social Insects." *Science* 191 (4224): 249–63. https://doi.org/10.1126/science.1108197.

Tsapelas, I., H. E. Fisher, and A. Aron. 2010. "Infidelity: When, Where, Why." In *The Dark Side of Close Relationships II*, edited by R. W. Cupach and H. B. Spitzber, 175–96. New York: Routledge.

Tucker, C. J., D. Finkelhor, A. M. Shattuck, and H. Turner. 2013. "Prevalence and Correlates of Sibling Victimization Types." *Child Abuse and Neglect* 37 (4): 213–23. https://doi.org/10.1016/j.chiabu.2013.01.006.

Tucker, C. J., and K. Updegraff. 2010. "Who's the Boss? Patterns of Control in Adolescents' Sibling Relationships." *Family Relations* 59 (5): 520–32. https://doi.org/10.1111/j.1741-3729.2010.00620.x.

Valderrábano-Ibarra, C., I. Brumon, and H. Drummond. 2007. "Development of a Linear Dominance Hierarchy in Nestling Birds." *Animal Behaviour* 74 (6): 1705–14. https://doi.org/10.1016/j.anbehav.2007.02.034.

Van Lawick, H., and J. Goodall. 1971. *Innocent Killers*. Boston: Houghton Mifflin.

Velando, A., and C. Alonso-Alvarez. 2003. "Differential Body Condition Regulation by Males and Females in Response to Experimental Manipulations of Brood Size and Parental Effort in the Blue-Footed Booby." *Journal of Animal Ecology* 72 (5): 846–56. https://doi.org/10.1046/j.1365-2656.2003.00756.x.

Velando, A., R. Beamonte-Barrientos, and R. Torres. 2006. "Pigment-Based Skin Colour in the Blue-Footed Booby: An Honest Signal of Current Condition Used by Females to Adjust Reproductive Investment." *Oecologia* 149 (3): 535–42. https://doi.org/10.1007/s00442-006-0457-5.

Velando, A., R. Beamonte-Barrientos, and R. Torres. 2014. "Enhanced Male Coloration after Immune Challenge Increases Reproductive Potential." *Journal of Evolutionary Biology* 27 (8): 1582–89. https://doi.org/10.1111/jeb.12416.

Velando, A., H. Drummond, and R. Torres. 2006. "Senescent Birds Redouble Reproductive Effort When Ill: Confirmation of the Terminal Investment Hypothesis." *Proceedings of the Royal Society B: Biological Sciences* 273 (1593): 1443–48. https://doi.org/10.1098/rspb.2006.3480.

Velando, A., H. Drummond, and R. Torres. 2010. "Senescing Sexual Ornaments Recover after a Sabbatical." *Biology Letters* 6 (2): 194–96. https://doi.org/10.1098/rsbl.2009.0759.

Velando, A., J. C. Noguera, H. Drummond, and R. Torres. 2011. "Senescent Males Carry Premutagenic Lesions in Sperm." *Journal of Evolutionary Biology* 24 (3): 693–97. https://doi.org/10.1111/j.1420-9101.2010.02201.x.

Velando, A., R. Torres, and C. Alonso-Alvarez. 2008. "Avoiding Bad Genes: Oxidatively Damaged DNA in Germ Line and Mate Choice." *BioEssays* 30 (11–12): 1212–19. https://doi.org/10.1002/bies.20838.

Velando, A., R. Torres, and I. Espinosa. 2005. "Male Coloration and Chick Condition in Blue-Footed Booby: A Cross-Fostering Experiment." *Behavioral Ecology and Sociobiology* 58 (2): 175–80. https://doi.org/10.1007/s00265-005-0911-0.

Viñuela, J. 1999. "Sibling Aggression, Hatching Asynchrony, and Nestling Mortality in the Black Kite (*Milvus migrans*)." *Behavioral Ecology and Sociobiology* 45 (1): 33–45. https://doi.org/10.1007/s002650050537.

Waal, F. D. 2007. *Chimpanzee Politics Power and Sex among Apes*. New York: The Johns Hopkins University Press.

Waal, F. D. 2019. *Mama's Last Hug*. New York: Norton.

Wagner, R, H. 1998. "Hidden Leks: Sexual Selection and the Clustering of Avian Territories." *Ornithological Monographs* 49: 123–45.

Walker, R. S., K. R. Hill, M. v. Flinn, and R. M. Ellsworth. 2011. "Evolutionary History of Hunter-Gatherer Marriage Practices." *PLoS ONE* 6 (4): 2–7. https://doi.org/10.1371/journal.pone.0019066.

Walter, K. V., D. Conroy-Beam, D. M. Buss, K. Asao, A. Sorokowska, P. Sorokowski, T- Aavik, et al. 2020. "Sex Differences in Mate Preferences across 45 Countries: A Large-Scale Replication." *Psychological Science* 31 (4): 408–23. https://doi.org/10.1177/0956797620904154.

Westneat, F. D., and C. R. Sargent. 1996. "Sex and Parenting: The Effects of Sexual Conflict and Parentage on Parental Strategies." *Trends in Ecology & Evolution* 11 (2): 87–91.

Westneat, D. F., and I. R. K. Stewart. 2003. "Extra-Pair Paternity in Birds: Causes, Correlates, and Conflict." *Annual Review of Ecology, Evolution, and Systematics* 34 (1): 365–96. https://doi.org/10.1146/annurev.ecolsys.34.011802.132439.

Williams, C. G. 1966. *Adaptation and Natural Selection a Critique of Some Current Evolutionary Thought*. Princeton, NJ: Princeton University Press.

Wilson, E. O. 1975. *Sociobiology: The New Synthesis*. Cambridge, MA: Harvard University Press.

Wilson, M., and M. Daly. 1992. "The Man Who Mistook His Wife for a Chattel." In *The Adapted Mind: Evolutionary Psychology and the Generation of Culture*, edited by J. H. Barkow, L. Cosmides, and J. Tooby, 289–322. New York: Oxford University Press. https://doi.org/10.1136/practneurol-2015-001124.

Wilson Saires, M. A., and K. D. Makova. 2011. "Genome Analyses Substantiate Male Mutation Bias in Many Species." *BioEssays* 33 (12): 938–45. https://doi.org/10.1002/bies.201100091.

Wingfield, J. C., G. Ramos-Fernandez, A. Nuñez De La Mora, and H. Drummond. 1999. "The Effects of an 'El Niño' Southern Oscillation Event on Reproduction in Male and Female Blue-footed Boobies, *Sula nebouxii*." *General and Comparative Endocrinology* 114 (2): 163–72. https://doi.org/10.1006/gcen.1998.7243.

Wittig, R. M., and Ch. Boesch. 2003. "Food Competition and Linear Dominance Hierarchy among Female Chimpanzees on the Taï National Park." *International Journal of Primatology* 24 (4): 847–67.

Wolf, J. B., and E. D. Brodie. 1998. "The Coadaptation of Parental and Offspring Characters." *Evolution* 52 (2): 299–308. https://doi.org/10.1111/j.1558-5646.1998.tb01632.x.

Wolke, D., N. Tippett, and S. Dantchev. 2015. "Bullying in the Family: Sibling Bullying." *The Lancet Psychiatry* 2 (10): 917–29. https://doi.org/10.1016/S2215-0366(15)00262-X.

Wright, J, and A. P. Cotton. 1994. "Experimentally Induced Sex Differences in Parental Care: An Effect of Certainty of Paternity?" *Animal Behaviour* 47: 1311–22.

Yu, Q., Q. Zhang, Q. Xiong, S. Jin, H. Zou, and Y. Guo. 2019. "The More Similar, the More Warmth: The Effect of Parent-Child Perceived Facial Resemblance on Parenting Behavior." *Personality and Individual Differences* 138: 358–62. https://doi.org/10.1016/j.paid.2018.10.027.

Yu, X., X. Li, and Z. Huo. 2015. "Breeding Ecology and Success of a Reintroduced Population of the Endangered Crested Ibis *Nipponia nippon*." *Bird Conservation International* 25 (2): 207–19. https://doi.org/10.1017/S0959270914000136.

Yu, X., N. Liu, Y. Xi, and B. Lu. 2006. "Reproductive Success of the Crested Ibis *Nipponia nippon*." *Bird Conservation International* 16 (4): 325–43. https://doi.org/10.1017/S0959270906000499.

Zajonc, R. B. 1980. "Feeling and Thinking: Preferences Need No Inferences." *American Psychologist* 35 (2): 151–75. https://doi.org/10.1037/0003-066X.35.2.151.

List of Common and Systematic Names

Alligator snapping turtle (*Macrochelys temminckii*)
American crocodile (*Crocodylus acutus*)
Arabian babbler (*Turdoides squamiceps*)
Atlantic central American milk snake (*Lampropeltis polyzona*)
Bald ibis (*Geronticus eremita*)
Barn owl (*Tyto alba*)
Barnacle goose (*Branta leucopsis*)
Black eagle (*Ictinaetus malaiensis*)
Black grouse (*Lyrurus tetrix*)
Black guillemot (*Cepphus grille*)
Black-legged kittiwake (*Rissa tridactyla*)
Blue-footed booby (*Sula nebouxii*)
Boa constrictor (*Boa constrictor*)
Bonobo (*Pan paniscus*)
Brahminy blind snake (*Indotyphlops braminus*)
Brown booby (*Sula leucogaster*)
Brown noddy (*Anous stolidus*)
Brown pelican (*Pelecanus occidentalis*)
Caledonia crow (*Corvus moneduloides*)
California sea lion (*Zalophus californianus*)
Canvasback duck (*Aythya valisineria*)
Cat (*Felis catus*)
Cattle egret (*Bubulcus ibis*)
Chimpanzee (*Pan troglodytes*)
Clark's spiny lizard (*Sceloporus magister*)
Common kingfisher (*Alcedo atthis*)
Cormorant (*Phalacrocorax spp.*)
Coyote (*Canis latrans*)
Crested ibis (*Nipponia nippon*)
Crow (*Corvus spp.*)
Dog (*Canis lupus familiaris*)
Domestic fowl (*Gallus domesticus*)
Dunlin (*Calidris alpina*)
Elephant cactus (*Pachycereus pringlei*)
Eurasian oystercatcher (*Haematopus ostralegus*)
Eye-gnats (*Liohippelates spp.*)
Fox (*Vulpes vulpes*)
Galapagos fur seal (*Arctocephalus galapagoensis*)
Garlic pear tree (*Crateva religiosa*)
Garter snakes (*Thamnophis spp.*)
Gorilla (*Gorilla spp.*)
Great blue heron (*Ardea herodias*)
Great egret (*Ardea alba*)

Great tit (*Parus major*)
Greater ani (*Crotophaga major*)
Green iguana (*Iguana iguana*)
Greylag goose (*Anser anser*)
Heermann's gull (*Larus heermanni*)
Hermit crab (*Coenobita compressu*)
Japanese quail (*Coturnix japonica*)
Junglefowl (*Gallus gallus*)
Magnificent frigatebird (*Fregata magnificens*)
Mallard (*Anas platyrhynchos*)
Mauritius kestrel (*Falco punctatus*)
Meerkat (*Suricata suricatta*)
Mouth brooding cichlid (*Geophagus spp.*)
Nayarit coral snake (*Micrurus proximans*)
Noddy tern (*Anous stolidus*)
Osprey (*Pandion haliaetus*)
Paradise fish (*Macropodus opercularis*)
Peacock (*Pavo cristatus*)
Pied flycatcher (*Ficedula hypoleuca*)
Raven (*Corvus corax*)
Red deer (*Cervus elaphus*)
Red grouse (*Lagopus lagopus scotica*)
Red head duck (*Aythya americana*)
Red warbler (*Cardellina rubra*)
Red-billed tropic bird (*Phaethon aethereus*)
Red-footed booby (*Sula sula*)
Sandhill crane (*Grus canadensis*)
Scrub jay (*Aphelocoma californica.*)
Siamese fighting fish (*Betta splendens*)
Soft bodied tick (*Ornithodoros spp.*)
Sooty tern (*Onychoprion fuscatus*)
South polar skua (*Stercorarius maccormicki*)
Spinytail iguana (*Ctenosaura pectinate*)
Spotted hyena (*Crocuta crocuta*)
Western capercaillies (*Tetrao urogallus*)
Western Mexican whiptail (*Aspidoscelis costatus*)
Western terrestrial garter snake (*Thamnophis elegans*)
White pelican (*Pelecanus erythrorhynchos*)
Wild turkey (*Meleagris gallopavo*)
Willow grouse (*Lagopus lagopus*)
Wolf (*Canis lupus*)
Yellow baboon (*Papio cynocephalus*)

Index

For the benefit of digital users, indexed terms that span two pages (e.g., 52–53) may, on occasion, appear on only one of those pages.
Note: Figures are indicated by f following the page number